DATE DUE

OCT 1 8 1991	
NOV 6 1991	
MAY 2 2 1995	
MAR 2 4 1997	

The Longest Cave

by
Roger W. Brucker
and
Richard A. Watson

With a New Afterword

Southern Illinois University Press
Carbondale and Edwardsville

Grateful acknowledgement is made to Ronald Sukenick for
permission to quote from his book *98.6*, a novel. Copyright
© 1975 by Ronald Sukenick; and to Shapiro, Bernstein &
Co., Inc., for permission to quote lyrics from the song
"The Death of Floyd Collins" by Rev. Andrew Jenkins and
Mrs. Irene Spain. Copyright 1925 by Shapiro, Bernstein
& Co., Inc., copyright renewed. Used by permission. All
rights reserved.

90   89   88   87   4   3   2   1

Library of Congress Cataloging-in-Publication Data

Brucker, Roger W.
  The longest cave.

  Reprint. Originally published: New York :
Knopf, 1976. With an afterword.
  Bibliography: p.
  Includes index.
  1. Mammoth Cave (Ky.) I. Watson, Richard A.,
1931–   . II. Title.
GB606.M35B78   1987   551.4'47'09769754   86-15537
ISBN 0-8093-1321-9
ISBN 0-8093-1322-7 (pbk.)

Dedicated to the Memory of

James W. Dyer
and
E. Robert Pohl

It's not much of a house, but we've got one hell of a basement.
      —Traditional Flint Ridge greeting

If you cannot understand that there is something in man which responds to the challenge of this mountain and goes out to meet it, that the struggle is the struggle of life itself upward and forever upward, then you won't see why we go. What we get from this adventure is just sheer joy. And joy is, after all, the end of life. We do not live to eat and make money. We eat and make money to be able to enjoy life. That is what life means and what life is for.
      —George H. Leigh-Mallory

In the late afternoon after he's done with his work George often gets his board and goes surfing. If you ask George why he likes surfing so much he gets a thoughtful look on his face strokes his blond beard squints gazing into emptiness and after a while says Guess I just like it.
      —Ronald Sukenick

Cave Conservation

Caves are fragile in many ways. Their features take hundreds of thousands of years to form. Cave animals such as blindfish are rare, and always live in precarious ecological balance in their underground environments. Cave features and cave life can be destroyed unknowingly by people who enter caves without informing themselves about cave conservation. Great and irreparable damage has been done by people who take stalactites and other flowstone features from caves and by those who disturb cave life such as bats, particularly in winter, when they are hibernating. Caves are wonderful places for scientific research and recreational adventure, but before you enter a cave, we urge you first to learn about careful caving by contacting the National Speleological Society, Cave Avenue, Huntsville, Alabama 35810.

R.W.B.
R.A.W.

Contents

Contents

Page

Illustrations

Photographs

Maps

Preface

This is a story of some cave explorers who are among the thousands of people whose lives have been intertwined with the great caves of Mammoth Cave National Park. The earliest prehistoric explorers of 4000 years ago mined the caves for gypsum and mirabilite, but there is evidence that they also explored just to see where the caves went. During the settlement of Kentucky in the late 1700s, the caves were mined for saltpeter to make gunpowder. Then as the frontier moved on westward in the 1800s, Mammoth Cave became a tourist attraction.

Discoveries made by the black slave guide Stephen Bishop, beginning in 1838, started the quest for enlarging Mammoth Cave. This led eventually to the discoveries made by Pete Hanson and Leo Hunt 100 years later. It also inspired others to search for new caves in Flint Ridge, north of Mammoth Cave Ridge across Houchins Valley. Lute Lee and Henry Lee found Colossal Cave in 1895. Edmund Turner explored Salts Cave, and in 1916 found Great Onyx Cave. Floyd Collins discovered his Great Crystal Cave in 1917, and then died in 1925 exploring Sand Cave in a narrow connecting ridge between Flint Ridge and Mammoth Cave Ridge.

At the beginning of the twentieth century, Horace Hovey and E.-A. Martel had inferred on geological grounds that the caves of Flint Ridge and Mammoth Cave Ridge were connected. During Floyd Collins' lifetime there were rumors that he had connected some of the Flint Ridge caves with one another. Some people said Floyd died trying to connect the Flint Ridge caves with Mammoth Cave.

Jim Dyer inherited Floyd's legacy. He managed Floyd Collins' Crystal Cave in the late 1940s, and introduced Bill Austin to caving. In 1955, with Jack Lehrberger, Bill made the first connection between two major Flint Ridge caves, Unknown Cave and Crystal Cave.

Jim, Bill, and Dr. E. Robert Pohl encouraged a group of cavers who incorporated in 1957 as the Cave Research Foundation under the presidency of Phil Smith. A few of these explorers became connection fanatics who used this organization to integrate the caves of Flint Ridge, and to make the final connection in 1972 between the Flint Ridge Cave System and Mammoth Cave.

The story of exploration up to 1953 is told in Appendix I, Historical Beginnings. We urge everyone who wants to start at the beginning to read it first. (We also recommend that you glance at the Glossary of Cave Terminology and the Picture Glossary of Cave Maneuvers—Appendixes III and IV—before you begin.) The body of this book contains the story from 1953 to the final connection in 1972.

This is the story we tell. It is a vastly one-sided story in which the impressive scientific contributions of the Cave Research Foundation are mostly ignored. It overstresses connection also at the expense of those hundreds of cavers who contributed primarily to enjoy the fun, fellowship, and adventure of caving. Nevertheless, connection is a major theme in the history of these caves. It led to the integration of the Flint Mammoth Cave System—the longest cave.

That was the big goal. During the years of caving described in this book, the dream of connecting the Flint Ridge Cave System with Mammoth Cave began to be referred to as the Everest of speleology. Everest was not climbed by one man or by one team of climbers, but by generations of devoted explorers. For many climbers, one season on Everest was enough, just as one trip or expedition underground has been enough for many cavers in Flint Ridge. Once can be just enough, the adventure of a lifetime, to be neither missed nor repeated. It took many teams of explorers to establish routes and base camps on the slopes of Everest. Similar teams worked for years to establish routes in Flint Ridge out to and beyond Candlelight River, so that finally a team could move rapidly and securely from the Flint Ridge Cave System to make the final connection with Mammoth Cave.

Nevertheless, although the skills and endurance required in mountain climbing are similar to those required in caving, a mountain is very unlike a cave. A mountain is in view and presents a precise goal: the peak. Possible routes can be examined through binoculars or from a helicopter. Photographs can be used to plan routes. Tourists can watch climbers. Rescuers often can see climbers in trouble, and at least know where to go. The stages of a climbing expedition are often precise, from planning through build-up to triumph or failure, and then retreat. None of this is true for big-cave exploration.

The routes in a cave are hidden. A caver may know where he wants to go, but he has no way of knowing beforehand how to get there, or whether it is possible to get there at all. Usually the goal is just the unknown. Cavers explore caves to see where they go. And caving is the antithesis of spectator sports. The only way to discover that cavers are in trouble is to notice that they have not come back. Finally, the hidden nature of caves necessitates that planning be piecemeal and often less than precise. When you do not know what lies ahead, you can plan only in general and one step at a time. It would be embarrassing to tool up for a month-long expedition, and then find that the cave ended just around the next bend in the passage.

Also, the mountain or the big wall ends. To be sure, other routes can be pioneered—straight up, in winter, at night. Or a party can traverse Everest. But even for those who come after, there is a set goal. And once you have made a hard climb, you can conclude your climbing career there at a well-defined point. You know when you are on a mountain peak, and you know when you have climbed a mountain. But in a cave, you can never be certain that the cave does not go on. If you conclude your caving career by thoroughly exploring a hard cave, likely as not someone else will come along to discover new passages you missed.

This, then, is the most striking difference between caving and mountain climbing. When you climb a mountain, you can see your progress.

Often you can see where you began, and often you can see your goal. As you climb, your route spreads out below you. The peak up ahead looms closer. The sky opens up and the lowlands stretch far below. Finally, you can go no higher. The horizon circles you. You have climbed the mountain peak, and from there you can often see the whole way along which you have come. And there, tiny in the distance, is the base camp to which you must return.

It is utterly different in a cave. Within seconds you lose sight of your starting point. The sinuous passages twist and turn. Always you are confined by walls, floor, and ceiling. The farthest vistas are seldom more than one hundred feet—along a passage, down a pit, up at a ceiling. You are always in a place; you never look out from a point. The route is never in view except as you can imagine it in your mind. Nothing unrolls. There is no progress; there is only a progression of places that change as you go along. And when you reach the end, it is only another place, often a small place, barely large enough to contain your body. It is conceivable that you have missed a tiny hole that goes on. You may not have reached the end at all. The only sign that you have reached the end is that you cannot go on. And there is no view. You certainly cannot see the base to which you must return.

Even when exploring a cave, often you do not know where you are, but only where you started from and (you hope) how to get back. A mountaineer can usually see where he *is*. Cavers usually have to wait until their surveys are made into maps to know where they *have been*.

There is one close analogy, however, between caving and mountain climbing. Once a pioneer team sets a route, others can follow. This does not mean that lesser physical skills are needed by the followers, but that they have less mental tension because they know it can be done. A lot of caving has to do with finding out what can be done.

You can "do" a cave by going through all its known passages. But goals are almost impossible to set or reach. The goal of connecting the great caves in Mammoth Cave National Park is not typical of world caving. In Hölloch of Switzerland—for many years the longest known cave in the world—the goal was simply to find and to map 100 kilometers of passages. The reaching of that goal was an anticlimax. Momentum carried exploration onward, and more passages are discovered in Hölloch each year.

Everest is the highest mountain. Flint/Mammoth is the longest *known* cave. But what cave is *really* the longest? No one knows. It is hidden underground where exploration never ends.

What lures explorers to big caves is the possibility of discovering many miles of passages where no human being has ever been before. Through even the smallest virgin passage, a caver *might* find marvels. It happens in the best-known caves. Mammoth Cave has been known since the late 1700s, yet explorers have found several miles of virgin passages in Mammoth Cave since 1970.

Caves also lure with their personalities. Some people feel that all caves are hostile, and they are uneasy in a cave the moment they step inside. Others feel their spirits rise when they enter a cave, and thus believe that all caves are friendly. Caves close in on some people and open up for others. Objectively, as a hole in the ground, each cave has distinctive features that make it difficult or easy to traverse. Dark or light walls and large or small passages reflect light and sound differently. Temperature, humidity, stone texture, dust or sand or mud or water, bats and rats, salamanders and crickets, crystals, and flowstone decorations combine to affect different people in different ways.

Unusual noises in caves may raise hackles or superstitions. Trickling water, breezes, carbide lamps, and echoes often give the impression of unintelligible voices almost out of hearing. Some cavers can make themselves half believe in "little people." In Floyd Collins' Crystal Cave in Flint Ridge, cavers have heard behind them in the darkness the faint call: "Wait for me." It jumps to your mind that the voice belongs to Floyd— whose body is in a coffin in the Grand Canyon of Crystal Cave—that Floyd is calling, wanting to come along. Depending on your temperament, you wait and yell back for Floyd to get a move on, or you scurry on in hopes that he will not catch up. Floyd was one of the greatest cave explorers who ever lived, so most Flint Ridge cavers would like him to come along, if only in spirit. So far, he has not shown up in person.

For experienced cavers, caves often generate another kind of feeling. Besides seeming hostile or friendly or indifferent, caves can also seem to be dead or alive. Several Flint Ridge cavers feel that Great Onyx Cave is a dead cave. Most of the passages are dry, and some are incredibly dusty. Water drips into a pool at the bottom of a deep pit, but explorers who waded, poked, and floated corks in the pool could locate no outlet. The best exploration route—out the drain of the deepest pit—is closed. Great Onyx Cave thus seems stagnant and dead.

A live cave, on the other hand, is a going cave. In it you feel the pulse of possibility, and there is mystery and allure around every corner. You just know that this cave goes, if only the right lead can be found. Floyd's Lost Passage in Crystal Cave felt that way in November of 1954. In Floyd's Lost Passage the air moves, water drips, and there is shifting mud on the floor. These are signs of live cave. Where water and air move, passages go, and you might be able to follow. Even though more than 30 people had explored there for a week only a few months earlier and had not found much new passage, cavers still knew that somewhere out of Floyd's Lost Passage they would find more cave. And they did.

Why do people go caving? Mallory said it best: "What we get from this adventure is just sheer joy." As you read this book, remember that for every fanatic who had his eyes on the goal of the big connection, there were two dozen cavers who helped just because they loved caving. Even the connection fanatics did it to enjoy life: "That is what life means and what life is for."

Roger W. Brucker and Richard A. Watson

Acknowledgments

Hundreds of people contributed to the adventures described in this book. Some of their names are listed in Appendix V. Inevitably in a list so long, we have missed some people who helped. To us, those absent names are symbolic of the several hundred thousand hours of largely selfless underground exploration and surveying that is *not* described in this book. That work is the essential foundation on which the connection efforts were built.

Among these people, we thank particularly *for providing materials from which we wrote portions of this book:* Sarah G. Bishop, William P. Bishop, John F. Bridge, Thomas A. Brucker, Denver P. Burns, Patricia P. Crowther, Joseph K. Davidson, David W. Deamer, Frederick J. Dickey, Burnell F. Ehman, Jacob H. Elberfeld, P. Gary Eller, Charles B. Fort, John P. Freeman, Russell H. Gurnee, Charles F. Hildebolt, Carol A. Hill, David A. Huber, Mark D. Jancin, John J. Lehrberger, Barbara MacLeod, Roger E. McClure, Arthur N. Palmer, Cleveland F. Pinnix, Thomas L. Poulson, L. Greer Price, Jack C. Reccius, David B. Roebuck, Robert H. Rose, Stanley D. Sides, Gordon L. Smith, Philip M. Smith, William J. Stephenson, Norbert M. Welch, Stephen G. Wells, M. Spike Werner, William B. White, John P. Wilcox, and Richard B. Zopf; *for reading the manuscript and providing editorial assistance:* Jacqueline F. Austin, William P. Bishop, John F. Bridge, Bobbie Bristol, Joan W. Brucker, Thomas A. Brucker, Denver P. Burns, Angus Cameron, Patricia P. Crowther, Joseph K. Davidson, Frederick J. Dickey, Dennis E. Drum, Mary E. Drum, Burnell F. Ehman, P. Gary Eller, Donald Finkel, John P. Freeman, James A. Hedges, Carol A. Hill, Peter Kurz, Janet E. Levy, Harold Meloy, Arthur N. Palmer, Thomas L. Poulson, James F. Quinlan, Sally Rogers, Jerome P. Schiller, Linda Kay Sides, Stanley D. Sides, Philip M. Smith, Constance Urdang, Patty Jo Watson, Bethany J. Wells, Stephen G. Wells, M. Spike Werner, Willam B. White, and John P. Wilcox; *for assistance with the photographs, maps, and drawings:* David J. DesMarais, Patricia P. Crowther, Richard B. Zopf, Ellen Brucker, Diana O. Daunt, Donald E. Coons, William R. Crowther, John P. Wilcox, Tomislav M. Gracanin, Robert Halmi, and Bolt Beranek and Newman Inc.; *and for cooperation and assistance in the field:* the National Park Service of the United States Department of the Interior; Mammoth Cave National Park Superintendents R. Taylor Hoskins, Thomas C. Miller, Perry Brown, Paul McG. Miller, John A. Aubuchon, Robert Bendt, and Joseph Kulesza; Mammoth Cave National Park Chief Naturalists Willard Dilley, George Olin, R. Alan Mebane, Edwin Rothfuss, and Leonard W. McKenzie; and the many other interested and helpful officials and employees at Mammoth Cave National Park.

We especially thank the following: John P. Wilcox, who brought to his many examinations of successive versions of the manuscript the same diligence and persistence that led him to make the final connection; Philip M. Smith, Denver P. Burns, and Joseph K. Davidson for many years of attention to detail; Harold Meloy, Stanley D. Sides, and Patty Jo Watson for guidance in our interpretations—with which they do not always agree —of history and prehistory; Patricia P. Crowther and William R. Crowther for map making; Linda Kay Sides for compiling the List of Participants and biographical data; Joan W. Brucker for many readings and for compiling the index; and our parents, who taught us fortitude.

During the years we were involved in this adventure, we were always grateful recipients of the cordiality of Dr. and Mrs. E. Robert Pohl at Mammoth Onyx Cave and of Jim and June Dyer in Columbus, Ohio. We also appreciate more than any of us ever said the hospitality of Jacque Austin and Bill Austin, who opened their home to cavers on Flint Ridge.

And what are we to say about Bill Austin and Jack Lehrberger? As historians, we are greatly frustrated at their decision to provide us with very little information about the caving they did. (Despite this, they helped us clear up some serious inconsistencies.) As cavers, we understand their desire to keep to themselves what were intensely private experiences. And as authors of this book, we admit that we are delighted at their reticence, for it was the legend of Austin and Lehrberger that led us on, and we expect that legend to lead others on as long as anyone cares about exploring caves.

Finally, we dedicate this book to the memory of James W. Dyer and E. Robert Pohl. Without Jim's enthusiasm and inspiration, there would never have been a tribe of Flint Ridge cavers. And if Dr. Pohl had not provided the access underground and the foundation in science to keep our explorations going, these adventures would never have taken place. We are grateful.

R. W. B.
R. A. W.

The Longest Cave

The First Connection

Caves and cavers
have secrets. Unknown Cave and Crystal Cave
are connected.

1 Bill Austin unrolled the map. Roger Brucker looked at it blankly, for it was unlike other cave maps he had seen. He had expected to see a map of the maze of intertwined passages which he knew were under his feet, underground in Kentucky below the Austins' house on lonesome Flint Ridge. But this map was a simple line plot, a pattern of straight lines connecting pinpricks. Only every fifth survey station was marked, and thus only the general trend of the passages was shown. There was no detail at all about dimensions, twists, and turns.

"What do you think of that?" Bill asked.

Roger had known Bill for two years, but he was still uncertain what to think. He looked at Bill, noncommittal. Bill gave him the Flint Ridge smile, a broad, face-filling grin that said he knew things others wanted to know and could learn, if they worked hard.

"You see this passage here? There are several pits along it with leads out of them. Now about here"—Bill pointed to a blank spot on the map—"is the upper end of Eyeless Fish Trail in Crystal Cave." He spanned the distance with thumb and forefinger. "It's about this far. Now what we'd like you boys to do is come back next weekend and go in Crystal to this point. Jack and I will go in . . . another way . . . and pound on rocks in the pits. If you hear us, you come on through." Bill quickly rolled up the map. That Roger and Jack Lehrberger—a friend who had done a lot of caving with Bill—would agree was assumed. Roger would bring Red Watson. Red had been caving for less than a year, but he had learned fast and was an obvious choice for a trip to the limits of Crystal Cave.

Bill and Roger stepped outside into the bright sunlight of that beautiful September day in 1955. Bill nodded good-by and then walked purposefully toward a car full of tourists that had just driven up. He was manager of the cave underfoot—Floyd Collins' Crystal Cave, isolated in an island

of private property on Flint Ridge within Mammoth Cave National Park—and fall tourists were rare. These would not escape.

Late Friday afternoon the next weekend, Roger and Red started the eight-hour drive from Ohio, where they lived, to Crystal Cave in Kentucky. Like Bill, they were in their early twenties. Roger worked for a small industrial-advertising agency in Yellow Springs, and Red was an Aerial Photography Officer in the Air Force, stationed in Columbus. They were pleased to be leaving all that behind.

"It's got to be Salts Cave," Roger said.

Red had never been in Salts Cave—nor had Roger, for that matter—but he knew that it was a big cave about a mile south of Crystal Cave in Flint Ridge.

"This is that secret work Bill and Jack have been doing?" Red asked.

"Sure. It must be. Salts Cave!" Roger said, pounding the steering wheel. "We'll connect Crystal with Salts!"

It was a pleasant thought. Red settled back as Roger guided the car. The tires hummed on the highway. "No matter where you are in this country," he muttered, "you have to take a long car ride before you can do anything."

"It's part of outdoor adventure," Roger said. "Here we are, roaring down this magnificent highway, out for adventure." He sang and prolonged the word "adventure."

They were quiet for a while, then Roger cocked his head and sang (a bit off key) lines from an old Kentucky ballad:

> *Oh come all you young people and listen while I tell*
> *The fate of Floyd Collins, a lad we all knew well.*
> *His face was fair and handsome, his heart was true and brave.*
> *His body now lies sleeping in a lonely sandstone cave.*

As dusk deepened, the travelers crossed the Ohio River and drove up the escarpment onto the limestone plateau. On the horizon to the south, a row of knoblike hills rose through the blue haze, announcing cave country. Darkness fell. The strip of blacktop highway unrolled beneath the car. Roger and Red were silent.

Goin' cavin'.

After a while Red said, "But old Floyd *died* in that sandstone cave."

Some hours later as they turned into the mile-long access road to Floyd Collins' Crystal Cave on Flint Ridge, they thought of Floyd again and of how he had died, trapped in Sand Cave in 1925. In a few months it would be thirty-one years from the date of his death. The next day, Roger and Red would walk past Floyd's coffin on their way through Grand Canyon to the lower levels of Crystal Cave.

Floyd had died, it was said, after fifteen days in the hole. Later, a shaft was sunk to take out his body. Pictures were taken of him, and Roger and Red had only recently seen them. They would never forget the frozen look on that mud-dripped face. Part of an ear had been eaten away, by cave rats or crickets.

Roger stopped the car in front of a small house at the end of the road.

"Y'all looking for something?" Bill Austin had come silently up to the car. "Glad you could make it," he said. Behind him sat Max, the Austins' large Weimaraner dog, attentive in the midnight air.

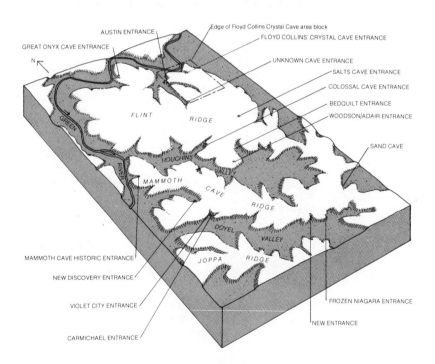

Block diagram showing the location of principal ridges and cave entrances. Block measures 6 miles north and south, 4.5 miles east and west.

As they entered the house, Jacque Austin handed Red a plate with a large piece of apple pie on it. He took the pie gratefully and sat down on the floor with his back against the wall. The dog walked over to sit beside him. Jacque enjoyed feeding her caving friends, and she made wonderful apple pie. Still, Red was apprehensive of Jacque, whose disapproval could be formidable. He remembered one occasion vividly.

"Wait a minute," Jacque had said on catching sight of Red on one of his early visits to Flint Ridge. "Red, honey, come over here," she ordered with mock little-girl sweetness.

Red walked over to her. She squinted, crouched down, and looked up at him. Then she put her arm around him.

"Hold still," she said, "I think there is *something on your face. A bug!*" With her open palm, she hit him resoundingly across his new mustache.

"Got 'im!" she said with great satisfaction. Then she turned and walked briskly back to the house, her back shaking with suppressed laughter.

Red dozed while Roger discussed the next day's plans with Bill. Roger was wide awake. He had brought the map he had been preparing of Crystal Cave. Bill helped him orient it with relation to the line plot Bill had made of the "other" cave he and Jack had been exploring. Roger was wildly curious to know what cave the line plot represented.

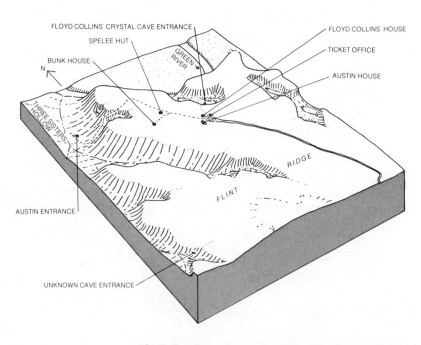

Block diagram showing the location of principal features in the Floyd Collins' Crystal Cave area. Block measures about one mile square. Vertical scale is exaggerated.

"That'll really be something when we get Crystal and Salts connected," Roger tried.

Bill remained silent for a moment. Then he said, "You guys can sleep late. You needn't go into the cave until four tomorrow afternoon. Jack and I will come in later and start pounding around ten p.m. All you have to do is get out there on time, listen, and pound back."

Roger and Bill synchronized their watches.

"You can sleep out in the Guide House," Bill said.

"He won't tell us a thing!" Roger exploded, almost before Red had closed the door behind them. "But it *must* be Salts Cave. Won't that be something!"

When they got back to the Guide House, several hundred yards behind the Austins' house, Roger said, "I'm really tired of all this secrecy. To-morrow they'll take us on a long rinky-dink to confuse us, and then out through Salts Cave." Thinking of Salts Cave, he cheered up again. He was actually quite happy with the mystery, and enjoyed the secrets as much as Bill did. There had been rumors about connections between Salts Cave and Crystal Cave in Flint Ridge for thirty-eight years, ever since Floyd had discovered Crystal Cave.

"Tomorrow we'll make the connection!" Roger said, hitting Red on the shoulder.

Red crawled into his sleeping bag and was soon asleep again. Roger prepared for bed slowly. He blew up his air mattress, placed it on the floor, and at last slid into his sleeping bag. All was quiet, but for a long while Roger lay wide-eyed in the night.

Bright sunlight soon filled the Guide House. Red awakened slowly. Eventually he groaned and crawled out of his sleeping bag in a faint haze of dust and down. Red's freckles and carrot-red hair blazed in the sun-light. Despite his mustache, he looked much younger than his twenty-four years, with a smooth baby face, the always defiant hair, and bright blue eyes. Roger was sitting upright in his sleeping bag, his shirt rumpled from having been slept in.

"Good day," Red murmured.

"It'll be a good day if you prepare for it," Roger replied. Roger was twenty-five, with a burr-head haircut and happy eyes under droopy eye-lids. He was a medium-sized caver, five foot ten, weighing about 170 pounds. Red verged toward the small end of the cavers' scale, five foot eight and 145 pounds. Bill and Jack fitted the classic model—set by themselves, naturally—of around six feet and 160 to 170 pounds of bone and muscle. There were debates about the proper size for a caver, but the only agreed-upon conclusion was that a caver cannot be fat.

Red put on a pair of Levi's that were so small he could not button them up. Next came a T-shirt full of holes. He then put on a buttonless red flannel shirt, pinning it together with two large safety pins. Finally, he pulled on an enormous pair of khaki trousers that were ripped at the knees and the rear. Red tied them securely around his middle with a piece of rope, then sat down on the floor to search in his pack for his boots.

"Are you dressing for a clown act?" Roger asked.

Red put on his boots without answering. Roger got dressed in clothes nearly as ragged, and then they walked toward the Austins' house, wondering whether this caving weekend was special enough for Jacque to prepare breakfast for them.

Roger knocked tentatively on the door.

"The latchstring is out!" came a yell from inside.

"She means come in," Red said.

"I s'pose y'all want some breakfast," Jacque said. "Sit down." She took some biscuits from the oven.

"What *is* that outfit you've got on?" Jacque asked Red as she pushed the butter plate toward him.

"I like caving because I can wear out my old clothes," Red replied.

"Red, honey," Jacque said, laughing, "you have succeeded beyond your wildest dreams."

Later, Roger and Red sat on the porch of the Crystal Cave Ticket Office, listening to wizened, lame, one-eyed Kanah Cline tell stories about early exploration in the lower levels of Crystal Cave. Years earlier he had shown Jim Dyer the way to the mile-long passage known as Floyd's Lost Passage, and Jim had introduced Bill to caving. Bill had introduced Roger and Red to the big cave, so in the caving family Kanah was their great-grandfather. He relished the role, leading them to understand that the old-timers were as tough as any cavers ever had been or would be.

It was a slow day. Jack Lehrberger arrived in time to eat the early supper Jacque had prepared. Jack was tall and lean like Bill, but gave the impression of being broader. He was majoring in mathematics at the University of Louisville. With classic Latin features, dark deep-set eyes, chiseled nose and chin, perfect teeth in a flashing smile, and a brooding courtly manner, Jack was surely the handsomest and most self-assured young man in Kentucky. He taught dancing for Arthur Murray, and with his fiancée had won ballroom-dancing contests.

As a caver, Jack was a loner. He had gone on exploration trips of twenty hours or more alone, deep into virgin passages where he might not have been found for a long time had he lost his way. He had discovered the lower levels of Salts Cave, penetrating deeply into Flint Ridge in the area south of Crystal Cave. He had slept alone beside Floyd's coffin in Crystal Cave.

The caving exploits of Bill and Jack were already beginning to be legendary. On twenty winter nights, on trips lasting twelve—sixteen—twenty—twenty-four hours they had, it was whispered, discovered more passages underground in Flint Ridge than anyone ever had before. Not even Floyd Collins' discoveries matched their miles. Roger and Red viewed them with awe. This feeling was tempered, however, with considerable self-respect. Roger and Red had caved with Bill and Jack several times, and

*Typical cave pack contents include equipment
and supplies for up to twenty-six hours underground. Spare lamp parts usually include a
jackknife, a small coil of wire for repairs, and
other items of personal utility.*

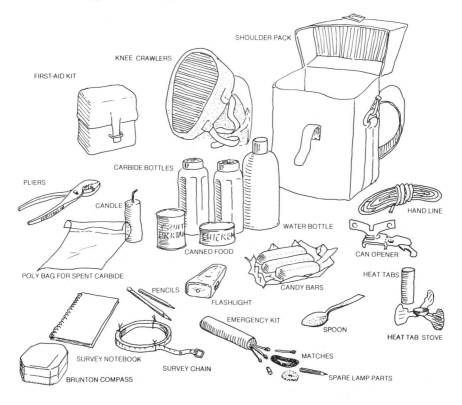

knew they could keep up. And they had been invited to share what would
surely be a great adventure.

Around 3:30 p.m., Roger and Red each prepared a small canvas bag.
In the bags they put plastic bottles of water and carbide for their lamps,
some candles and matches in waterproof containers, and some small cans
of fruit and meat. It was time to start at last.

Bill and Jack were still dressed in sport shirts. They had not even
begun to prepare for their trip underground. They would be taking the
easier, quicker way in that "other" cave that Roger and Red would see only
if a connection with Crystal Cave were made.

At the Crystal Cave Entrance, Roger and Red filled their carbide
lamps. Pieces of carbide about the size of your small fingernail go into
the bottom half of the lamp. The top half is filled with water. Then the
two halves are screwed together tightly and sealed with a rubber gasket.
A valve allows water to drip into the carbide chamber, one drop about every
two seconds. Acetylene gas is generated and escapes under pressure from
a nozzle or "tip" on the front of the lamp. The gas burns with a bright

yellow-white flame, and provides light directed as far as fifty feet by a 2.5- or 4-inch diameter reflector. The lamp is held firmly on your hard hat by a bracket, the hat is held firmly on your head by a chin strap, and thus light is provided wherever you turn.

Red closed his hand over the reflector to trap a pocket of gas. Then he struck his palm downward to spin the serrated wheel mounted on the reflector. Steel grated against flint, hot sparks flew out, and the gas burst into flame with a loud pop. Roger leaned over to light his lamp from Red's flame.

And so Roger and Red started along the classic route in Floyd Collins' Crystal Cave. First they walked down the steep stone steps in the sinkhole entrance. Once through the wooden door, they smelled the sharp odor of burning gas from their lamps mingling with the dank odor of clay and limestone. They walked the few hundred feet through a narrow passage to the steep trail down into the Grand Canyon.

"Come along, Floyd!" Roger shouted as they walked past the coffin at the bottom of' the Grand Canyon.

Red said nothing. He disapproved of levity with Floyd's ghost, but agreed that if Floyd wanted to come along, he had a right.

The main tourist route in Crystal Cave trails up and down through a long passage with dimensions up to thirty feet wide and high. Huge piles of breakdown blocks cover the floor, and gypsum crystals glitter on the walls and ceilings. Roger and Red walked quickly through this passage, having seen it many times before. Their minds were on the caves beyond.

After several thousand feet they came to a huge boulder that leaned across the passage, Scotchman's Trap.

"And why is it called Scotchman's Trap?" Roger asked gleefully.

"Because it's so tight," he answered himself.

After ducking under the boulder, Roger and Red paused to put on miners' knee crawlers—thick rubber pads, each held in place with two straps. Then they slid feet first about eight feet down through a hole off the side of the main passage. At the bottom they got on hands and knees to crawl a few tens of feet through a dusty passage to an intersection. There they turned right and crawled down the Crawlway 1200 feet to the Keyhole.

The Keyhole is about two feet wide, ten inches high on one side, and four inches high on the other. It is the only known way from the upper to the lower levels of Crystal Cave. The Keyhole is a tough physical and psychological dividing point between the commercial cave above and the wild cave below. Roger and Red slipped smoothly through the Keyhole, having almost—but not quite—forgotten what a formidable test it had been for them the first time through. It had receded in their minds as a difficulty because they had gone on to pass through so many similar passages deeper in Crystal Cave.

Two thousand feet beyond the Keyhole they came to a different kind of test, Bottomless Pit. They shuffled cautiously across a narrow, sloping ledge, barely glancing into the darkness below.

After Bottomless Pit, they came to Formerly Impossible. Here a buttress of rock bulged out into a pit some thirty feet deep. The passage goes on beyond, but the buttress had made it impossible to get around to it without fancy rope work. Then Bill had figured out that a well-placed charge of dynamite would dislodge a wafer of limestone about a foot thick and two feet wide from the buttress. This opened a horizontal slot, just large enough to slide through to the passage beyond. Because you are wedged in the slot, the traverse is probably less dangerous than it always seems to be.

Then there is the drop through the Crack in the Floor. It is a very narrow one. Roger wedged himself down into it, his chest being almost too large to fit. Then his head sank out of sight. He held on briefly, and then dropped several feet down into Floyd's Lost Passage. Red followed easily down through the hole. It had taken them about an hour to reach the huge passage that Floyd had discovered after years of exploring. It had been "lost" on Floyd's death in 1925, and was visited regularly again only after Jim Dyer and Bill started exploring the lower levels in 1948. Now, in 1955, it was difficult for Roger and Red to conceive of Floyd's Lost Passage as having once been the farthest limit of Crystal Cave. Yet, just the previous year a party had been lost for fourteen hours on the route from Scotchman's Trap to the Lost Passage. For some cavers this part of the cave could still be at the limit of endurance and difficulty.

Roger and Red did not pause. They hurried down the long crawlways

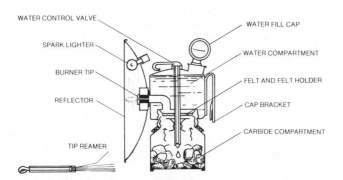

A carbide lamp produces acetylene gas by dripping water on lumps of calcium carbide. The carbide compartment is filled one-third to one-half full, the drip is adjusted, and the lamp is screwed together. Gas escapes from the burner tip. The spark lighter contains a flint, and when the lighter wheel is spun with the heel of the hand the gas ignites and burns with a yellow-white flame. The tip reamer contains fine wires to ream clogged burner tips; wires retract into the case for carrying.

of B-Trail to Bogus Bogardus Waterfall. Here they climbed down to enter a tight tube twelve inches high and two feet wide, Fishhook Crawl. This was very familiar to them because they had been on the parties that had discovered these passages.

Red pushed his head out into the ten-foot-deep Black Onyx Pit at the end of Fishhook Crawl. He reached far out for a handhold to swivel his body out of the narrow passage to slide down into the pit. Roger followed across the pit into another belly-crawl that led after a few feet to another pit, which they crossed on a ledge. Then for several hundred feet they walked and crouched, and sometimes straddled a canyon twenty feet deep. They walked across the Balcony over the black void of the Overlook. Ducking beneath a spray of water, they crouched along, soon to drop down into Storm Sewer, a long passage with a rectangular cross-section that starts as a walkway but becomes a 1000-foot crawlway leading down to Eyeless Fish Trail.

"I'm sure Bill and Jack are trying to connect Crystal with Salts Cave," Roger complained happily. He crawled down the lower end of Storm Sewer on his hands and knees, then rose to lope along on all fours like a bear. Red followed with a loose and rolling "four-footed" amble. They could sustain this pace for hours, but it had to be learned. All this talk of man's backbone being fundamentally designed for horizontal and not vertical orientation of the body, Red thought, might be true. But there is nothing like horizontal travel to expose the weakness of your back, and to disrupt the contents of your stomach, until you learn to hold yourself in. Have you ever tried to crawl the length of a football field? Try it some-time, on that nice soft grass. That is 100 yards. So far, Roger and Red had crawled 1000 yards, on sand and rocks, and there might be another 1000 in mud and water before they could turn around to crawl back. Or maybe they would connect with that "other" cave, and go with Bill and Jack out an entrance different from the one they had entered.

At 9:50 p.m. they were moving rapidly up Eyeless Fish Trail, walking in a stream between mud banks that slanted up ten feet to the stone ceiling. The water was cold.

Then they were stopped dead in their tracks by an alien sound:
Thud . . . thud . . . thud.
"Did you hear that?" they asked together.
"Get a rock!" Roger said.

Their hearts raced as they looked wildly around for a stone with which to pound back. There were only pebbles on the stream floor. They raced ahead, searching for a rock. Finally, Roger found one and slammed it against the ceiling.

Splat!

"Yech!" Roger said. He and Red clawed the mud that had sprayed into their expectantly uplooking faces. Then Roger scraped the layer of mud off the ceiling to reach bare rock, and pounded again. The limestone hammer was not effective. It shattered on impact. But whether Roger's pounding was getting through or not, the other pounding had stopped. Roger and Red listened, heard nothing, and hurried on.

Ducking under a low projection, they stood up again in the passage and started running upstream. Every few minutes they stopped, scraped mud from the ceiling, and pounded. Red collected more rocks to use as others crumbled under the blows.

"Why didn't we bring a hammer?" Roger asked.

They moved through the water, stooping, crawling, and running. After about 800 feet they came to a junction where a tributary stream trickled in. There they saw the initials "B.A." and "P.M.S." drawn in the mud. This was the farthest point of penetration by Bill Austin and Phil Smith on their discovery trip of the previous year.

BEAR-WALKING

CRAWLING

Red crawled up the tributary branch. "It goes as a crawl, but it's full of water," he said.

"Go on, go on," Roger insisted.

"Okay, okay," Red said. "I'm up on my hands and feet. In a minute I'll have a crawl in the water . . . oops."

Splash!

"All right, now I'm crawling, but it's a belly-crawl ahead. I can see only about six inches of air space." Red had stopped.

"All right, come on back," Roger said reluctantly.

Before turning, Red shouted up the passage. The water swallowed the sound. It would be a cold crawl, but it was clearly passable. And it could easily be the drain of a pit in the passage in that "other" cave where Bill and Jack were. Red shouted again. There was no answer, not even an echo in the dull passage. He could not pound because he could find no rock in the stream bed. There was one final thing he could do. With the flame of his carbide lamp he wrote the initials "R.W." and "R.B." on the wall where Bill and Jack would see them if they came through. Then he backed up to the junction where Roger was pounding.

The bed of the tributary stream was a fine, light-colored sand, with leaf fragments and acorns littering the bottom. The main passage was larger, almost high enough to stand up in, and seemed to offer the best possibility for connection, except for the fact that they had already traveled so far upstream in Eyeless Fish Trail.

"We've come maybe a thousand feet since we heard them," Roger said, "and Bill's map shows only about three hundred feet between that point and his pits."

"Could we have missed a side passage?" Red asked.

"It's always possible, but I don't think so," Roger replied.

They went slowly on up the crouchway. Now they were in virgin cave.

"Hey, pits!" Roger shouted.

The passage opened up. It was perfect. They had come into the bottom of a vertical shaft complex, surely one of the main inputs for the river in Eyeless Fish Trail. Roger picked up a solid piece of breakdown rock from the floor and pounded heavily. There was no reply.

An hour later they were dejected. They had climbed carefully up thirty feet over knife-edged flutes of limestone that had been sharpened by falling water. There were several pits in the cluster, but they had found no way out of them, and no sign of Bill and Jack. They continued to pound periodically, although it was well past 11:00 p.m.

They gave up at last. They left dye-marker bags in the bottoms of pits where an explorer might look down and see them, or find them if he came in from some unnoticed side passage. They also left a note describing where they had searched. Then they began the long haul back toward the entrance of Crystal Cave.

Roger's light had been dim for some time, and just before they were about to climb the mud bank out of Eyeless Fish Trail at the junction with Storm Sewer, his flame sputtered out. On a projection of mud and stone that bridged the stream they stopped to change their carbide. They were tired, and disappointed that they had not connected Crystal Cave with that "other" cave.

"Why did we hear only three thumps?"

"Maybe the time was wrong."

"No, it was right," Roger said, glancing at his watch. "Bill and I synchronized our watches. We heard those thumps just about ten o'clock."

"Maybe they were back the other way, downstream, and we ran on upstream out of range."

"Maybe. It doesn't seem likely, though. The map showed Bill's passage upstream."

"I don't suppose we imagined it?"

"I heard it clearly, didn't you?"

"Yes. It stopped us cold."

They thought about it for a while.

"I don't really think we could have missed a side passage," Roger said, "but there might have been a little hole out of the ceiling in the pits."

"Yes, but then we should have heard more."

"I know. Well, acoustics are funny in caves. Maybe conditions were just right where we heard those thumps, but nowhere else."

They had exhausted the topic.

"Damn!" Roger said with annoyance. He held up his lamp. As the gas burns, carbon tends to clog the hole in the tip. Roger had thrust a strand of wire through the hole to ream it out. The wire had broken off in the tip, sealing it shut so no gas could pass through.

He worked silently, trying to extract the piece of broken wire.

"My turn," Red said impatiently. Roger handed over the lamp. Red pried the tip itself out of the lamp. He peered at the small piece of brass-bound porcelain, about a quarter of an inch in diameter and half an inch long.

"Don't drop it, or we *will* be up a creek," Roger warned.

The wire seemed to be permanently emplaced. The tip, and thus the lamp, were useless.

"What else have you got for light?" Red asked.

Roger rummaged through his bag, turning out two stubs of candle and a small, rusty flashlight. The beam of the flashlight was dim, good for about half an hour. Even at top speed, they were four hours from the entrance.

"Well, we can do it with one lamp, but some of the climbs will be interesting," Red said. "I suppose you can carry a candle part of the way."

"What have you got for extra light?" Roger asked.

"Nothing but some candles," Red replied. "I never thought a flashlight was worth its weight to carry, it burns out so quickly."

"Okay," Roger said, "we can do it easily enough on one lamp." He left unspoken the admonition that they had better be careful with that one lamp. Their safety factor had been more than cut in half by the loss of one of their two lamps. If they lost the other, despite the candles, they would probably just have to wait until Bill and Jack came looking for them. The embarrassment and loss of confidence would be tougher to take than the discomfort of waiting in the dark cave.

"Let's get moving, I'm getting chilly sitting here," Red said.

"Eat something first," Roger said. "We've been in here for eight hours with nothing but a couple of candy bars, and these early morning hours are worst on a long trip. It'll lighten our loads, too."

Taking out his food, Red came across something in his pack that he had carried without using on so many trips that he had forgotten it. He pulled out a heavily taped cylinder about an inch in diameter and three inches long. Cutting into it, he revealed a spare-parts kit for a carbide lamp. There was a new tip in it. Red looked up to see Roger watching. Their eyes met and Roger nodded without comment.

When they had finished their canned meat and fruit, Roger took the spare tip and carefully pressed it into the gaping hole in the front of his lamp. He adjusted the water drip, screwed the two parts of the lamp together, and felt for the pressure of gas escaping out the new tip against his tongue. He spun the steel on flint.

The passage was suddenly bright with new light. Roger and Red looked at one another with uncontrollable glee. They closed up their bags and started out of the cave. Sixteen hours after they had entered, they emerged into early morning sunshine, to trudge up the hill to the Ticket Office.

"Where were you?" Bill asked, looking up from some paper work. He was clean and neatly dressed, and looked well rested, a complete contrast to the two tired and muddy cavers he was questioning.

"We heard three thumps," Roger said.

"What time?" Bill asked.

"Ten to ten," Roger replied.

"Hmm," Bill mused.

"You *did* pound, didn't you?" Roger asked.

"Sure. Did *you* pound?" Bill asked in turn.

"Sure," Roger answered. "Did you hear us?"

"Nope," Bill replied. "Too much water falling. We quit pounding at ten after ten."

They commiserated with one another about the difficulty of communicating in caves. Something always seems to go wrong. Sometimes individuals who on the surface are reasonable, precise, and reliable will leave notes underground that are classics of ambiguity. A typical example is: "Have gone on out." That one was left at a junction of three passages, making it entirely unclear whether the explorer had taken the right, left, or front passage, or, possibly, gone out of the cave itself. Another gem is: "Meet me here in one hour." That was found by explorers several years after it had been placed by a person or persons for a person or persons still unknown.

"Never mind," Bill said. "What did you see?"

He listened carefully to Roger's description of the pit complex, but quizzed Red most closely about the tributary passage containing leaves and acorns.

"The drain of the pit we pounded in," Bill said, "had a light-colored sandy floor, and there were oak leaves and acorns in the water."

"How high was the water?" Red asked.

"It would be a wet one," Bill replied, breaking into a grin. "You'd get the family jewels wet on that one."

Bill's quiet pleasure cheered Roger and Red.

"Why didn't I crawl on?" Red asked, as he and Roger walked back to the Guide House to change clothes. "I bet that's it."

"Yeah," Roger said exuberantly. "Next trip we'll make the connection."

Two weekends later, Dave Jones and Roger went to Flint Ridge to test a radio transmitter Dave had developed. Bill and Jack went into that "other" cave while Dave and Roger remained on the surface. The transmitter did not work, but they had arranged to do some pounding as well. This time Bill and Jack could hear Roger, who was pounding on the surface, but Roger and Dave could not hear Bill, who was pounding underground.

On the way out of the cave, Bill and Jack dropped down into the pit where they had been pounding two weeks before. They crawled through water one to two feet deep for 100 yards in a passage that was never more than three feet high. They came to a fork and turned left. Within fifty feet they found the initials "R.W." and "R.B." They added the initials "J.L." and "B.A." and an arrow with a point on each end, plus the exclamation: "WE DID IT!" Then they went out Eyeless Fish Trail, Storm Sewer, the Overlook, Fishhook Crawl, B-Trail to Floyd's Lost Passage, and up to the Crawlway through the Keyhole to Scotchman's Trap and on out the entrance of Floyd Collins' Crystal Cave, where they greeted Roger and Dave with beaming faces.

The first connection had been made, but it was not with Salts Cave as Roger had thought it would be. Bill and Jack had entered Unknown Cave, a nearly forgotten hole in Flint Ridge about midway between the Crystal Entrance and the Salts Entrance. It was in Unknown Cave that Bill and Jack had discovered miles of virgin passages on those long winter nights. Now Unknown Cave and Crystal Cave were connected.

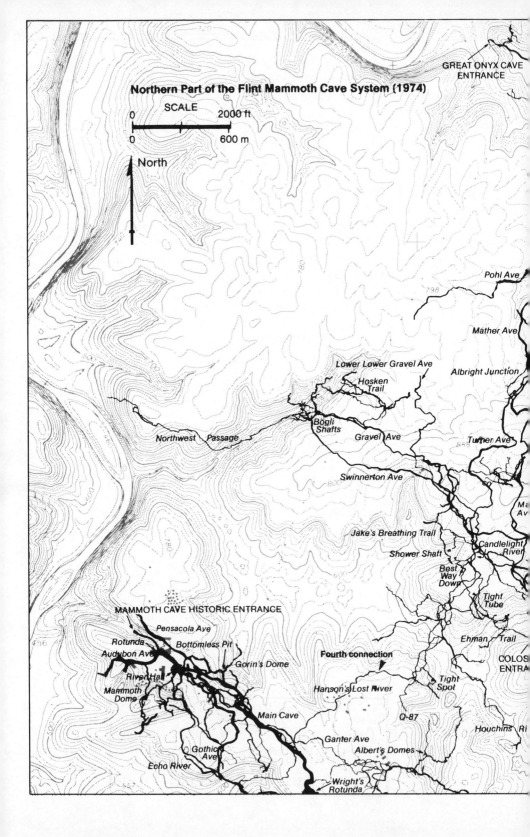

Northern Part of the Flint Mammoth Cave System (1974)

SCALE

0   2000 ft

0   600 m

North

GREAT ONYX CAVE ENTRANCE

Pohl Ave

Mather Ave

Lower Lower Gravel Ave

Albright Junction

Hosken Trail

Bögli Shafts

Gravel Ave

Turner Ave

Northwest Passage

Swinnerton Ave

Ma Av

Jake's Breathing Trail

Candlelight River

Shower Shaft

Best Way Down

Tight Tube

MAMMOTH CAVE HISTORIC ENTRANCE

Pensacola Ave

Rotunda

Bottomless Pit

Ehman Trail

Audubon Ave

Gorin's Dome

Fourth connection

COLOS ENTRA

River Hal

Mammoth Dome

Hanson's Lost River

Tight Spot

Main Cave

Q-87

Houchins Ri

Gothic Ave

Ganter Ave

Albert's Domes

Echo River

Wright's Rotunda

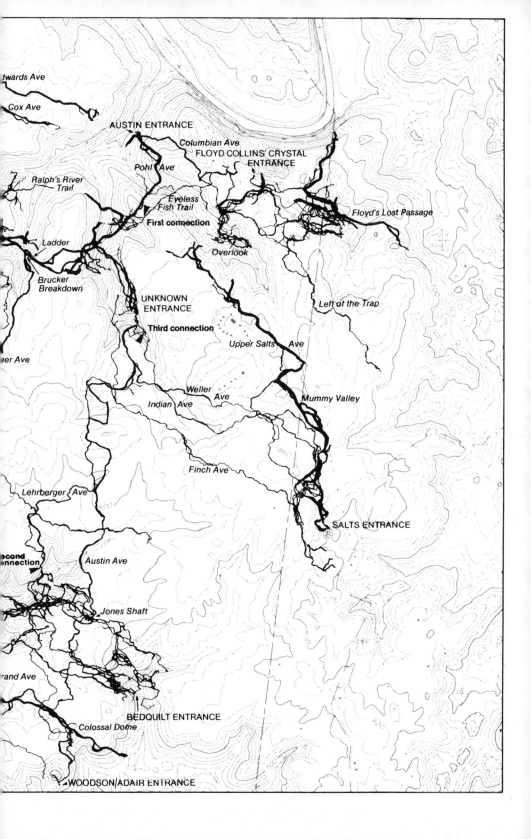

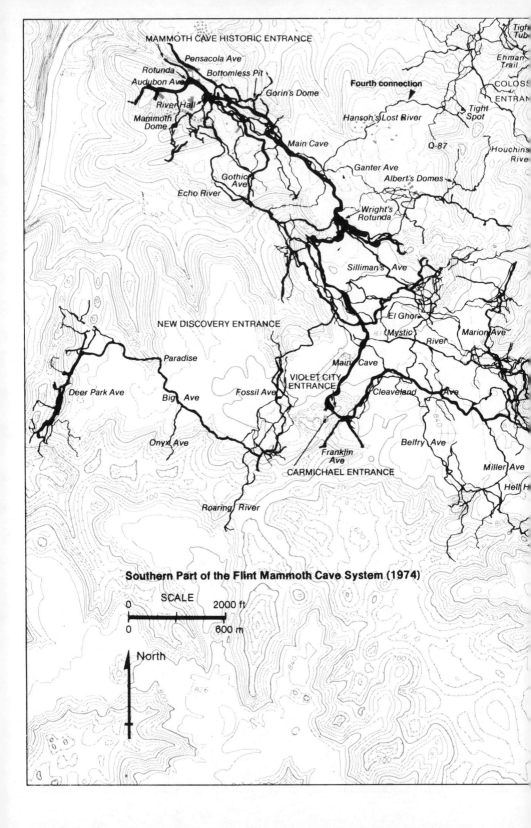

MAMMOTH CAVE HISTORIC ENTRANCE

Pensacola Ave

Rotunda

Audubon Ave

River Hall

Mammoth
Dome

Bottomless Pit

Gorin's Dome

Fourth connection

Hanson's Lost River

Main Cave

Gothic
Ave

Echo River

Ganter Ave

Albert's Domes

Wright's
Rotunda

Silliman's Ave

NEW DISCOVERY ENTRANCE

Paradise

Deer Park Ave

Big Ave

Fossil Ave

Onyx Ave

El Ghor

Mystic River

Main Cave

VIOLET CITY
ENTRANCE

Cleaveland Ave

Marion Ave

Belfry Ave

Miller Ave

Franklin
Ave

CARMICHAEL ENTRANCE

Hell H

Roaring River

Tigh
Tub

Ehman
Trail

COLOSS
ENTRAN

Tight
Spot

Q-87

Houchins
River

Southern Part of the Flint Mammoth Cave System (1974)

SCALE

0   2000 ft

0   600 m

North

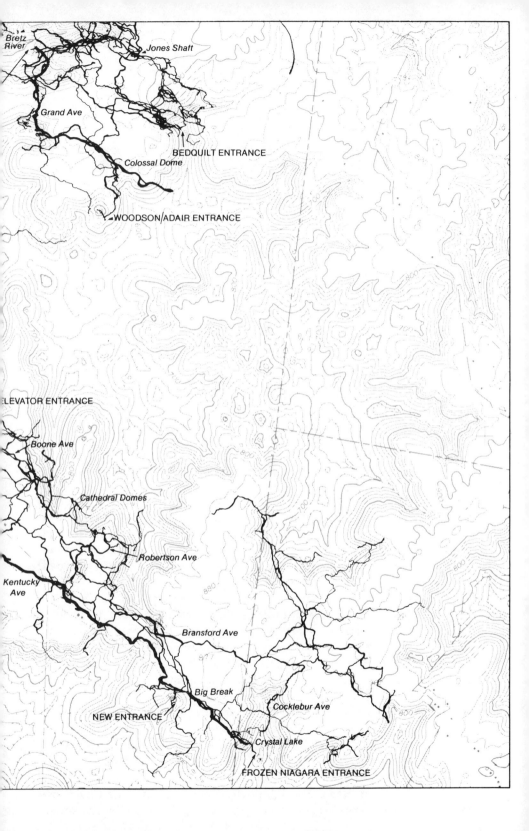

The Flint Ridge Con

Jim introduces Roger
to the Kentucky caves and to
the idea of connection.

2 Two years before that first Flint Ridge connection, Joan and Roger Brucker were adapting to life with Tom, their three-month-old son. He was a sociable baby who could be carried around like a football. So they packed camping gear and drove to Mammoth Cave National Park in Kentucky to attend the field trips of the April 1953 National Speleological Society Convention. Arriving late at night, they pitched their tent in the campground and went to sleep, despite Tom's being cheerfully wide awake.

The next day Roger went on a field trip guided by Jim Dyer, who by then had retired as manager of Floyd Collins' Crystal Cave. They drove to a churchyard, where they parked, walked across a graveyard, and proceeded to a small cave on the Northtown Ridge, northeast of Crystal Cave and Mammoth Cave National Park. Jim was a lean, clean-cut man in his early forties, and although not a native, he spoke with a faint Kentucky drawl. There was something sly and conspiratorial in his voice.

Jim ushered his charges to the cave with a steady stream of extravagant talk and just a bit of a leer in his eyes. Roger had heard the baloney of cave guides all his life, but there was something more here. It clicked into place when he learned that Jim had worked in a carnival. When Jim invited someone to enter a cave, the innuendo in his voice suggested that the delights inside might be more than a strong man could resist.

Jim took the group into the cave's walk-in entrance. This little Kentucky cave had a passage twenty-five feet wide and ten feet high. After 300 feet it ended in a massive light-brown wall of flowstone that resembled a waterfall. Several members of the party scrambled up over it to check for leads, but there were none.

On the way out, Roger asked Jim why a cave like this one would just end after such a short segment of large passage. Jim's eyes narrowed, and lines squinched out at their corners. He lowered his voice and confided, "Maybe it does go on. What we need are some experts to really look at these things."

Of course, Roger thought, an expert is just what is needed. He made his firm, expert way out of the cave.

Thereafter on other field trips Roger was in the forefront of those who scrambled around at the end of each cave visited, verifying the conclusion that the cave did not go on. He nodded knowingly when the young man who had dug into Cub Run Cave pointed out that he had known there was a cave behind the limestone cliff because a spring emerged from it. Roger recognized that Short Cave must be a part of a much larger underground drainage system, and he watched with envy and approval some cavers who were digging through a pile of breakdown blocks that closed a passage.

One evening they assembled at Mammoth Onyx Cave for a private showing of Kentucky's most colorful cave. Mammoth Onyx Cave lived up to the description in a promotional brochure that contained an odd combination of ballyhoo and science. As Roger walked out of Mammoth Onyx Cave in the early night darkness, he was caught by the glassy eye of a slight man in a crisp, short-sleeved white shirt and bow tie who was staring out abstractedly from under a bare lightbulb in the Mammoth Onyx Cave Ticket Office.

Dr. E. Robert Pohl acknowledged Roger's presence with a barely perceptible nod, not enough to include him, but enough not to discourage. Dr. Pohl had been Jim Dyer's employer when Jim was manager of Crystal Cave. Roger had passed from the hands of the tent-door barker into the inner sanctum. He was standing in the presence of the medicine man himself.

"This is where they'll be tonight. I expect they'll be there all night." Dr. Pohl tapped the map with a well-manicured, bony finger. "Bill Austin, Jim Neidner, and perhaps several others. They'll go in by way of Scotchman's Trap," he continued, moving his finger with a slow sweep along some braided lines of a map of Crystal Cave. "Then they'll double back beneath the Crawlway. They'll go to Bottomless Pit to explore." Dr. Pohl indented the map with a sharp fingernail and turned his glittering but expressionless gaze onto Roger. "Perhaps you could join them," he said.

Roger was mesmerized. It was like discovering that you could make love in the afternoon. Caving at night! In the year he had been exploring caves it had not occurred to him that one might explore at night. Cavers went into the caves in the daytime. In the evenings they sat around campfires, and slept at night, like normal people. At that time it was inconceivable to him that a cave could be so large that one would cave at night because it might take all the next day to get out.

Roger did not go on that long trip in Crystal Cave that night. He met Bill Austin for the first time the next day. As manager of Floyd Collins' Crystal Cave, Bill was all business as he conducted the National Speleological Society tour group down to the Grand Canyon in Crystal Cave.

"Is it really true that this is the largest underground room in Kentucky?" Roger whispered to Jim Dyer, who had appeared at his elbow soon after he had entered the cave.

"Nah," Jim said, "but we've got more unexplored passages in Crystal Cave than any other cave in Kentucky."

Jim showed Roger the tiny hole that was the start of the fabulous "Sixteen Hour Trip," a belly-crawl through hundreds of yards of gypsum crystals. Roger found it hard to imagine spending sixteen hours crawling in a cave. How many candy bars would you need for such a trip?

The next day Roger and Jim, by now old friends, climbed 100 feet down wooden stairs to see the "river" in Great Onyx Cave—another privately owned commercial cave in Flint Ridge surrounded by Mammoth Cave National Park. They came to a wooden dock built over a pool of water. A trickle of water pattered down one wall, but there was no sign of an exit for the water. It must go somewhere, Roger thought, for the pool was only about eight feet wide and thirty feet long. Jim said he thought the pool had been dammed artificially.

Back in the main passage, Jim suggested that Roger look carefully along the walls. Great Onyx Cave was in the northern part of Flint Ridge, not very far west of known passages in Crystal Cave. Perhaps a connection between these two Flint Ridge caves could be found.

"See anything funny?" Jim asked.

Roger strained to give an expert answer, but he could see nothing that looked funny. Finally he said, "Well, it's hard to see. They've piled up sand and rocks against the wall in making the trail."

"That's it, boy!" Jim said, slapping Roger on the back with delight. "Even if there were a side lead, how would you ever find it?"

Roger walked on again, thunderstruck. These people in Kentucky had so much cave they could close off passages without even thinking of it. Back in New York, people would dig for weeks to go into a muddy hole fifty feet long.

Interrupting Roger's thoughts after a long silence, Jim said, "Maybe they don't *want* those passages explored." Roger nodded. It certainly looked to him as though there ought to be a connection between Great Onyx Cave and Crystal Cave.

As they continued to walk through Great Onyx Cave, Roger mulled over a question he was bursting to ask. When the National Speleological Society group was being taken through Mammoth Cave, he had heard numerous wild tales. The wildest was a whispered comment that Jim Dyer was known to have gone into Crystal Cave and come out in Mammoth Cave. He was said to be the best cave explorer since Floyd Collins. Finally, Roger asked Jim about it.

"Nope!" Jim said, grinning broadly as he put on the conspiratorial air that was the essence of the man. "I've heard that story myself. It hasn't happened yet, but it might be true someday."

By this time Roger was far from being the expert he had been earlier in the week. Now he was a disciple. He wanted to be a caver like Floyd Collins, Jim Dyer, and Bill Austin. He, Roger Brucker, wanted to explore and connect these big Kentucky caves.

Jim Dyer, 1953.

The Rinky-Dink

Bill takes the Ohio cavers on a run-around
that makes them dizzy with the vastness of Crystal Cave.
They are filled with the desire to explore it.

3 Phil Smith had also been at the April 1953 National Speleological
Society Convention that had inspired Roger. Phil, however, was not
surprised at what he learned in Kentucky. With a childhood friend,
Roger McClure, he had been exploring caves since high-school days. In
1952 they had hitchhiked from their home in Springfield, Ohio, to Kentucky to see Crystal Cave. In Columbus, Phil and Roger McClure had
started the Central Ohio Grotto of the National Speleological Society. Jim
now arranged for Phil, Roger McClure, and Roger Brucker to take a trip
on the 1953 Thanksgiving weekend with Bill Austin in Crystal Cave.

Jim was shaky with flu, and had a sleepless night. He accompanied
them down the commercial route of Crystal Cave until they came to the
duck-under at Scotchman's Trap. Here Jim said, "You boys just go along
with Bill. He'll take care of you."

Bill nodded, and plunged down into the hole. Without time for
thought, Phil and the two Rogers followed. Bill was out of sight, but they
could hear him moving down the Crawlway ahead as they scurried along.
The abrupt start had them out of breath already. Before they had gone
1000 feet, the Velveeta cheese and crackers they were carrying were
crushed and impregnated with sand. This food seemed inedible already,
but there had been talk of a twelve- or fourteen-hour trip, so they clung
doggedly to the crumbling mess.

Bill paused long enough at the Keyhole to allow them a view of his
body entirely filling it, and then they heard him rocketing down the
passage on the other side. They had to follow immediately if they were
to keep up. Bill was about the same size as Phil, but seeing Bill at the
Keyhole, Phil realized that his own shoulders were probably wider than
Bill's. They were, but before he had time to worry, Phil had scraped

through the ten-inch-high hole. On they went, taking the upper and then the lower of passages that diverged vertically, the right and then the left of passages that diverged horizontally. The neophytes looked around rapidly at every intersection, remembering the first maxim of caving— that a passage always looks new to you when you go back through it from the other direction. Down, around, up, over, back, under, and on. They were not sure that they could find their way back on their own.

Bottomless Pit is reached through a crouchway that leads to a high canyon. They could feel the open space out there. Below is sixty feet of blackness. A large boulder rests on the edge of Bottomless Pit. Bill stood waiting for the Ohio cavers between the boulder and the right wall. When the three could see him, Bill crossed over.

He stepped onto a mud-slick ledge a foot or so wide that slants thirty degrees down into the pit from the right wall. In the same movement, he leaned over the open pit to place the palms of his hands against the opposite wall, three feet away. Without pausing, he shuffled sideways along the ledge, ten feet to the other side. There he scrambled on hands and knees six feet up a steep slope to a crawlway that leads away from the pit.

Jim Dyer first crossed Bottomless Pit this way with Luther Miller in 1948. In 1953 there was a thin steel cable across the top of Bottomless Pit at hand level along the wall opposite the ledge, but it was rusted and looked insecure. Phil held it under the palms of his hands as he pressed against the wall while crossing the pit. What if he slipped off to hang on the cable over the pit? It is a long drop to the floor. Phil wondered whether —if he slipped—he could climb hand over hand on that thin cable with knuckles scraping against the rough limestone wall. Following quickly in Bill's footsteps, Phil and the others went across.

Safely in the crawlway, they doubled back down a slope to the bottom of Bottomless Pit. The passage out of the pit led between two enormous limestone blocks.

"You're meat in the sandwich," Bill said as they slid through.

Then they came to the brink of a pit about fifteen feet deep.

"Watch carefully," Bill said. He sat on the edge of the pit and then turned over with his stomach to the pit wall. Then he slid down backward into the pit until he stood on a tiny nubbin some five feet down on the wall of the pit. The side of his face against the wall, he bent at the waist and stuck his rear out into the pit, lowering his upper body while maintaining friction with his hands on the wall. Phil could not imagine what Bill was going to do next. The bottom of the pit was still at least ten feet below Bill's feet.

"This is where you put your hands where your feet are," Bill said tersely. He moved his feet backward just as his fingers reached the nubbin he had been standing on, and slid down with the toes of his boots scraping the wall until he was hanging by his hands. Then he dropped

The method of putting your hands where your feet are to climb down Bottomless Pit.

the remaining few feet into the pit. Phil and the two Rogers stared with disbelief, but each performed the bold maneuver exactly as Bill had shown them. They wondered how they would get back up.

The next obstacle was the Crack in the Floor. Again, Bill had arrived well before them, and he was already jammed halfway down into the Crack. It was only eight to ten inches wide. With a wild grin, Bill worked his way down, his body jerking sideways back and forth as he slipped out of sight. This was not the only way to Floyd's Lost Passage, he said, but it was the quickest.

"Come on, come on," a muffled voice rose with laughter up through the Crack.

Phil would not fit through the Crack. So he and the two Rogers followed Bill's instructions on how to go through an alternative route to Floyd's Lost Passage.

When they arrived, they found Bill very much at home. Now there was no haste. The Ohio boys had been angry at Bill, but now they found him to be communicative and friendly. Ignoring their crackers and cheese, he prepared a spaghetti dinner for them, the finest meal they had ever had in a cave. Then they looked at gypsum flowers and walked the length of the mile-long Lost Passage. Bill showed it with pride. They forgot the annoyances of the trip in, even forgot that none of them could possibly have memorized the way on that fast trip. It did not matter. They were not going back that route anyway.

They followed Bill up a muddy avenue, and listened in awe as he showed them Floyd's Jump Off. However, they did not climb the pile of rocks Floyd had collected so he could get back up into the passage out of which he had rashly jumped. Bill wanted instead to go to a place that had not yet been explored. He started chimneying up the wall opposite Floyd's Jump Off. Sixty feet up, he paused for a moment on a ledge, and then jumped lightly three feet across the gap into a passage that opened in the other wall. Bill crouched there to secure the others as one by one they made the leap. Later, they noticed that Jim Dyer turned white when Bill told him he had taken them that way. However, Bill had watched them carefully, and the thrills of fear the Ohio cavers had felt merely added spice to their confidence that they could do it.

They explored awhile, but now, after nearly fourteen hours in Crystal Cave, the new Flint Ridge cavers were beginning to tire. Bill seemed ready to go on as long as they would follow. He was already suggesting another place that they might explore.

The word "rinky-dink" cropped up in Phil's mind. Were they being given a run-around? He looked sharply at Bill. Bill's face often had a stony expression, but he seemed continually amused at the antics of the Ohio cave explorers. It was obvious that Bill had a wild sense of humor. This appealed to Phil. It was a great trip.

Roger Brucker had also been thinking deep thoughts. First it was Jim Dyer, and then Dr. Pohl, and now here sat the third magician, Bill Austin, who smiled when the going got tough, but was dead serious over trifles. Bill was making dry comments as the three ate raisins from small, crushed boxes. It was an initiation ceremony. A sorcerer and three apprentices sat in the cold yellow light of carbide lamps, deep in Floyd Collins' Crystal Cave.

Roger McClure was painfully aware of the spectacle he and the two other Ohioans made. They were desperately tired, and their clothes had been torn to shreds in the tight crawlways. He tried to make more saliva so he could lick a thin layer of dirt from his lips. Bill sat relaxed, neat and unwearied in tight-fitting Levi's pants and jacket, ready to show them more of Crystal Cave. Bill fingered a shoulder bag. He had been able to carry not only his carbide, water, and spare gear, but also the ingredients for the spaghetti dinner in that bag without difficulty.

"I'm going to get one of those bags," Roger McClure muttered.

Somewhere they had got back onto the route by which they had entered the lower regions.

"Now that we're back where you've been," Bill said, "you lead."

Phil started ahead. He *did* recognize the passage, and he remembered that it was only about 1000 feet long. He was not sure where they had gotten back into it, but he thought the Keyhole had to be coming up soon. The Keyhole was obscure, but he found it easily. It was not far from there to Scotchman's Trap. He began to estimate the distance they were traveling. The Crawlway went on and on. When he was certain that they had gone too far, Phil stopped. Where was Scotchman's Trap? Looking

back, he saw Bill's questioning face. Phil
went on. When he stopped to look back a
second time, he saw that the two Rogers
were with him, but Bill was not in sight.
Rather than trying to pass one another in
the narrow passage, each caver turned
around in place, with Roger McClure now
in the lead and Phil bringing up the rear. They started back.

"You fellows lose something out this way?" Bill asked when they
reached him.

Roger McClure grunted and looked around carefully, particularly right
under the ledge on which Bill was sitting. There was nothing there, so
he continued to crawl back down the passage. Bill dropped in line behind
Phil. After 100 or so feet of backtracking, Roger McClure found an opening in the wall. It is obvious, if you are looking for it. On the way in they
had been so rushed that they had not noticed—and Bill had not bothered
to tell them—that the Crawlway is as big to the left as it is to the right
where they had turned to follow Bill after entering the Scotchman's Trap
passage. There were no obvious signs, and it was clear to them now that
to find the way, you simply had to know where you were.

Bill said nothing. This was his way of teaching. He provided the situation and let others draw the obvious conclusions. Although a sign could
easily be put up so that no one would miss the Keyhole, for example,
there were too many junctions for signs. In a big cave, you cannot depend
on artificial signs. You have to learn to recognize the natural signs so that
you can maintain a sense of where you are, just as a woodsman finds his
way around in a dense forest. At first a cave—like a forest—looks all the
same, but if you pay careful attention, you can notice thousands of subtle
differences from passage to passage. Bill had taught the Ohio cavers
a lesson: To explore in Floyd Collins', you simply had to learn the
cave.

Roger McClure remembered the short crawl to Scotchman's Trap.
Soon they all stood with relief on the tourist trail in the upper levels of
Crystal Cave.

It had been a classic Flint Ridge rinky-dink. When the three walked
into the Grand Canyon, Bill, who had gone on before, sprang up from a
bench and bounded up the steep slope toward the entrance. "Just lock the
door behind you," he shouted as he strode out of view.

Bill's cavalier exit was the last straw. Roger Brucker said he did not
want to go into any cave again. Roger McClure thought it was bad policy
to run off and leave the party, even this close to the entrance. Phil was
breathing heavily, stumbling with fatigue. But beneath these feelings in
each of them was a surging excitement and joy. They were filled with the
mysteries of Floyd Collins' Crystal Cave. They had never had such an enjoyable trip as Bill's rinky-dink. They would be back to explore Crystal
Cave.

Attempt to Extend Crystal Cave

National
Speleological Society
cavers try to extend
Floyd Collins' Crystal Cave.
Lessons from the C-3 expedition lead to new techniques.

4 Joe Lawrence, Jr., and some of his caving friends from Virginia were also introduced to the lower levels of Crystal Cave by a Bill Austin rinky-dink. Joe was now planning to lead a National Speleological Society expedition into Crystal Cave in February 1954, with Bill as assistant expedition leader. Jim Dyer had agreed to be exploration leader. A number of other veterans were going to be on hand. Phil Smith, Roger Brucker, and Roger McClure were invited. Nothing could have kept them away. If Joe's explorers could extend Floyd Collins' Crystal Cave, the Flint Ridge cavers might find themselves in the longest cave in the world.

The story of the National Speleological Society C-3 (for "Collins' Crystal Cave") expedition of 1954 is told in *The Caves Beyond* by Joe Lawrence, Jr., and Roger W. Brucker. The expedition was modeled on European caving and mountain climbing. It took place at a time when adventure was king. Hillary and Tensing had just climbed Mount Everest, and the French were struggling with Annapurna. The voyage of *Kon-Tiki* was fresh as a modern seafaring saga, and Cousteau was telling fascinating stories of undersea adventure. In the Pyrenees, the French were going deeper into the earth than cavers had ever gone before. It was time for adventure underground.

Most of the planning for the C-3 expedition took place in Philadelphia and New York, far from Crystal Cave. The C-3 expedition grew like a rolling snowball until the final call was for thirty people above ground to supply another thirty people who would stay underground for a week. Joe assigned a crew to install phone lines to Floyd's Lost Passage. Arrangements for photography and publicity took hours of long-distance telephoning. There were endless problems of supply, finance, and personnel. When the expedition finally took to the field in mid-February of 1954, Joe was

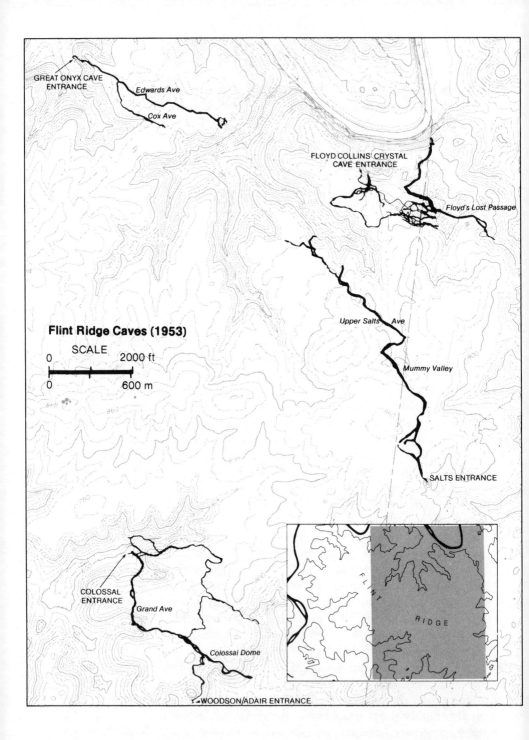

Map of Flint Ridge caves as they were known in 1953.

nearly worn out before the main force had entered Crystal Cave. A sleeping pill was prescribed, and after ten hours of sleep Joe was ready again to take command.

For a couple of days people dribbled down the long crawlways into Floyd's Lost Passage. They were laden with sleeping bags and supplies packed into torpedo-shaped sheet-metal cans, invented by Russ Gurnee. Telephone parties laid wire and made circuit hook-ups. When Jim Dyer arrived in Floyd's Lost Passage, he began assigning people to exploration parties.

Joe's strategy for finding new passages was: Look everywhere. When a passage goes, survey it. Bill and Jim translated this into: Explore out B-Trail and Bogardus Waterfall Trail. These were places Bill and Jim knew had potential because they had explored them themselves. These passages also led toward Salts Cave, with its huge main passage a mile to the south of Crystal Cave in Flint Ridge.

Jim sent Jack Lehrberger, Russ Gurnee, and Roy Charlton out as the spearhead. After twelve hours they reported that they had found a new connection between B-Trail and Bogardus Waterfall Trail. Surveys of these passages were begun.

One day a supply party, traveling without a guide, got lost in the crawlways on the way to Floyd's Lost Passage. The normal three- or four-hour supply trip stretched into a fourteen-hour nightmare. The mental strain of being lost was harder on some members of the party than the physical rigors of the cave. It was a reminder that even those passages that were beginning to be routine to some of the cavers imposed a barrier for others.

Phil Smith had been assigned to the supply force. He and other pack-horses made six or seven supply runs from the surface to Floyd's Lost Passage in the lower levels of Crystal Cave, hauling Gurnee cans full of canned goods, rice, oatmeal, raisins, and chocolate bars. By the end of the C-3 expedition, Phil was frustrated because he had not been able to participate in any of the exploration.

Roger McClure had better luck. First he had helped survey around Floyd's Lost Passage. Then he went on some exploring parties. But he, too, felt disappointed, because he had been disoriented in the big cave system.

Roger Brucker was assigned to cook meals in Floyd's Lost Passage for supply teams and explorers. When he decided that it was time to get some sleep, he found that all the sleeping bags were taken, so he joined a climbing party. From then on—to his joy—it was assumed that he was a caver, not a cook, so he helped Roy Charlton and Roger McClure survey the Bogardus Waterfall Trail. The survey data were phoned to the surface, where they were plotted, and later a map was returned to the cavers. It showed a braided set of lines—"like copulating snakes," Russ Gurnee said—but there was no tie to other mapped passages in Crystal Cave. The new map was just a set of lines floating unattached on a sheet of paper.

About halfway through the C-3 expedition, Earl Thierry, Roy Charlton, and Roger Brucker got together to grumble about how little they were finding out about Crystal Cave. Despite their surveying, they had no understanding of how one passage related to another in the total cave pattern. Roger pulled out his photostat copy of the topographic map that showed the contours of the Flint Ridge surface terrain overlying Crystal Cave.

"Where are we?"

"Who knows? Crystal Cave."

Roger had seen *the* map of Crystal Cave just before the C-3 expedition. Bill had unrolled a vellum plot of survey stations connected by straight lines. There was neither scale nor north arrow on that map. What Roger had seen was of no help now.

"Here," Roy said. He drew three parallel lines about two feet long in the sand on the passage floor. "Luther Miller described Crystal Cave to me this way: The top line is the commercial route to Scotchman's Trap and beyond. The second line is the next level down, the Crawlway. The bottom line is the lowest level, Floyd's Lost Passage." Roy poked his finger at a place one-third of the way along the Lost Passage line, and drew a perpendicular through the other two lines. "That," he said, "is the C-3 Waterfall Route by way of C-Trail, connecting the three levels."

They knew that Luther had helped Bill survey Crystal's lower levels, so they trusted the sketch. But what did it show?

"We ought to just survey our way out of the cave," Earl said.

"That'd be a waste of time," Roger said, "even if we had the time and manpower to do it."

"What about the tourist map of Crystal Cave?" Earl asked. He unfolded a copy he had been carrying.

"It looks right," Roger said, "but Bill told me that it was inaccurate."

"It says right on it, 'This map was prepared from an accurate instrumental survey,' " Earl said.

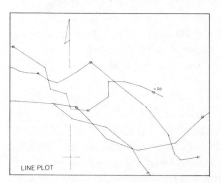
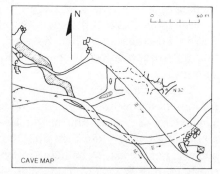

A line plot of a cave passage survey contrasted with a more detailed cave map from the same survey.

"It may have been prepared *from* an accurate instrumental survey, but how accurate is the prepared map?" Roger asked.

"I suppose if the cave did go outside Crystal Cave property under Mammoth Cave National Park land, Bill and Dr. Pohl wouldn't want people to know," Earl replied.

"Wait a minute!" Roger said, pointing to the map. "Look, here is Bottomless Pit. If we can survey to Bottomless Pit, we can hook our map to the tourist map. Then we would know where we are."

Suddenly the three complainers were conspirators. First they surveyed Floyd's Lost Passage, tying it into the initial part of the B-Trail survey. This showed them how the main passages fitted together. They knew Bill Austin had already surveyed Lost Passage, but they guessed that he probably would not give them access to the map.

Bill was irritated when he heard the news. "Why survey Lost Passage?" he asked. "We've already done it."

"I guess we didn't know," Roger said.

Then they surveyed from Floyd's Lost Passage along a passage that went within 300 feet of Bottomless Pit. Without telling anyone, they extended their survey over to Bottomless Pit. Now they were ready. Roy took the tourist map, compass, and tape up into the tourist route to figure out the map's scale and orientation. Earl took the expedition survey notes home to plot at 100 feet per inch. Roger enlarged the tourist map to the same scale. When these maps were put together, the conspirators would know more than anyone else on the expedition about how the passages of Crystal Cave were connected with one another. They might know even more than Bill Austin himself.

Weeks went by. Finally, Roger received the C-3 expedition map from Earl. Roger slid the translucent drawing of the upper-level tourist route over that of the lower-level passages and gazed at the result. He was at last beginning to know Crystal Cave. Then he enlarged the topographic map of the surface of Flint Ridge above Crystal Cave to the same scale, and traced all three maps together on one piece of vellum. The composite map showed a variety of surface and underground relationships.

Roger could see clearly that the piles of breakdown blocks that terminated upper-level passage segments occurred beneath surface valley walls. Deepening surface valleys had cut down through these passages, dividing them and leaving piles of breakdown blocks to terminate the passage segments. Underground vertical shafts—as Dr. Pohl had pointed out—were always located beneath the edges of valley walls where the beds of impermeable sandstone and shale that protect the ridge tops had been eroded away to allow water to seep down. This seeping water had dissolved out the cylindrical vertical shafts, sometimes to diameters of as much as forty feet and heights of over 150 feet. They provided connections between different horizontal levels. Just by putting the cave map and the topographic map on one sheet, Roger had illustrated the solutions

*Part of the survey route to the Bottomless Pit
during the C-3 expedition.*

to some geological and exploration problems, but had raised many other questions.

In April 1954, the National Speleological Society Convention was held in Pittsburgh. Roger took his new map. Bill looked at it carefully in silence. A muscle twitched in his jaw. Then his expressionless face broke into a grin.

"Okay," he said. "You win."

Roger heaved a sigh of relief.

"But I'm disappointed that you guys did a bootleg survey," Bill said. "Why didn't you just ask me, if you wanted to see the map?"

"We didn't think you would show it to us," Roger replied.

Bill laughed. "You may be right."

Then Bill explained why the Crystal Cave map had to be kept secret. If the National Park Service knew of the existence of any extensions of Crystal Cave beyond the private-property boundary lines into Mammoth Cave National Park, they might prohibit trespassing under Park land from the Crystal Entrance. This might stop the exploration of Crystal Cave.

"None of us wants that," Bill said.

Roger realized that Bill had said "us." Bill was including Roger in the inner circle.

"We certainly want to continue the work," Roger said. "But aren't you exaggerating about the Park Service?"

Bill told him that almost immediately after Ed Comes had plotted the B-Trail and Bogardus Waterfall Trail surveys during the C-3 expedition, one member of the team, Bill Stephenson, had delivered a copy of the map to the Park Superintendent. The Superintendent had looked at the blue-line dittoed map of these lower-level passages and asked how it connected to the map of the upper levels of Crystal Cave. Stephenson did not know, but promised to try to find out.

Roger was shocked—not so much by the deceit, for he had used deceit to make his map—but by Stephenson's seeming desire to do harm to his Crystal Cave hosts. Years later Bill Stephenson explained why he had given the map to the Park Service. He believed then that if the National Speleological Society cooperated with the Park Service, the Society would obtain access to Mammoth Cave and other caves in Mammoth Cave National Park in which exploration was forbidden at that time. He had not considered the possibility that the Park Service might instead curtail exploration even through the privately owned Crystal Cave Entrance. In fact, he did not gain access to Park caves. On the other hand, during the seven years between 1954 and 1961, the year the Park Service bought Crystal Cave, Park Service officials never explicitly forbade exploration out under Mammoth Cave National Park land from the Crystal Entrance.

Although nearly three miles of passages were surveyed in Crystal Cave on the C-3 expedition, most of them had been known before. The lack of new exploration resulted in part because so much effort was spent just in running the expedition. Time was wasted, for example, because there was

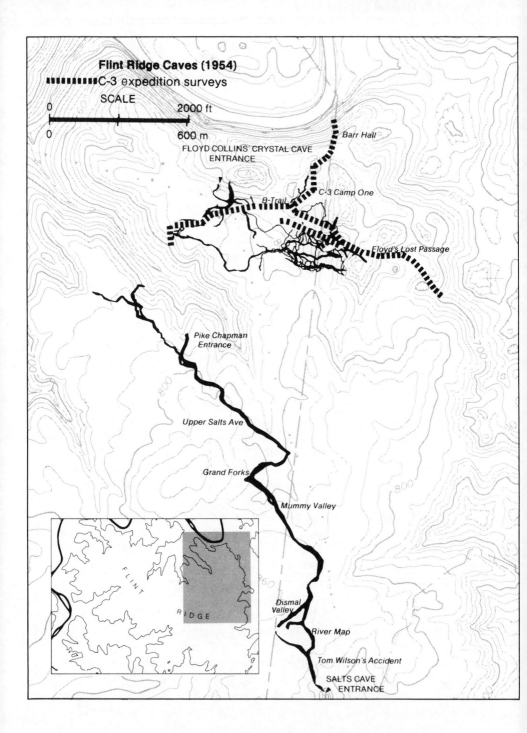

Map of Flint Ridge caves showing the addition of C-3 expedition surveys in February 1954.

great interest in setting up a second base camp, just as mountain climbers do. Consequently, some of the explorers were moved to Camp 2, only about twenty minutes' travel time and 1200 feet through canyons and crawlways from Camp 1 in Floyd's Lost Passage. The explorers from Camp 2 fumbled around in a baffling maze of canyons. They mapped nothing and they had no systematic way of communicating with one another about where any party had been, so they had no way to follow up discoveries.

When the C-3 expedition was over, Phil Smith, Roger Brucker, and Roger McClure agreed that it had been a failure. Exploration was hindered by lack of maps, and important expedition needs were subverted by a preoccupation with publicity. It was an arrogant conclusion by three newcomers, but they were convinced that it was accurate. With the passage of time, their judgment on the C-3 expedition became less harsh. Joe Lawrence had learned—and had taught the three critics—a great deal about how to run, and how not to run, an expedition in a big Kentucky cave. No one before had tried mounting a major cave expedition in the United States, and Joe had made a magnificent effort. How else was one to find out what worked and what did not?

One thing was certain: It was not worth the effort to support people underground for a long time. Phil was adamant that the use of supply teams was a waste of time and of cavers' energy and enthusiasm. The two Rogers agreed that they had been seriously fatigued by their long stays underground. There had to be another way.

By participating in the C-3 expedition, and by analyzing it together afterward, Phil Smith, Roger Brucker, and Roger McClure formed a close bond. During the C-3 expedition a dream emerged in their minds. They would work with Bill Austin to explore all of Floyd Collins' Crystal Cave. Then they would help Bill connect Crystal with the other big Flint Ridge caves. Then they would find a way under Houchins Valley between Flint Ridge and Mammoth Cave.

Someday—as Jim Dyer said—they would connect with Mammoth Cave.

Big Discoveries Expand Unknown/ Crystal Cave

An effort to connect
Crystal Cave with Salts Cave fails, but opens
the heart of Flint Ridge.

5 Soon after Roger Brucker unveiled his map of the lower levels of Crystal Cave in Flint Ridge, Bill Austin recognized from the maps and descriptions that Jack Lehrberger, Russ Gurnee, and Roy Charlton—who had been sent out to explore around Bogardus Waterfall—had actually gone to another place, subsequently named Bogus Bogardus Waterfall. This left a passage still to be explored at Bogardus Waterfall. If it went, it might connect Crystal Cave with Salts Cave. So in July 1954 Bill, Jack, Roger, Phil, and Dixon Brackett took a grinding eighteen-hour trip into the Bogardus area to straighten out the ambiguities of the C-3 expedition map.

The explorers proceeded to the real Bogardus Waterfall. Nearby was the entrance to a small passage over which the words "A NICE CRAWL" had been written in soot with the flame of a carbide lamp. Bill said he had written them in 1948 when he was exploring with Jim Dyer, and that the passage went a long way. He had turned back at a pit. There would be plenty for everyone out there, he said, so he, Jack, and Phil crawled on their bellies into the small passage. Roger, after his long drive from Ohio to Kentucky, was beginning to feel the effects of lack of sleep, so he decided to take a nap while the others explored. Dixon stayed with him. They slept for several hours, awakening when they heard a shout. The others had found their way to the bottom of Bogus Bogardus Waterfall by a new route. Roger fired up his carbide lamp and crawled down through a muddy, tight tube to the others. Dixon followed. By this time the advance party was ready for a meal.

Bill and Jack had taken a long crawl leading to a small elliptical tube with a sandy floor that took them to a cluster of dry vertical shafts. Then they followed another crawlway, but abandoned it to crawl through a small side passage back into the Bogardus Waterfall Trail. The crawlway they had abandoned was still unexplored. Roger asked how to find it. Jack pointed across the stone lily pads at the entrance to an ocher-colored tube just big enough to slide into. While the others were eating, Roger crawled off to explore.

Out of sight and hearing of the others, Roger found the passage getting smaller. Grape and sharp crystals on the walls, ceiling, and floor tore at his clothing—which is why they named it Fishhook Crawl—and he tensed as he realized that he would have to back out of the passage if he could not find a place to turn around. He decided that he must have missed the side passage that led back to Bogardus Waterfall Trail. There were no scuff marks on the floor. The passage ahead was virgin.

He decided to go around one more corner. Wham! As he stuck his head around the corner, a cold breeze struck him squarely in the face. It came from a black hole that had appeared at the end of the tunnel. The passage was even smaller ahead. Roger's adrenalin surged as he pushed rapidly forward on his belly. The breeze felt like a gale. The hole, however, was a deception. It was just another bend in the passage. So were the next several "holes." Then, after about 400 feet of very tight belly-crawling, Roger thrust his head out a tiny windowlike opening into a pit about ten feet in diameter, with a basin floor ten feet below. The pit was decorated with brownish-black flowstone, so Roger named it Black Onyx Pit. Above the window the walls ballooned out.

Roger could not turn around, or even roll over onto his back in the tiny crawlway. But if he could get out of the window and up onto a ledge a few feet to the left on the wall of Black Onyx Pit, he could see what he had found, and he could turn around for the long crawl back. He had to push his body out into empty space until he could bend sideways at the waist. Then, holding on to slick cave onyx projections at the level of the hole, he slid his legs out and down the wall into a layback against the wall. The traverse across the side wall to the ledge was not difficult. He did not think getting back into the window would be easy, but he was sure he could do it.

The ledge was a delight. It led to an extension of Black Onyx Pit that was nearly thirty feet long, eighteen feet wide, and fifty feet high, with its floor fifteen feet below Roger's perch. Peering into the gloom, he could see several leads, including the entrance to a passage five feet in diameter that departed from one wall. He could do nothing more now, however, for he had overstayed his time and the others would be expecting him. He climbed back over to the window.

The window in the wall of Black Oynx Pit leading back into Fishhook Crawl was not easy to enter. Roger put his arms into the opening and hung there, his feet swinging free, but the hole was not large enough for him to pull himself up by his arms and at the same time bend his head and shoulders into it. Finally he had to reverse his entrance procedure, pulling his body up in a layback on the slippery wall until he could slide his head and shoulders sideways into the hole. Then, with his arms out-

stretched down the passage and his shoulders wedged, he flailed his legs
in the open air of Black Onyx Pit until he managed to slide forward into
Fishhook Crawl on his belly. Then he slid forward like a snake in a hurry
toward Bogus Bogardus Waterfall. When he was nearly there, he heard
Jack calling him. He answered elatedly. Jack listened to his account, and
then went off down Fishhook Crawl. He checked Roger's Black Onyx Pit,
and came back smiling.

The breakthrough into a new passage complex of Crystal Cave in
the heart of Flint Ridge had been made. Later it was learned that during
the C-3 expedition a crucial crawlway opening along the line of discovery
had not been found because during an exploration-trip rest stop Russ
Gurnee had sat his six-foot-five-inch frame on the ledge above it, hiding
it completely. Now, however, the way was clear. They planned a Thanksgiving expedition to explore the passages leading off Black Onyx Pit. It
would be a big job, and there was some discussion about personnel.

Red Watson had joined the ROTC during the Korean War to avoid
the draft. The war was over by the time he got his commission. He spent a
year in Denver going to an Air Force aerial-photography school, and then
was sent to an air base outside Columbus, Ohio. Searching for outdoor
adventure, he looked through the membership list of the National Speleo-

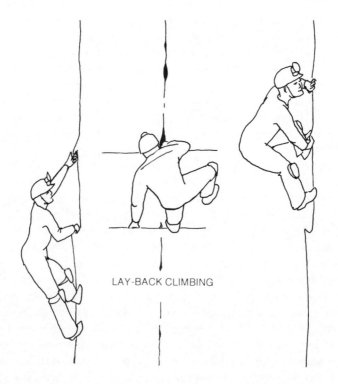

LAY-BACK CLIMBING

logical Society for the member in Columbus who had the lowest NSS number, and who thus would have been an NSS member longer than any of the others.

Jim Dyer answered the phone and after listening awhile said, "You call Phil Smith."

Phil was working on his thesis for an M.A. degree in educational psychology at Ohio State University, but he told Red to come on over anyway. Red climbed up to a room under the eaves of an old house where slanting ceilings forced tall Phil to stoop except in the center of the room. The room was piled with books, papers, and camera equipment. On the walls were cave maps, and photographs of caves and of scenes from Shakespearean plays.

Red and Phil talked, sizing each other up. Phil stretched out lazily in a chair, his legs crossed at the ankles, his hands clasped, flesh spare on his lanky frame. Red roamed the room, looking at everything, his compact body full of nervous energy.

"Have you done much caving?" Phil asked.

"Some," Red said.

"Our caving is pretty strenuous," Phil went on.

Red changed the subject to mountain climbing. He had done some climbing, but did not want to mention that he had been in only four caves in his life, one of which was about 100 yards long, and another of which was Carlsbad Caverns on a commercial tour. The other two did not amount to much.

Four hours later Phil and Red had discussed if not solved most of the world's major problems, and Red left with an invitation to go with Phil to his home in Springfield for Thanksgiving dinner. Then they would go to Floyd Collins' Crystal Cave for a big trip in Flint Ridge. Red stumbled down the stairs, glancing at the magazines Phil had given him. He read through the rest of the night. In the *Louisville Courier-Journal Magazine,* an extensive feature article emphasized the danger of getting lost while exploring caves. Red wondered whether Phil knew Crystal Cave well enough not to get lost. "Seven Days in the Hole" by Robert Halmi, reporting in *True Magazine* on the C-3 expedition, filled Red with visions of horror of the Keyhole, Bottomless Pit, and Formerly Impossible. He slept fitfully, cave darkness terrifying his dreams. He woke around noon, and never thereafter was he so frightened of caving as he was before ever really going caving at all. The reality of those dread places came as a welcomed relief when he actually experienced them in Crystal Cave.

Phil outfitted Red with caving gear, and lectured about caving during the long ride from Ohio to Kentucky. Red listened carefully, cramming, a bit anxious because he thought everyone took him to be an experienced caver. In fact, he had not fooled Phil, and he was wrong in thinking that he had to be an experienced caver to join this Flint Ridge crew. When Jim Dyer had suggested to Phil the need to develop teams of explorers,

Jim had also remarked that desire and fitness were as important as previous caving experience. Red was Phil's first green recruit.

In turn, Red had fastened on Phil as the one he should follow to learn. However, when he reached Flint Ridge, Red began to get confused. He had thought Phil was the organizing force behind this expedition. Then he met Roger Brucker, who seemed to view the exploration trip as part of his cartography program. However, Bill Austin, manager of Crystal Cave, was directing operations. And underground, on the long crawl out to the Bogardus area, Red found some of the cavers deferring to Jack Lehrberger, who charismatically projected the image of being the biggest daddy cave explorer of them all. The amazing thing was that there was no conflict in these crisscrossing lines.

Phil was busily managing, an occupation that led him three years later to found the Cave Research Foundation. Roger Brucker was constructing maps, and would become the first Chief Cartographer for the Flint Ridge Cave System. Bill was an heir to the finest cave in the world—Floyd Collins' Crystal Cave—and he was protective of his property. Jack, most interested in exploration, was quietly eager to go beyond the limits of previous penetration in Flint Ridge. Roger McClure, Dave Jones, Jack Reccius, Bill Hulstrunk, Dixon Brackett, Bob White, and Red Watson—only seven Indians for four chiefs—followed in the low passages on hands and knees, obeying orders. Red crawled blindly, because he soon saw that he could not memorize the complicated route back as he had read in a book that he should. He was dependent on the others.

During the C-3 expedition, heavy-duty phone wire had been laid from the Austins' yard where the Headquarters tent was located on the surface of Flint Ridge to Floyd's Lost Passage in Crystal Cave underground. The heavy wire and field telephones had taken so much energy to install that Dave Jones, who was recording engineer, adapted lightweight, sound-powered telephones for use in Crystal Cave. One could both listen and speak through a single crystal earphone that could be clipped anywhere along a thin, plastic-covered wire. A thousand feet of this wire weighed only a few pounds. At Floyd's Lost Passage, Dave spliced one of the rolls into the heavy telephone line. Transmission was loud and clear. Jacque Austin on the surface expressed suitably serious interest in the continuously reported accomplishments as the explorers unrolled the wire toward Bogardus Waterfall.

When the end of the previous survey was reached, Roger Brucker assigned segments of the passage ahead to various teams of surveyors. After five hours the new route had been surveyed all the way to Black Onyx Pit.

Meanwhile, Phil and Bill had gone beyond Black Onyx Pit to make a stupendous discovery. Instead of going up onto the ledge Roger had found to explore the high leads, they decided to continue through a crack at the bottom of Black Onyx Pit. It was Bill's experience that such drain passages often lead to larger passages. Within twenty feet they came to another drop, a much larger and deeper part of Black Onyx Pit. Phil un-

Dave Jones holding a Jones Phone, 1960.

coiled a shining new white nylon rope that contrasted with the muddy explorers. He tied the rope to a rock projection to use as a hand line for the climb down.

The passage continued on the other side of the pit as a canyon thirty feet high and five to ten feet wide, with possible routes of travel on several levels. Phil explored a small side passage while Bill went on down the main canyon. Coming back from the side passage, Phil decided to go back to the roped pit to leave a note for the surveyors who would be coming along. Bill and Phil were only fifty feet beyond previously known limits, but already there were so many leads to choose from that the follow-up teams would need clear guidance.

Then Phil followed Bill's scuff marks over large boulders jammed in the canyon passage.

"Come up here and look at this!" Bill shouted.

Phil rushed on to find Bill standing on the edge of the biggest pit either had ever seen. They called the wide platform they stood on the Balcony, and the pit itself the Overlook. Their lights reached neither the ceiling nor the floor of the pit in front of them. Its far wall was nearly fifty feet away. They threw in a rock. Several seconds later it splashed in a deep pool. They estimated that the Overlook had a total height of about 150 feet. Phil and Bill were standing on the Balcony about sixty feet above its bottom.

This discovery spurred them on. There were at least three obvious leads off the Balcony of the Overlook.

They chose a passage opening behind a spray of water on the same level as the entry passage. A few minutes after ducking through the shower, they came to a junction. In one direction a canyon led down to a rectangular passage four feet wide and six feet high. It offered more of the same wet and muddy prospect. The passage straight ahead was somewhat smaller, but it showed a sprinkling of gypsum on the walls, a sandy floor, and a change in the texture of the rock itself.

The choice was obvious. At Bogus Bogardus Waterfall the explorers had gone down to lower, wetter cave. Fishhook Crawl, Black Onyx Pit, and the passages beyond to the Overlook were also low. They had gone deep into the depths of Crystal Cave. To find big, dry passages high up under Flint Ridge, perhaps without any direct surface entrance, they would have to climb back up again. Cave passages on higher levels are often characterized by growths of gypsum crystals on their walls. When Bill saw the sparkling of gypsum in the passage before him, he was as pleased as he had been when he found the Overlook. Behind him, Phil was certain that *this was the way* to connect Crystal Cave with Salts Cave.

They walked rapidly into the passage, then crouched. Finally, when they were forced to their bellies by the lowering ceiling, Bill, in the lead, turned a disgusted look back to Phil.

"Oh, hell!" Phil said, and started backing out.

Back at the junction again, they climbed down into the wet passage. Phil was overheated from the crawlway. He was also already somewhat uneasy about having turned back, because the gypsum crawl did go on. It still does, unexplored.

The wet passage Phil and Bill now ran down got to be ten feet wide and seven feet high in places. It kept its rectangular shape, and it trended distinctly downward. They had very early along named it Storm Sewer.

After several hundred feet the ceiling closed down until they were crawling through mud and water in a passage three feet high and ten feet wide. The persistent rectangular cross-section was so striking that the tunnel seemed man-made. A thin layer of mud covered ceiling, walls, and floor. They knew they were getting down near the base-level of the Green River, which was known to backflood from the surface into the cave passages for thousands of feet and as high as fifty or sixty feet above normal river level.

"What's the weather like outside?" Phil asked.

"I haven't the foggiest," Bill replied.

This base-level caving was something new to Bill and Phil. They had never gotten this low under Flint Ridge before.

After a quarter of a mile Storm Sewer opened into a large passage making a T-junction. It was as though they had now entered the master sewer, with a vaulted limestone ceiling stretching six feet over a mud-bank floor twenty to thirty feet wide. A few hundred feet to the right, the room closed down to a muddy crater with no opening at the bottom. To the left, the big passage contained a small stream that flowed in a mud-walled canyon. Passage walls were twenty feet apart, with a fifteen-foot ceiling. Bill slid fifteen feet down a mud bank to the water on his feet, twisted back, and almost fell, crying, "Blindfish!"

All of these discoveries seemed to offer names immediately. The passage was named Eyeless Fish Trail, and the river it contained, Eyeless Fish River. The river was a major discovery that culminated years of exploration in Floyd Collins' Crystal Cave. This small stream was surely a part of a large underground river system that extended near base level beneath all the caves of Flint Ridge. If they could follow it, they would surely connect all the Flint Ridge caves into one system. Eyeless Fish River might even extend under Houchins Valley, which separated Flint Ridge from Mammoth Cave Ridge. The blindfish in Eyeless Fish Trail of Crystal Cave might be able to swim all the way to Mammoth Cave. But even if they could, this would be no promise that cavers—obeying the rule that you should follow the water to find new cave—actually could follow. Underground rivers have a tendency to siphon—that is, the ceilings of the passages that contain them sometimes meet the surface of the water, making it virtually impossible for cavers to go on. Siphons are formidable barriers to exploration even for cavers with special diving equipment, and there is never any guarantee that the ceiling will rise into open air beyond.

Downstream, the water of Eyeless Fish River flowed into a narrow black opening. At this level, so close to that of Green River outside, it would probably soon siphon. Even if it did not siphon, it would be very difficult to follow. Bill and Phil decided to explore upstream. Eyeless Fish Trail might intersect the bottoms of pits up which they could climb into unknown higher-level passages.

They sloshed upstream until the roof of Eyeless Fish Trail dipped down so that further progress would have required crawling in the water. Here they traced their initials, "p.m.s." and "b.a.," in the mud, and turned back. In a few hours they had found more new cave passages in Crystal Cave than half a hundred explorers had during the previous thirty years. On the way out, Phil sank to his crotch in the mud trying to climb up the bank of Eyeless Fish Trail. He noted that the mud out that way constituted a hazard.

It was not only the mud that was dangerous. A few years later Bill, Roger Brucker, and Dave Jones took two top adventure journalists—Coles Phinizy and Robert Halmi of *Sports Illustrated*—to see Eyeless Fish Trail. The water was up, and at one point a foot-wide bridge of mud crossed the river with a dark pool of water lying on either side. Bill walked across, with Phinizy and Halmi following. Roger, however, had slung around his neck a steel ammunition case full of Bill's camera equipment.

Bill Austin, Dave Jones, and Roger Brucker strap on knee crawlers at Floyd Collins' coffin in the Grand Canyon of Floyd Collins' Crystal Cave, 1955.

He decided, for safety's sake, to crawl on his hands and knees across the bridge. Sliding slowly down the mud bank on all fours, he slid smoothly off and head first into a pool about two feet in diameter on one side of the bridge. He went completely under, turned over, and came up banging his head solidly on a ledge underwater. The river ordinarily flowed on a level about six feet lower than it was now, and Roger knew there were passages down at that level into which he could be swept. He floundered wildly in a long moment of panic. He could not breathe. He was totally submerged in cold water, suspended in total darkness. There was nothing for his hands and feet to get a purchase on. Then his arms and head emerged back out the opening.

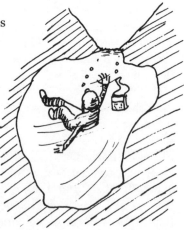

"Help!"

Bill had already raced back down the mud bank and was crouched on the bridge, extending a hand to Roger. "Taking a bath?" he asked, as he grabbed Roger firmly by a wrist.

The steel ammunition case also popped to the surface and was retrieved. Bill showed great satisfaction with the fact that the camera equipment he had packed emerged from the experience safe and dry. Roger emerged safe, but he was very wet. At a supply dump on the way out of the cave, Dave ran a butane blowtorch over Roger, raising clouds of steam from his clothing. Despite this, it was a long, cold trip for Roger from Eyeless Fish Trail to the Crystal Entrance.

On the original discovery trip, Bill and Phil returned from Eyeless Fish Trail to Black Onyx Pit to find Jack Lehrberger and Dixon Brackett exploring leads in that vicinity. Roger Brucker was supposed to meet them there, but the surveying had taken longer than they had expected. Always impatient with delays, Phil was furious. There was so much to be done. Then the surveying parties began catching up with the explorers. Everyone was so excited at the descriptions of the new finds that Bill and Phil immediately guided some of the surveyors on to the Overlook. The rest of the party laid the telephone line into a side pit, a dark, elongated room six or eight feet wide with a forty-foot ceiling. Only the lights made it seem like a camp. This dismal place was fittingly named Camp Pit.

At the Overlook there ensued a great rock-throwing orgy. In a few minutes all the larger loose rocks on the Balcony—no more than ten or twelve—had been pitched over, with enormously satisfying results. Each time, a long silence was followed by a deep, heavy *ker-chunk* from the pool of water far below. Nothing was left except a huge block of limestone about six feet long, three feet wide, and three feet high, seemingly just teetering on the edge of the Balcony. By this time Jack Reccius, a caver from Louisville who had done a lot of exploring in Salts Cave and Unknown Cave with Jack Lehrberger, was wild with enthusiasm. He rushed to the edge of the pit without thought of life or limb to strain at the block of limestone.

"Help me, help me!" he yelled. Several people tried, gingerly, to pry the block over the edge without going over themselves. Others were not about to join the madmen. Red crouched back against the wall of the Balcony, well away from the edge of the Overlook. He had gone along with everything so far, but he did not want to get close to the edge of that deep pit. The boulder won. Twenty-two years later it is as solidly attached— and as precarious-looking—on the edge of the Overlook Balcony as it has probably been for hundreds, if not thousands, of years.

The pitch of excitement could not be subdued. Bill and Phil led some of the party on out Storm Sewer to see the blindfish in Eyeless Fish River. On the way back they stopped at a walking lead, a broad opening ten feet high and five feet wide in the side of Storm Sewer.

"Red," Phil shouted, ."this one's for you! Explore some virgin cave."

The others stood back expectantly as Red ducked forward. He had done all right so far as part of a survey team. Now the lies had caught up with him, and he felt as though he had told a group of climbers that he could climb when he never had before, and then found himself forced into leading up a vertical wall.

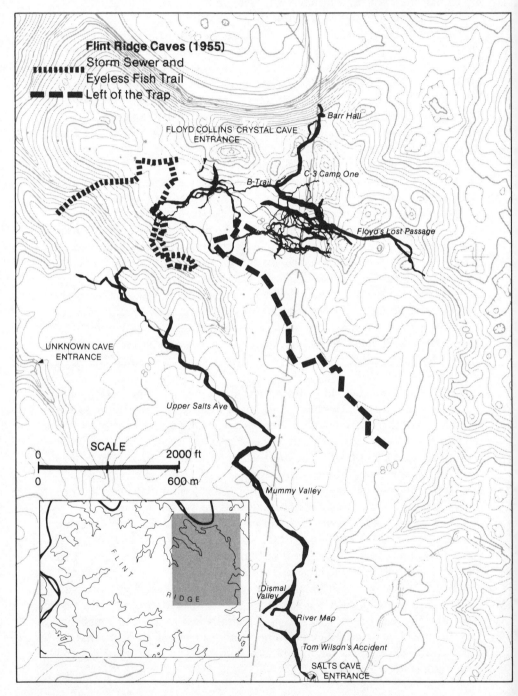

Map of Flint Ridge caves showing the breakthrough discoveries of Storm Sewer, Eyeless
Fish Trail, and Left of the Trap in 1955.

Flint Ridge Caves (1955)
Storm Sewer and
Eyeless Fish Trail
Left of the Trap

Barr Hall

FLOYD COLLINS' CRYSTAL CAVE
ENTRANCE

C-3 Camp One

B-Trail

Floyd's Lost Passage

UNKNOWN CAVE
ENTRANCE

Upper Salts Ave

SCALE

0   2000 ft
0   600 m

Mummy Valley

FLINT

RIDGE

Dismal
Valley

River Map

Tom Wilson's Accident

SALTS CAVE
ENTRANCE

"We'll wait here," Phil said, slumping down to sit with his back against the wall of Storm Sewer. The others sat down, too.

Not only had Red never explored virgin cave before, he had to go alone. His body carried him along. A few feet into the passage, he stepped down and turned a corner. There in front of him was a blank wall. It was not a pile of breakdown blocks, nor a mud fill plugging the passage. It was a solid stone wall. For a moment he thought he had been tricked, but, no, he had noticed quite carefully that there had been no tracks on the floor. Phil had not set it up for him.

"It ends!" Red shouted back.

There were sounds of disbelief and disgust from Storm Sewer. Groaning, the others got up to set the newcomer straight. They walked around the corner and stared for a moment in disbelief. So much new cave had been discovered on this trip that everyone expected every passage to go and go. Then they all burst out in laughter. Phil could say nothing because he was choking, so he just pounded Red on the back. Back out in Storm Sewer, Phil wrote over the top of the arched opening in big letters with the flame of his carbide lamp: "WATSON'S FOLLY." That seemed a little hardhearted to Red at the time.

It had been a hard trip for everyone. Some people in the party had complained about the difficulty on the way in, and had several times suggested that the group turn back. After reaching Camp Pit, one person refused to go a few hundred feet farther down the passage to see the Overlook. Three of the party members never entered Floyd Collins' Crystal Cave again after this trip.

For others, this discovery trip was the beginning of Flint Ridge caving. Jim Dyer's hypothesis and hope that Floyd Collins' Crystal Cave was the heart of a large Flint Ridge Cave System had been confirmed. For Phil and Roger, the discoveries proved the value of a kind of organized cave exploration they were developing in the Flint Ridge Reconnaissance, and that would lead to the founding of the Cave Research Foundation. For Red, it was the grandest adventure he had ever had. He had fallen in love with Floyd Collins' Crystal Cave.

Blindfish (Amblyopsis spelea).

Refining the Cavers and Their Methods

The cavers again try to connect Crystal Cave
with Salts Cave. They fail, but
again improve their caving abilities.

6 There was much to do beyond the Overlook. However, Jim Dyer had often spoken of a mysterious passage in Crystal Cave that posed a formidable problem in Flint Ridge exploration. When you drop down through the tight hole just past Scotchman's Trap off the commercial route, you are funneled down into a small canyon. To your right are the Crawlway, the Keyhole, and the route to Floyd's Lost Passage. To the left is another passage, the one Bill Austin had let visiting cavers get lost on when he had introduced them to Crystal Cave. During the C-3 expedition everyone had been warned not to go Left of the Trap, because that crawlway went on forever. No one had ever been to its end.

"Who knows what might be out there?" Jim mused.

Phil Smith and Roger Brucker decided in 1955 to undertake a real test of intensive caving Left of the Trap. It was practical work, of course. But it was the mystery that had fired everyone's enthusiasm. Jim told of a Spaniard who once had come to Crystal Cave. Supposedly he had gone underground Left of the Trap for days, marking his way with blue chalk. He had come back, Jim said, raving of crystal palaces, sparkling lakes, and enormous rooms. Jim knew for a fact that the gypsum displays Left of the Trap were spectacular. He also remarked that the passage headed straight down Flint Ridge from Crystal Cave toward Salts Cave.

The first team could not have been more ready. Left of the Trap begins as a canyon, fourteen inches wide, with walls covered by a tan gypsum crust. The explorers walked along sideways, disarmed already by the easy canyon. Then the canyon changed abruptly to an extremely disagreeable belly-crawl. The floor is a fine, easily inhalable dry sand. On every trip you must push sand aside to make way, for your dragging feet pull sand back in to fill the trail behind you. After two football-field lengths of seemingly interminable flat belly-squirming, you come out into another narrow canyon. This is the Cigar Box, named for a wooden cigar box left by the old-timers who had explored there. You can stand up here.

Jim had warned that, so far as anyone knew, the Cigar Box was the last place anyone could stand up Left of the Trap.

Phil and the others tried, as some had before them, to chimney up the walls of the narrow fifteen-foot Cigar Box. It was too tight at the top to go on. The only way on from the thirty-foot-long room was another dusty crawlway. Every surface was covered with crunchy white and ocher gypsum encrustations. There were billions of gypsum crystals of every description. The explorers sneezed, and crawled on southeast, hoping for a bend straight south toward Salts Cave, now only half a mile away in Flint Ridge from this long passage in Crystal Cave. However, the passage went straight on, and all possible side leads turned out to be mere pockets in the wall. Some are just large enough for you to crawl inside. There you find yourself in a circling bower of glistening white gypsum flowers, a pristine nest very remote from all the rest of the world.

It was a long crawl. Phil and the others had very sore backs and knees even before they turned back. On the first trip they had crawled more than ten hours out Left of the Trap, and it was more than a ten-hour crawl back to Scotchman's Trap again.

Jim ignored their aches and pains. "There's lots of cave out there," he said at his Ancient Mariner best. "And the old-timers who tried to explore it were greenhorns or drunk. You guys could really hit it big Left of the Trap."

Phil and Red, Roger Brucker and Roger McClure, Dave Jones, Burnell Ehman, Dave Huber, the entire new team, hitched up their survey gear and packs, to push out on the new frontier. It was the ultimate challenge. The Left of the Trap crawlway trends in one direction, southeast down Flint Ridge. It stays the same size. It goes on and on and on and on and on and on. . . . Long stretches are too low for hands and knees, but seem too high for belly-crawling. Every time you try to rise, your collar or belt scrapes a gritty deposit from the low ceiling down your neck or into the back of your pants. Gypsum sand gets behind your knee-crawler straps, which saw into your tender flesh. There is no way to keep the crumbling gypsum out of your shoes, clothes, and hair, or eyes, nose, and throat.

Special care with equipment is required Left of the Trap. You carry plenty of water, because the passage is dry as a salt mine. Most importantly, you cultivate the feeling that it will be all right if you do not stand up for eighteen to thirty-six hours. There are no options. You stay horizontal, and you survey. Your goal is to connect Crystal Cave with Salts Cave so that the Flint Ridge Cave System will be the longest in the world. But secretly you know that the hope of making that connection is not *really* why you are there with gypsum grit in your eyes, under your arms, and between your toes: You have put your body on the line Left of the Trap because you are a caving monomaniac and *you want to know where that passage goes!*

The painfully slow surveying on several trips Left of the Trap only tantalized the explorers. With the extension of the passage, they decided that they should stock an underground base with food, water, and carbide at the farthest point of penetration. Then two of the toughest cavers, lightly laden, could move rapidly to the supply point to pick up supplies for a long exploration push.

Jack Lehrberger and Dave Jones were obvious choices for the exploration. The others set up a supply dump. Then Jack and Dave rushed far out to where the passage became slightly damp. They limped back after thirty-six hours to tell wearily of having found some old tin cans where they stopped. Jack thought at first that possibly he had left the cans there on a solitary trip in Salts Cave. That would mean that they had made the connection between Crystal Cave and Salts Cave. But then Jack concluded that the cans must have been from an earlier exploration trip he had taken alone Left of the Trap. He was sure of it. The connection between Crystal Cave and Salts Cave had not been made.

The results of the team-supported exploration push left Phil and the others dejected. Even with the help of a supply dump, and after all the work required to get those supplies out there for them, Jack and Dave had gone no farther Left of the Trap than one man—Jack himself—had previously gone *alone*. The new team had done an important job Left of the Trap, surveying it for more than a mile down Flint Ridge, but several all-out efforts had not yet taken them past the point of Jack's previous penetration.

The explorers had reached an endurance barrier for single-trip exploration underground in Flint Ridge. Jack and Dave had gone for thirty-six hours without sleep, but it was clear that twenty-eight to thirty-two hours was about all of this kind of caving anyone could tolerate and remain halfway efficient. So why not set up camps and sleep in the cave? Phil still opposed setting up big base camps in the cave, but temporary traveling camps might be okay. He located some paper sleeping bags that could be folded small enough to be carried into the tight passage. The next team out Left of the Trap went prepared to camp. Unfortunately, the paper bags did not allow enough body moisture to escape, and the sleepers woke drenched in cold sweat after only thirty to forty-five minutes. Down-filled bags were too expensive for destructive cave use. Camping Left of the Trap was thus abandoned.

Phil's experiments with supply dumps and camping Left of the Trap had a profound effect on caving in Flint Ridge. It is extremely difficult to haul supplies through long, low crawlways. It is as tiring to carry a small bag with ten pounds in it while crawling in a cave as it is to carry forty pounds in a backpack along a mountain trail. A second problem was that supportive manpower did not exist. Unlike some of the European caving groups, the new Flint Ridge explorers did not have a large group of pack-horse cavers on hand. Phil had done support work on the C-3 expedition, and was adamantly opposed to doing more. It was frankly doubted that

Americans ever could set up an expedition in which many self-sacrificing cavers carried supplies deep into a cave so that a few elite cavers could explore and survey. Fred Bögli ran such a program for exploration in Hölloch, the Swiss cave that was soon to surpass Mammoth Cave as the longest cave in the world. But almost every American caver was interested in doing his own thing, not in supporting someone else while he did his thing. Consequently, when the single-trip endurance barrier was reached Left of the Trap, exploration there was abandoned, and interest turned to other parts of Crystal Cave. Since then, only one camp has been set up underground in Flint Ridge, and it has not been used.

Exploration and surveying techniques refined on the trips Left of the Trap became standard elsewhere in the cave. Before, it had been thought that mapping would be inefficient unless surveyors were preceded by pathfinders. The early directors of the Flint Ridge Reconnaissance discussed at length the qualifications required of the pathfinder. He was ideally to have all the virtues of an Eagle Scout plus the great stamina and resourcefulness of Daniel Boone. Bill and Phil were pathfinding when they discovered the Overlook and Eyeless Fish Trail in Crystal Cave. However, it turned out that it was about as efficient to survey while exploring as to send out pathfinders. This was in part because often what the pathfinders reported could not be understood until the passages were surveyed.

One technique developed Left of the Trap is the system of leapfrog surveying. One party moves ahead, counting twenty or thirty bends in the passage where survey points are required. Then that party starts surveying. The rear party surveys the points the first party had counted off, and then the rear party leapfrogs the forward party for another twenty or thirty stations, and so on, as each party in turn surveys twenty to thirty stations down the passage.

Another surveying procedure is attractive, but is not always to be recommended. A survey party explores forward counting fifty or sixty survey points, and then surveys back. This permits free exploring. And while surveying back, the surveyors are inspired by a known objective and movement toward the cave entrance. The danger is that the explorers now turned surveyors might lose interest, or for some other reason not complete the survey. Hanging surveys—those not tied into the passages on the map—are useless.

The policy was soon established that explorers were expected to survey the passages they explored, preferably while exploring them. Some cavers said that this destroyed the spontaneity of exploring virgin cave, and others were not interested in making maps. Those who became Flint Ridge cavers accepted surveying as an integral part of exploring, and for many of them, making the map is one of the central pleasures of caving.

However, not everyone appreciates the maps. Some cavers view the underground as a world totally other from that of the surface. For them the cave experience is spoiled if they know the relation of the cave passages—beyond the entrance—to surface features. Surveying and mapping bring the cave world sharply into quantitative focus. Some cavers prefer to keep the experience entirely qualitative. In the cave they do not want to know where they are in relation to the surface world. Jack Lehrberger may have felt this way. But a major goal of most Flint Ridge cavers —however they may appreciate the caves aesthetically—is to make the caves known scientifically.

Even those who appreciate the value of maps often prefer while actually caving to enjoy the mystery of not knowing exactly where they are in a cave. While surveying Left of the Trap in Crystal Cave, Roger Brucker once left a note in the center of the passage: "Greetings Sucker—you are 270 feet from the Cigar Box." On a later trip Burnell Ehman came upon the note, and howled, "I don't *want* to know how far I am from the Cigar Box!"

Roger McClure and Phil Smith consult maps,
1953.

Cave War

Political obstacles confine the explorers,
but harden their determination to
explore the caves of Flint Ridge.

7 Phil Smith, Roger Brucker, and Roger McClure had organized the Flint Ridge Reconnaissance as a project of the Central Ohio Grotto of the National Speleological Society, for the exploration and scientific study of all the caves in Flint Ridge. This was bold, because the only Flint Ridge cave open to them in 1955 was Crystal Cave under its island of private property inside Mammoth Cave National Park. West of Crystal Cave on Flint Ridge the owners of Great Onyx Cave forbade exploration. Finally, the National Park Service forbade cave exploration in Mammoth Cave National Park, and would have to be approached if entry were to be gained legally to Salts Cave, Colossal Cave, Unknown Cave, and other Park caves in Flint Ridge.

Thus, while Roger, Bill Austin, and others organized exploration and mapping trips in Crystal Cave, Phil visited university scientists, local Kentucky organizations, the National Park Service, and the U.S. Department of the Interior, seeking both political and monetary support for the exploration of the caves of Mammoth Cave National Park.

Always in the background of Phil's activities was Jim Dyer. Jim became a second father to Phil, urging him on to levels of accomplishment that Jim himself had not attained. Phil also spent long hours talking to Dr. E. Robert Pohl, who represented the owners of Crystal Cave. Dr. Pohl became Phil's main adviser on science and local politics. Phil made friends with Cal Miller, the current Superintendent of Mammoth Cave National Park. Cal was a caver who had been active in the early exploration of Carlsbad Caverns. He enjoyed hearing about the continuing discoveries in Flint Ridge, and gave Phil high hopes for a cooperative agreement with the National Park Service. Then, in one of those swift transfers for which the Park Service is famous, Cal was replaced. The new Superintendent, Perry Brown, seemed to have a negative interest in caves. He apparently believed that anyone who went into a cave voluntarily was crazy.

In the spring of 1955 Roger Brucker wrote the new Superintendent a long, earnest letter asking permission to survey Salts Cave. Superintendent Brown asked for additional information, which Roger provided. The Superintendent replied that there would be a problem because Jack Lehrberger was a member of the Flint Ridge Reconnaissance. Brown had heard rumors that Jack had already entered Salts Cave in Mammoth Cave National Park illegally, and he remarked that Jack Lehrberger and his friends were *persona non grata*. More letters gave rise to another demand by the bureaucracy. The group had to be sponsored by a national organization. Within weeks Roger changed the project from that of the Central Ohio Grotto to that of the National Speleological Society.

Brother Nick Sullivan and Bill Davies of the Society then lobbied in Washington. It was hoped that pressure from the National Speleological Society would force the Park Service to open Salts Cave to exploration. The officials at Mammoth Cave National Park, however, stood firm. They had strict, long-standing policies against exploration. They made more demands. The cavers obtained physical examinations and shots for tetanus and typhoid. Then, just as it appeared that Salts Cave would be opened, Superintendent Brown withdrew permission at the last minute. His stated reason was that Salts Cave was too controversial. Probably he thought that a connection between Crystal Cave and Salts Cave might boost the price when Floyd Collins' Crystal Cave was finally sold to the National Park Service.

Instead, Brown offered permission to survey in Colossal Cave. After discussions with Phil, Jim, and Bill, Roger declined. The cavers were determined to work in Salts Cave. They were convinced that they could outlast Park Service reluctance.

The antagonism between cavers and the managers of Mammoth Cave already had a long history by 1941, the year the National Speleological Society was formed and Mammoth Cave National Park was established. Because organized cavers knew the restrictions against exploration, they made few serious attempts to gain access to Park caves until the 1950s. This opposition to exploration had origins in local circumstances that had nothing to do with modern sport caving or speleology.

What primarily led to the lock-out was the bitter memory of a long period of cave robbery and vandalism in the Mammoth Cave area, known as the time of the cave wars. During the 1920s, when more than twenty commercial caves were open, cave onyx, stalactites, and other cave deposits were sold at many houses along the main roads. These deposits were fair game anywhere. If clandestine mining diminished the attractions of a rival cave, tough luck! If a rival cave could be closed, more tourists would be available for the others. This cave warfare reached its peak with the stealing of Floyd Collins' corpse from Crystal Cave.

The land for Mammoth Cave National Park was purchased parcel by parcel by a local holding company. By 1941, nearly 500 families had been moved off their home places. Ill feeling and smoldering anger followed these evictions, and the high incidence of poaching of deer and ginseng in Mammoth Cave National Park today can be traced in part to the feeling by many old-timers that Park land should be a wide-open local resource.

For more than 100 years, rivalry over access to Mammoth Cave had been vicious, and had even led to armed encounters. All previous owners had jealously guarded entry to Mammoth Cave, and the National Park Service was to be no exception. After the transfer of Mammoth Cave to the Service, most of the same guides, management, practices, routes, and myths—such as the falsehood that Mammoth Cave was 150 miles long, when only about forty-four miles of passages had been mapped—were maintained. Most important, the Service continued to charge a fee for entry to Mammoth Cave, as had always been done. If other Park caves were opened for free entry to cavers, this might diminish the number of tourists willing to pay to enter Mammoth Cave. The lockout seemed eminently rational to the Park Service bureaucrats.

Either the desire to protect the caves from vandalism or the desire to run a commercially sound operation was reason enough for rigid entry restrictions, but there was a third practical reason. Park Service officials over the years have sometimes loved the caves and sometimes hated them, but whether they entered the caves only when absolutely necessary or liked to poke around off the tourist trails themselves, they knew that the dangers of cave exploration can be formidable. The nightmare of the abortive fifteen-day attempt to save Floyd Collins when he was trapped in Sand Cave in 1925 crept into every discussion of exploring. No Superintendent wanted to have a circus like that in his park. No, sir! Cave exploration was strictly forbidden. Two guided tours in Salts Cave were permitted, however. One was organized by Bill Stephenson for the National Speleological Society in 1950, and the other by Charlie Fort to prepare a picture article for the *Louisville Courier-Journal,* also in 1950.

If the National Park Service lockout had been maintained, the caves of Flint Ridge would not have been connected to one another in the 1950s and 1960s, nor would the big connection have been made between the Flint Ridge Cave System and Mammoth Cave in 1972. The change in policy was one result of the refusal of the owners of Floyd Collins' Crystal Cave and Great Onyx Cave to sell their properties when land was purchased for the Proposed Mammoth Cave National Park in the 1930s. These caves remained as islands of private property within Mammoth Cave National Park when it was established in 1941.

Inholdings in national parks are always a nuisance to administrators, but these inholdings were particularly annoying. As full-scale operating commercial caves within Park boundaries, they were in direct competition with Mammoth Cave. They ate up maintenance appropriations because by court order the Park Service had to grade and repair roads through the

Park to them. The successful operation of these two caves also had an effect on the salaries of employees in Mammoth Cave National Park. Money spent on souvenirs at Great Onyx Cave could have gone to concessionaires at Park headquarters. The more people who toured Floyd Collins' Crystal Cave, the fewer guides would be needed in the Park. Finally, the salary of Park Service employees was related to the total number of visitors to the Park each year. It is reasonable to pay administrators more for handling larger operations. This is part of the explanation why some national-park superintendents have worked to increase the parks' attractions, developments, and number of visitors.

Floyd Collins' Crystal Cave, owned by the Thomas family estate, and Great Onyx Cave, owned by the Cox family estate, were viable commercial operations, which is adequate reason for their owners' refusal to sell the caves to the National Park Service. Further, from the beginning there was a battle over the value of the two inholdings. The Park Service tried to show that the private caves were not worth much, and the private owners did everything they could to increase the value of their properties. The continual maneuvering caused bitterness on both sides.

Some private-cave owners—and not just those with inholdings inside the Park—set up official-looking booths along the road outside the Park. Here they stationed "cappers"—solicitors in khaki uniforms and Smokey Bear hats to direct tourists not to Mammoth Cave, but to other commercial caves in the vicinity. Partisan souvenir sellers who were interested in having tourists drive all the way into the Park countered with billboards in official Park Service colors that announced: "NO ONE HAS THE RIGHT TO STOP YOU ON THIS ROAD."

Suspicion, secrecy, and competition between the Park Service and the private-cave owners reached its peak in 1954 when the National Speleological Society was invited to conduct the week-long C-3 expedition in Floyd Collins' Crystal Cave. Rumors were spread that Crystal Cave was as dead as Floyd himself, and that thirty-two people may have gone into the cave, but they were doing nothing, for the expedition was just a publicity stunt. One newspaper reporter said that he had come to expose a hoax, but he stayed to write enthusiastic stories about the exploration. The publicity was nationwide, and daredevil Robert Halmi's color photographs and article in *True Magazine*, "Seven Days in the Hole," were sensational.

Floyd Collins' Crystal Cave had not received so much publicity since Floyd was trapped in Sand Cave. Naturally, Park Service negotiators were chagrined, for there was no question but that the value of the private inholding had been increased.

One incident cut through the rivalry and helped, finally, to gain Park Service approval of the Flint Ridge cavers. On May 7, 1955, Bill Austin received a phone call from the *Louisville Courier-Journal* asking him to take pictures of a man trapped in a cave near Sulfur Well, Kentucky. Bill promised no pictures, but he immediately threw his cave gear into his car. He and Dr. Pohl drove quickly to the cave where the man was trapped.

Some Park Service officials were there to direct a rescue, but no one seemed to know exactly what to do.

Bill took a look in the cave at once. The man was trapped in a breakdown room about twenty feet from the entrance. He had been trying to remove some columns from a cave wall. When he had pulled on a column, he had released about a ton of assorted boulders it had been holding fast. A piece of flowstone had slid over to pin the man's legs, but it had also held up the largest boulder just enough to let him live. Bill asked for assistance, and Henry Porter, from Knob Lick, volunteered. He and Bill jacked up the top rock, propped it up with two-by-fours, and then jacked up and propped the block of flowstone that was pinning the man's legs. Then he could be removed. The trapped man's legs were permanently paralyzed, and Bill took no pictures of his rescue operation for the *Louisville Courier-Journal.*

Park Service officials who had observed Bill's rescue operation thanked him, and reported favorably about him to Superintendent Brown at Mammoth Cave National Park. The episode represented one of the truce periods in the long era of cave wars. The cave people understood and respected one another, even when they were rivals. The Park Service managers recognized the integrity of Bill Austin. This created good will that helped the cavers eventually to gain permission to explore the caves in Mammoth Cave National Park under Park Service auspices.

Unknown Cave

Bill and Jack secretly explore
a big cave near Crystal Cave. Now they make
the Ohio cavers co-conspirators.

8 It was hot and dry during the late summer of 1955 in Kentucky. Roger and Phil were occupied Left of the Trap. Their heads were full of plans, programs, and experiments, and their files began to bulge with letters, directives, and results. Through the Flint Ridge Reconnaissance they were busily—even bureaucratically—taking over. Bill and Jack saw this, and seemed to be going along with it. They agreed even to such rules as that some cavers would be excluded from participation for undefined "conduct unbecoming to a speleologist." But Phil observed that Bill and Jack smiled a lot. Roger noticed that they said yes a lot. Obviously, something peculiar was going on. *What were Bill and Jack up to?*

They could not be exploring in Crystal Cave, for then secrets would be difficult to keep, and also unnecessary. So what were the mysterious hints about new discoveries? It must be, Phil and Roger decided, that Bill and Jack were exploring in the lower levels of Salts Cave. They were looking for a connection between Salts Cave and Crystal Cave in true old Austin-and-Lehrberger style, secretly, alone, and for themselves. Roger was curious. Phil was furious.

Then came the bombshell. In September 1954, Bill wrote to Phil and Roger to tell of a tremendous discovery. Bill did not say where, but Phil and Roger were sure that it was in Mammoth Cave National Park, somewhere in Flint Ridge near Crystal Cave.

Bill and Jack had done a nifty bit of pit work to arrive in a fair-sized passage. Twelve hours and more than three Lehrberger cave miles of walking passage (a Lehrberger cave mile contains a minimum of 6000 feet) later they had seen fantastic arrays of gypsum and mirabilite crystals in huge walking passages. There was so much cave that Bill finally, in a big walking passage, just sat down and refused to go any farther. He was a very happy caver.

Bill and Jack had made the greatest cave discovery—both in length

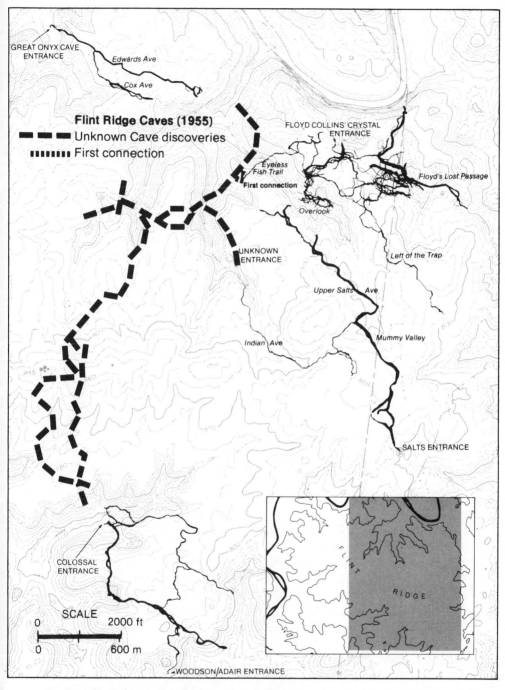

Flint Ridge Caves (1955)
- - - Unknown Cave discoveries
||||||| First connection

GREAT ONYX CAVE ENTRANCE

Edwards Ave

Cox Ave

FLOYD COLLINS' CRYSTAL ENTRANCE

Eyeless Fish Trail

First connection

Floyd's Lost Passage

Overlook

UNKNOWN ENTRANCE

Left of the Trap

Upper Salts Ave.

Mummy Valley

Indian Ave

SALTS ENTRANCE

COLOSSAL ENTRANCE

SCALE

0 ——— 2000 ft

0 ——— 600 m

F L I N T   R I D G E

WOODSON/ADAIR ENTRANCE

Map of Flint Ridge caves showing the discoveries that Jack Lehrberger, Jack Reccius, and Bill Austin made in Unknown Cave in September 1954. The first connection route between Unknown Cave and Crystal Cave was discovered by Bill Austin and Jack Lehrberger in September 1955. Within seven months two other Unknown/Crystal connection routes had been discovered.

and in beauty—in the Kentucky cave area since Stephen Bishop. But *where* was it, this wonderful new section of cave? Was it in Salts Cave? Or was it an entirely new cave? Phil and Roger knew that all they could do was wait. Having given in to the desire to tell, Bill and Jack would sooner or later want to show.

There are few records of exactly what Bill and Jack did. The following is a somewhat speculative version of how one of the greatest underground discoveries ever made took place.

Bill and Jack, with the help of Jack Reccius, carried a clandestine survey from historic Mummy Valley in Salts Cave through a tight belly-crawl to the lower level—Indian Avenue—that Lehrberger and Reccius had found in 1954. The survey led northwest through Indian Avenue for several thousand feet. It terminated in a set of very narrow canyons and a steep pile of breakdown blocks. When the survey was plotted and scaled off onto the topographic map, the end point was seen to lie almost exactly on a "Y," the symbol for a cave entrance. It was on Mammoth Cave National Park land, about a mile southwest of the Crystal Cave Entrance on Flint Ridge. They then checked on Park Service maps. On one of them someone had written the word "unknown" by the "Y." Unknown Cave was born. Bill and Jack lost no time in checking it out. They hoped that Unknown Cave would be a back entrance to Salts Cave.

On that first trip in Unknown Cave, they crept over the masonry sill of an ancient gate. The walking passage straight ahead pinched to a crawlway, then ended. A branch to the right led to the brink of a pit. They descended a twenty-foot pitch, then climbed farther down a steep slope into the bottom of the pit, some forty feet below the tiny window by which they had entered. On the wall of the crawlway that led out of the pit for several hundred feet they found the scratched names of the Collins family.

On a later trip Jack Reccius decided that there must be a lead under one wall of the entrance pit where the sound of rushing water could be heard. He and Jack Lehrberger moved rocks until they could squeeze on their bellies into a stream bed. From here they crawled into a small room. There were several possible leads, but they followed the water through a tight tube resembling a drainage culvert. The tube opened into a canyon, down which the explorers chimneyed forty feet to be stopped by a big pit.

Bill Austin and Jack Lehrberger then planned a trip. In early September of 1954—without ropes—they made the perilous descent of the newly found pit. After they reached its bottom, about thirty feet down, they climbed down a canyon, and then walked into a passage heading north. The way south—toward Salts Cave—was blocked by a massive pile of breakdown. They looked carefully, but if there were a way to the end of Indian Avenue in Salts Cave—less than 200 feet away—it was not obvious.

The explorers walked north for about 2500 feet, then turned to walk

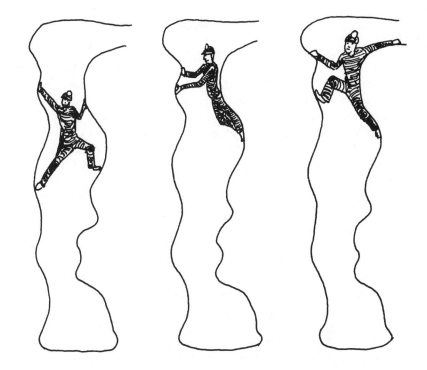

west for almost 3000 feet. Here they encountered another massive break-down pile, an enormous inpouring of rocks, probably the collapse of an outside valley wall. They climbed the unstable pile, searching for a way over. Near the top a comfortable walking passage led off. They ran south down this passage.

They were rather pleased with themselves. In about an hour they had found more than a mile of virgin walking cave in the heart of Flint Ridge, and they seemed to have entered a system of walking passages that would not end. There had not been a discovery like this since Salts Cave and Mammoth Cave had been walked into. Hour after hour they pounded on-ward, mostly in passages from twenty to more than 100 feet wide and from ten to twenty feet high. They did not take account of the point at which the size of their discovery exceeded that of the discovery of the main cave in Salts. They did not stop to consider the moment when they charged on past the imaginary point that would have marked the longest distance any explorer had pushed virgin cave in Mammoth. They did not even know that theirs was probably the longest cave-discovery trip ever taken. They were running, wild with joy and excitement.

At last a barrier loomed. It was another pile of breakdown, but it was fairly dry, and a crawlway led around it to more walking cave. However, Bill called a halt. His feet were blistered. Weary, sated with ever mounting excitement and now with fatigue, Bill and Jack placed their initials and the date on a rock to mark the day of the greatest discovery. Then they slowly retraced their steps for something more than three miles of awesome new cave.

Back at the entrance pits, the difficulty of the climb brought them back to ordinary reality. Reaching the Unknown Entrance at last, they looked out into the stark light of a beautiful fall day and discussed quietly whether they should wait until dark to emerge. They knew some Park Rangers who would love to catch them trespassing in a cave on Mammoth Cave National Park land. It would have been a long wait. Cautiously they slipped out of the Unknown Entrance and faded like shadows through the trees. Soon they were loping up a trail toward Crystal Cave property.

It was a classic Austin-and-Lehrberger cave trip. They caved hard and fast, and often separated for as long as four hours to explore in different directions. On one of these separations Jack had barely gotten out of earshot when he found a deep pit. A narrow ledge extended across one wall. Jack started across on his hands and knees. It was narrower than he had thought, so he glanced up to look for handholds so he could stand up. As he looked up, a drop of water landed on the tip of his lamp and extinguished the flame. He reached up to spin the steel on flint to light the lamp again, but the wheel spun free. The cap and spring holding the flint had come off and the flint was lost. Jack felt for his matches. They were gone. He did not trust himself to move either forward or backward on the ledge over the pit in the blackness.

About an hour later Jack heard scuffling. Then Bill's light appeared.

"Funny place to take a nap," Bill said, as Jack backed along the ledge to safety.

Another time, when Bill and Jack were clearing a narrow crawlway of breakdown blocks so they could get through, a huge chunk of one wall tipped slowly over to pin Jack by his chest against the other wall. Bill was behind him, and they discussed the possibility of Bill's being just able to slide on his belly underneath Jack to the other side, where he might be able to push back on the big block. Jack finally vetoed this plan, took a deep breath, shoved hard against the block, and slid backward as fast as a belly-scoot could take him. The block hesitated a moment, and then tipped heavily over into the space Jack had just vacated.

Soon after the big discovery, Bill and Jack began to survey Unknown Cave. Still trying to find out where they were, Roger told Bill that he was working on a new base map of Flint Ridge to cover all the area under which they might find cave extensions. He asked how much land south and west of Crystal Cave he should include. Bill said to include everything west to Great Onyx Cave and south to Mammoth Cave. Roger decided that Bill and Jack must have hit the jackpot in Salts Cave.

Bill may have shown the new discovery to the Ohio cavers when he did because there was a lot of surveying to be done fast. A rough survey had shown that the north end of Unknown Cave terminated under a valley wall very close to the surface on Floyd Collins' Crystal Cave prop-

erty. An entrance dug at that point would give easy, legal access to passages that were then miles and hours from the illegal Unknown Entrance.
Jack Reccius had been apprehended by Park Service Ranger Joseph
Kulesza at the entrance of Unknown Cave, so very precise surveying that
would enable a dig from the surface to hit the passage had to be done before the Service closed the Unknown Entrance.

But first there was the great possibility of making a connection between Crystal Cave and Unknown Cave. Bill and Jack held their cave
close to their chests, but they were not constitutionally greedy. Roger
Brucker and Red Watson knew Eyeless Fish Trail in Crystal Cave as well
as anyone, and had been on the discovery trip out there. So Roger and
Red it was. They were sent out to pound and to listen for pounding. And
what sport it would be to take them on a seven-mile rinky-dink (for that
was about the length of the new discoveries) if the connection between
Unknown Cave and Crystal Cave were made, and then to take them up
that horrendous climb out the Unknown Entrance! They would be expecting the Salts Cave Entrance, and so would be totally disoriented.

The story of that first connection is told in the opening chapter of
this book. After the connection between Unknown Cave and Crystal Cave
was made, the Ohio group was finally told that there was a brand-new
cave. The new discoveries were not in Salts Cave at all. In fact, although
it was surveying in Salts that had led Bill and Jack to Unknown Cave,
strenuous efforts on their part had failed to find a connection between
Unknown and Salts, even though passages in the two caves were separated by less than 200 feet near the Unknown Entrance.

Before anything else, of course, the Ohio cavers simply had to see
Unknown Cave for themselves. Then the need for a new entrance would
be obvious to them. There was no time to waste. Bill and Jack did not
know how much Reccius had told Park Service officials about Unknown
Cave, but they did know that he had told them it was fabulous. Twenty
years later, after a lot of caving, Reccius said that a seventy-two-hour expedition in Unknown Cave with Charlie Fort, Bill Walters, and John Key
in 1954 was his most memorable cave experience.

Phil loudly vented his annoyance that Bill and Jack had continued exploring alone despite being members of the Flint Ridge Reconnaissance,
and that they had shared their secrets with the Louisville cavers but not
with the Ohio group. But now fortunes were reversed. The Louisville
cavers—including Jack Lehrberger—were very much disapproved of by
the Park Service. It was not exactly planned that way, but it turned out
that in the eyes of the Park Service later on the Louisville group took the
rap as the bad guys (they got caught trespassing) and the Ohio group
gathered the benefits of being good guys (they did not get caught).

Just before dark on a cold day early in November 1955 the grand initiation began. Roger Brucker, Roger McClure, and Dave Jones had been
told to pack plenty of food for a long trip. One thing they had learned
from the Left-of-the-Trap experiments in Crystal Cave was what to eat on

long cave trips. They packed small cans of boned chicken and turkey, tuna fish, and vienna sausages, more small cans of peaches, pears, and pineapple, some larger cans of date-nut bread, and about five candy bars apiece. They joked about Red's having gotten married and gone back to Iowa for graduate school. Phil was in Greenland, navigating a traverse across the Ice Sheet. How could they miss this trip?

Bill led them down the slope behind his house on Crystal Cave property. They soon were walking on National Park land. Their route lay along a dry creekbed, over limestone outcrops and fallen trees. After about a mile of walking southwest, they climbed onto a terrace and walked to the base of a cliff where a black hole five feet wide and three feet high opened into the earth. Water from a spring splashed along the right wall to the entry chamber, and there an old, rusty gate stood ajar. They passed through the gate. Bill disappeared down a side passage for a few moments to return with a cable ladder wrapped around a four-foot length of tree limb. He pushed it into a right-hand lead, and the others followed. Bill readied the ladder for descent into a spacious pit.

One by one they climbed jerkily down the ladder, and then slid down the slope to the bottom of the pit. Bill and Phil cleared away the rocks concealing the stream passage, then the four bellied into the gravel-floored stream. The solid limestone ceiling pressed them flat into the cold water. After twenty feet the water plunged over a lip into a room three feet in diameter. One at a time the explorers crouched in the room, then slid sideways into a crack leading out of it. The crack narrowed progressively until it confined the explorers tightly on all sides. They moved forward by wiggling kneecaps, elbows, and toes. Then the floor of the gunbarrel passage dropped away into black nothingness below.

The traverse tested the nerves of the Ohio cavers. To keep from falling, they had to stretch and jam their arms and legs against the wall. Over the pit, the walls changed abruptly from being merely damp to being slick with wetness. It is thirty feet of ceiling walking for human flies.

Halfway across, Roger Brucker's light went out. The others had gone on ahead, so all was blackness. He took long, deep breaths to try to slow his beating heart. He shouted, but though he could hear the others, the passage contours were such that they could not hear him. Or they were too busy talking, he thought. Whatever the reason, he could not hold himself jammed into the crack over the pit for very long, and pushing on without a light was surely an invitation to a fall. Backing up would be as bad, if not worse.

Roger wedged the elbow of his right arm against the wall, and then carefully reached up to his hard hat to remove his lamp. He managed to jam his left elbow tightly against the other wall, so that he could hold the lamp in his right hand and spin the tiny wheel on the flint with the heel of his left hand. The sparks gave him the briefest of glimpses of the shiny black walls, but the lamp would not light. He shook it. It was out of water. Roger thought about that. He had handed his pack containing his water bottle on to the others before he started across the pit, and he supposed it was now lying somewhere in the passage ahead, out of reach.

Although he was beginning to feel the physical as well as the mental tensions of being suspended above the open pit—he did not know how far down the bottom was, but it was too far—Roger put the lamp back on his hard hat slowly. It would not do to drop it. Then he maneuvered his hands down to unbutton his pants. Retrieving the lamp, he urinated approximately into the small hole in its top. Pushing the lid firmly shut, he brought the lamp around to spin the wheel again. The flame shot out with a loud pop. Putting the lamp back on his hard hat, Roger completed the traverse, picked up his bag on the other side of the pit, and caught up with the others.

"What took you so long?" Dave asked.

Roger told him.

"Whew! I thought I smelled something," Dave said.

"Well, what are we waiting for?" Roger asked.

"Aren't you going to pour that out and put in water?" Dave asked.

"Why?" Roger asked. "It seems to be doing okay, and I always wondered how it would work."

"Well, you could at least button up your pants," Dave said, turning to follow Bill on down the passage.

The route had led down a forty-foot chimney between two muddy canyon walls a yard apart. At the bottom they stopped to examine a spectacular fossil protruding from the wall. It was a colonial coral about a foot in diameter with black crystallized stems radiating from a common base like a batch of fat pencil leads. The colony of animals had been fossilized in the calcareous ooze on the bottom of an ancient Mississippian sea, hundreds of millions of years ago. The silicious mineral matter was more resistant to solution than the surrounding limestone, and had been etched out by the water that had seeped down the crack to dissolve back the canyon walls.

Bill led the way to the far end of the canyon, where there was a manhole-sized opening in the floor.

"Next time we should bring a rope," Bill said. "Now watch. It's an interesting climb."

Bill crouched on his heels at the edge of the hole and bridged it with his arms. His legs swung free beneath him as he lowered his body into the abyss. In the hole up to his chest, his arms bent, elbows sticking out like a giant grasshopper's legs, he paused a moment. Then, pressing his chest against the rim, he kicked his legs out wide until his feet touched solid rock where the pit belled out below and behind him. Then he transferred his weight to his legs, keeping his feet secured against the wall by continuing to press in a long reach against the rim of the hole with his hands. With his head and lamp turned downward, Bill illuminated briefly

How Bill Austin climbed down the pit in
Unknown Cave.

the thirty-foot drop. Then he allowed his body to sink lower into the hole. Delicately he pushed off with his fingers, bent his knees, and tipped backward toward the wall of the pit, disappearing under the overhang.

"Okay, now keep watching," Bill said. "It's not over yet."

Three heads jammed into the hole to look down on a terrifying scene. Bill picked his way down the wall of the pit. Jagged pinnacles of water-cut limestone jutted upward, ready to rip apart whatever might fall down. The pit belled out to six feet in diameter, too wide to chimney. In turn, each caver climbed down. Dave and Roger Brucker later had anxiety dreams about the pit. Thereafter they used a safety rope, but no one ever relaxed there, for a fall in that confined place—even on a rope—would not be pleasant.

From the bottom of the pit the cavers climbed down a pile of large rocks to enter a passage six to eight feet wide and seven feet high.

"This is the Upper Crouchway. Look at this," Bill said. He had stopped at a place where the red sand floor changed to dark brown mud. In the mud were prints of naked human feet made by two prehistoric cave explorers who had walked along the passage thousands of years ago. Bill and Jack had built a protective fence of rocks around the footprints.

They walked on, concluding that it was more likely that the prehistoric explorers had come through some unknown connection with Salts Cave than that they had climbed down from the Unknown Entrance.

At a junction with a much larger passage, twenty feet wide and nine feet high, Bill turned left. In a few minutes they were in an oval chamber eighty feet long, forty feet wide, and fifteen feet high. Bill said that to the left—up a gleaming flowstone slide—lay some crawlways. Other big passages led off. They followed the low passage that continued in the direction they had been going. They had to crouch, then walk, then crouch again. The passage opened into a perfect oval cross-section, eight feet high by twenty-five feet wide, for many hundreds of feet. The floor was covered evenly with pebbles and slabs of uniform size. There was a sheen of gypsum in the dark gray limestone. The contours were beautiful, and the explorers were silent.

At one spot Bill led them to the right wall and gestured down into a hole. It looked as though someone had cut through the floor of an immense hotel corridor, revealing a slightly offset and even larger corridor just below. "Lehrberger and Reccius found a lot of cave down there," Bill said, "but that's another trip."

They continued along the upper level, squeezing to the left wall of a passage almost filled with great stone blocks for a couple of hundred feet until they came to a pile of muddy breakdown that appeared to seal off the passage. However, there was a black gulf upward between the end of the ceiling and the pile. Their lights flickered over the dim outline of the boulder slope.

"Now this is a bit touchy," Bill said. He did not have to tell them to watch carefully, although he did add, almost to himself, "One at a time."

The pile of wet, loose rocks extended upward out of sight at an angle of nearly forty-five degrees. It looked unstable, as though ready to ava-

Midway up the Brucker Breakdown. Explorer at the top lights up Turner Avenue.

lanche at the slightest touch. Later it was named the Brucker Breakdown.

Bill climbed. There were ominous rumbles and crashes of rock. "Try to keep in the path," he yelled. His voice drifted faintly down to the cavers who were watching from beneath the approach-passage overhang.

"Okay. Next victim," Bill shouted.

The beginning is fun. Dave moved sideways across the slanted face of a fallen ceiling block that is tipped so that a water-worn crack offers toeholds. Then he traversed up the soggy gravel matrix of the boulder pile. Back and forth, stepping gingerly, he held his breath as a boulder or the matrix slid slightly. Bill's tiny pinpoint of light at the top was the goal, and then there were two lights above, and finally three for Roger McClure to aim for as he climbed last. Even with three lights near the top, the room loses none of its immensity. The climb is eighty feet up, and the pile extends another forty feet upward in a nearly vertical mass that makes the ceiling look entirely inaccessible.

When Roger McClure arrived, they set off along a passage that opens into the big breakdown room like a balcony. They walked east, and then south as the passage turned. Its floor changed from mud to crunchy sand

Top of the Brucker Breakdown.

like golden-brown sugar. The walls were smooth and clean. The flat expanse of ceiling was marked by tiny joint cracks that extended a few feet, then disappeared, only to be paralleled by other cracks. The passage cross-section was a flat oval, thirty feet wide and seven feet high.

"This passage goes a long way," Bill said.

"Where does it go?" Roger asked.

"As near as we can figure, it goes right under Edmund Turner's grave in the Mammoth Cave Church cemetery."

"It should be called Turner Avenue, then," Roger said.

"Fine with me," Bill replied.

The passage narrowed and the gleam of gypsum appeared ahead. A shillelagh-shaped stalactite hung down two feet from the ceiling. Its bulbous tip, studded with large gypsum crystals, menaced the walkers, who stooped below it. Beyond were two columns, similarly encrusted with gypsum crystals.

"That's pretty unusual, isn't it—gypsum on onyx?" Roger McClure asked.

"Yup," Bill replied. "But wait a bit. You haven't seen anything yet." Bill walked on silently through a passage encrusted with gypsum. Before a bend in the passage he stopped. "Follow right in my footsteps," he said. "I'll clobber the first guy who gets out of line."

Ahead was a gleaming white column, eight feet in diameter and flared out at top and bottom where it met ceiling and floor. It stood in the center of a perfect flat ellipsoid passage, twenty feet wide. It was multicolored and sparkling from the facets of thousands of crystals. There were calcite helictites, stalactites, flowstone, and gypsum flowers and crust.

"But how can there be a flowstone growing on that one side and gypsum flowers all over, too?" Roger McClure asked. "I thought flowstone was only in wet cave and gypsum only in dry."

"Beats me," Bill said, "but there it is. Special atmospheric conditions, I suppose." There was considerable pride in his voice. It was a spectacular find, beautifully displayed with great restraint alone in the center of the passage. Mineralogically, the combinations were probably matched in few other places in the world.

The cavers walked carefully around the column, looking at delicate needles several inches long, tiny helictites like sprinkles of shredded coconut, and voluptuous globes of flowstone the color of buckwheat honey. Much of the column was powdered sugar coated with white gypsum dust. There were also helictites the size of tangled fingers, and sheets of flowstone drapery as thin as bacon, so that they transmitted yellow light from lamps held behind them. Brown, rust, rose, cream, and white, the New Moon rose into view around one bend in the passage and sank from view around the next bend farther along.

Charlie Fort had named this rococo grotesque the New Moon. He had been so impressed by his first climb down the Unknown Entrance pit that he had said he would not return until they found the moon down there. Jack Reccius told Charlie that they *had* found the moon, so they returned for their seventy-two-hour trip. Unknown Cave was then sometimes called New Moon Cave. The column is along the route later traveled hundreds of times by explorers looking for a way from Flint Ridge under Houchins Valley to Mammoth Cave, and so Flint Ridge cavers began referring to it affectionately as Old Granddad, the name it is best known by today.

"All right, if you were careful then, be supercareful now," Bill said as they moved on. He was a fanatic about preserving the caves as much as possible in their natural state, insisting that trails be narrow and that explorers have very good reasons if they got off them. He wanted the people to whom he was showing the cave to see it as a wild cave, as he first had. He continued to issue warnings as they moved into a fabulous area.

Every surface in the passage was covered with fibers, hairs, needles, crystals, cotton balls, and massive columns of gypsum, mirabilite, and even rarer minerals. Conditions of temperature and humidity are so critical for some of these hydrated minerals that mineralogists who later tried to take some out of the cave for X-ray diffraction analysis opened their specimen bottles in the laboratory to find nothing but a few globs of liquid.

The reflection of light was dazzling. Gypsum sand on the floor looked like snowdrifts. Gypsum crusts were six inches thick on the walls, and

Single-file foot trail down Turner Avenue.

Old Granddad Column in Turner Avenue.

littered the floor. A large piece of hollow gypsum stalactite had fallen and stood upright, so big that a carbide lamp could be placed inside to make the barrel-like sides glow yellow and cream. Cavers always talk of crystal paradises, and there are some spectacular ones in Texas. No one had ever expected to find such treasures in Flint Ridge in Kentucky.

They came to a junction in a large room. Turner Avenue continued straight ahead, while a lower level of the same size diverged slightly to the left. The party walked across the room into the continuation of Turner Avenue. The lower passage was later named Mather Avenue, after the first director of the National Park Service.

"From now on," Bill said, "I expect you to be *really* careful."

Soon they were kneeling, looking under a low ceiling back into an alcove off Turner Avenue. Totally transparent mirabilite stalactites a foot long dangled from the ceiling. They resembled glass more than stone. The mineral was clearer than ice, and looked artificial. Later Fred Benington analyzed similar material to find that it contained threads of sodium hemicalcium sulfate dihydrate, a mineral previously unknown in nature. It was an even more special world than that of the New Moon. Every surface was covered with sprays of mirabilite flowers. The crystals were so unstable that tiny, scintillating flakes fell from the ceiling. The crystal formations were so huge that later Bill was suspected by judges in a photography contest of using a miniature carbide lamp as scale to take a trick photograph of a mirabilite flower so it would look a yard in diameter.

"All right, let's move along," Bill said. "We can't spend all night here if we are to get any surveying done."

Leaving the mineralized area, they ran along the passage, having to stoop here and there, but mostly through high-ceilinged tunnels. These dark, drab, dull passages were more like the usual Flint Ridge cave. But only outsiders thought them really dull. For the Flint Ridge cavers, they offered the excitement of new discoveries ahead. The cavers leaped from rock to rock and trotted over gravel floors. In an hour they reached damp, cool mud floors which were an extreme contrast to the desiccated dryness behind them. From a pool in the bottom of a small pit they drank deeply and refilled their water bottles.

Bill led them into a crawlway, and then into a wide passage covered with mud. They chunked rocks into a deep pit at the left. Then picking their way down a long boulder slope, they came to the sandy floor of a large room with several passages leading out of it. Later cavers named it Argo Junction. Here Bill declared lunchtime, after which they started their survey.

The Ohio bunch had never surveyed like this before. The first compass sighting chained out to nearly 220 feet. Few shots were less than 100 feet. Surveying the twenty-by-twenty-foot passage thus went rapidly. Roger Brucker kept notes, and had trouble keeping the sketch of the passage details up to date with the pace that Bill and the others were setting. Hour after hour the repetitive process continued, setting the points, reading the compass, and chaining the distance. It was indeed real cave, as Stephen Bishop used to say.

"Where does that go?" Roger Brucker asked about a very large passage departing from the left wall.

"It don't go nowhere," Roger McClure twanged in a fake Kentucky accent. "It stays right there."

"Don't know," Bill replied, ignoring the joke that was as old as caving itself. "We haven't looked."

The Ohio cavers were still capable of being amazed. How could anyone just walk past a passage that big and not check it out?

"Maybe you can explore it someday," Bill said. "Meanwhile, *this* is the way to Crystal Cave." He turned down the main passage.

Around the tenth hour they totaled the footage they had surveyed. It was nearly 5000 feet. They were at a pile of breakdown that filled the passage from wall to wall and floor to ceiling.

"This is where we stopped the first time," Bill said.

"Is it the end?" Roger McClure asked.

"No, there's a crawlway around it." Bill ducked down.

"Are we going to survey on through it?" Roger asked.

"Yup," Bill replied.

Surveying in the crawlway after the big walking passages was a tremendous change of pace. It seemed almost an indignity to the Ohio cavers now to have to get down on their hands and knees. Survey stations could be placed only a few feet apart because of numerous bends. But there were compensations. Bill pointed out a nest of gypsum needles, some of them eighteen inches long.

Once out of the confining crawlway, they again racked up 100-foot shots. They climbed a slope to recognize the junction between Turner and Mather avenues that they had passed many hours before. They had surveyed about a mile of the loop. It had been a fantastic trip, and this was a fine place to end the survey.

However, as they relaxed and started rolling up the tape, Bill said, "We can't stop now. The day is young."

They ate another meal, then surveyed another 1000 feet. Some survey shots had to be repeated because Roger Brucker was so tired he forgot them before he got them written in the notebook.

"Only a few more stations and we'll be done," Bill said.

The survey booklet was filled with notes and the surveyors' eyes were bleary. After the tape was rolled up and the compass put away, Bill led them back past the snow-white gypsum, the New Moon column, and down the treacherous Brucker Breakdown. That was enough for one day.

Soon, however, they were back to see more. Not far inside the Unknown Entrance, Bill climbed through a window into a pit down a length of flexible cable ladder that was fixed there. It was an easy climb, with just a splash of water from above. They chimneyed carefully down the waterfalls of the drain into a second pit. Across this small pit was a climb over muddy rocks through a hole leading to another passage as wide as Turner Avenue, but low.

"This is the Lower Crouchway," Bill said, setting off in a duckwalk at a speed that would win races.

"Aren't we close to Salts Cave here?" Dave asked.

"Sure, right through that breakdown," Bill replied.

"Have you checked it for leads?" Roger McClure asked.

Bill's glance was adequate answer. He and Jack had looked rather carefully for a connection right there between Unknown Cave and Salts Cave, but had found none.

They stooped and ran down the passage. In some places they loped along on all fours like bears, with rumps high and heads lifted forward, sometimes scraping the ceiling painfully with their backs. It was tiring, and they tried all variations—sideways like crabs, even spinning around like a dog preparing to lie down. In a low place Roger Brucker tried rolling over and over like a log. Bill's approach was right. There was no comfortable way to cover the half-mile Lower Crouchway. It was best to go fast and get it over with.

"No, we haven't checked them all," Bill said, noticing that the others were looking at some leads behind slabs of rock the size of grand pianos on the floor of the passage. He and Jack had been exploring this new Unknown Cave for only a little over a year.

Then the passage opened up and all sighed with relief at being able to stand up again.

At a junction Bill nodded to a large passage on the left. "Good place to pick up an easy mile of survey if you ever want it," he said.

They bypassed the opening, walking northeast in a fine comfortable passage, thirty feet wide and fifteen to twenty feet high, later named Pohl Avenue. The floor was firm, damp mud. They were obviously very low in the cave system.

They came to some pits with waterfalls pouring in.

"Here's the place," Bill said, stopping at a pit and grinning at Roger Brucker.

"This is where you pounded?" Roger asked. "This is where the connection to Crystal leads off?"

"Yep," Bill replied.

Roger started to climb down. He was ready to check out this connection from Unknown Cave to Crystal Cave.

"No, no," Bill said. "There's nothing to see. Besides, it's up to here in water. Are you sure you want to get wet now?"

Roger grumbled. However, he followed Bill on past a spectacular waterfall flowing over a rim of rock and spraying into a deep pit.

Another walking lead to the left was bypassed as they continued north.

"This is it," Bill announced at last.

They had finally reached the end of Pohl Avenue. The passage terminated in breakdown under the edge of a hill, with water dripping in and flowstone cementing the pile.

"You're in Crystal Cave," Bill said.

"The Unknown Crystal Cave System," Roger Brucker said.

Whatever you called it, at this point Pohl Avenue was well within Floyd Collins' Crystal Cave property.

"Be nice to have an entrance here," Dave said.

"Exactly," Bill replied.

They made plans excitedly. Now they were at the end of the world, far extended in the depths of Flint Ridge. An entrance here would cut hours off each exploration-and-survey trip. It would also be on Crystal Cave property, so no Park Service Ranger could keep them from entering it. They could begin their work in the heart of Flint Ridge legally. Now they traced their long way back along Pohl Avenue to the Lower Crouchway and on to the Unknown Entrance.

Dr. Pohl, as manager of the Crystal Cave property, approved their plans for the new entrance. But before any digging could be begun, a closed survey from the end of Pohl Avenue underground to the Unknown Entrance and back on the surface to the site had to be completed. Another check was made by surveying from Pohl Avenue in Unknown Cave to Eyeless Fish Trail in Crystal Cave through a newly discovered route that was higher and drier than the original connection route. This made possible a second survey loop through the Crystal Entrance and back on the surface to the site of the dig. Each loop consisted of many miles of tortuous surveying. If the ends of the survey loops did not plot out very close to one another, there were errors in the surveys. Jacque Austin helped survey on the surface. The Ohio crew agreed to survey and to re-survey underground until Bill was satisfied with the closure accuracy.

Bill was amused. He had learned the Flint Ridge con from masters. But then, he also had an aptitude for it. Show them the situation so they can figure out what is needed, and then let them beg to help do it.

The next weeks were busy. The telephone wire was extended from Camp Pit in Crystal Cave through the Overlook, down Storm Sewer, up Eyeless Fish Trail, and through Columbian Avenue to the end of Pohl Avenue in Unknown Cave. This unwound ball of Tom Sawyer's string was supposed to be a communications outlet from the depths in case of emergencies, and for the transmittal of data to the surface. In fact, once it was laid, perhaps no more than a dozen parties ever took along with them the clip-on earphones that are required for talking and listening on the wire. The entire circuit has not been tested for breaks since 1957.

On many trips Bill and the Ohio cavers extended the survey and re-surveyed to correct elusive errors. At last Bill was satisfied. Dr. Pohl calcu-

*The Spelee Hut served as field headquarters
and dining area for the explorers through 1971.*

lated that the survey should be accurate within five feet. A stake in the ground near a small cedar tree in Three Sisters Hollow on Floyd Collins' Crystal Cave property marked the end point of Pohl Avenue in Unknown Cave below, the target for the new entrance tunnel. George Morrison, that old secret surveyor of Mammoth Cave, was probably grinning in his grave. The Kentucky cave-country tradition of clandestine underground survey was very much alive.

The Central Ohio Grotto had already staked their claim on Flint Ridge by erecting a small building—the Spelee Hut—on Crystal Cave property. Now the Grotto donated the first cases of dynamite for blasting open the new Austin Entrance to Unknown Cave. Of course, everyone called it the Austin Entrance to Crystal Cave, for the cavers were not supposed to be exploring in any other cave.

Bill bought a red Jeep truck, a farm tractor, a wheezing old Worthington vertical, water-cooled, single-cylinder, belt-driven compressor, a rock drill, wheelbarrows, and shovels. Dave Huber, Burnell Ehman, and Dave Jones organized a press gang to cruise the Antioch College campus in their home town of Yellow Springs, Ohio (where Roger Brucker lived, too).

"Say, how'd you like to spend a really great weekend?" they would ask.

Many non-caving Antioch students helped dig the Austin Entrance, some of whom never did and did not want to enter the cave.

Roger had obtained the total mileage Bill and Jack had surveyed in Salts Cave and Unknown Cave. The Floyd Collins' Crystal Cave mileage was added on, and a map-measuring wheel was run over the Colossal Cave map made by Vaughan and Marshall years before. On the basis of this, Roger and the others wrote a paper to present at the Atlanta meetings of the American Association for the Advancement of Science in December 1955. They wrote that Floyd Collins' Crystal Cave was the nucleus of a vast Flint Ridge Cave System. There were twenty-three miles of surveyed passages, and at least nine more miles of known passages. Most importantly, one passage crossed under a valley as deep as any on the Mammoth Cave Plateau. They concluded that there must be other passages at that low level, and thus that the main caves in Flint Ridge were connected. It was implied that the Flint Ridge caves could also be connected to Mammoth Cave through passages yet to be found under Houchins Valley, which separated Flint Ridge from Mammoth Cave Ridge. The conviction of the Flint Ridge cavers was that in time they would find the connections and survey them.

Brother Nick Sullivan, then president of the National Speleological Society, read the paper at the AAAS meetings, announcing that the Flint Ridge Cave System was the longest in the world.

This trial balloon was immediately shot down by Dr. Alfred Bögli, who for many years had been directing the exploration and mapping of Hölloch, a cave in Switzerland, which then had more than thirty-five miles of

Cross-section of the Austin Entrance.
Bill Austin broke an arm climbing up
to explore the Attic.

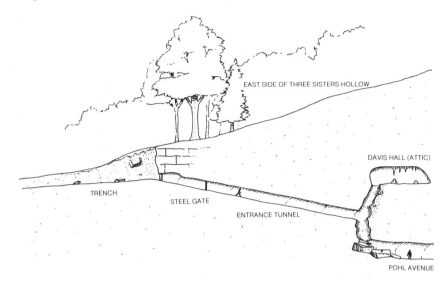

superbly surveyed passages. It did not matter. The Flint Ridge Cave System was definitely on the world map now, right there beside Mammoth Cave.

The paper was intended in part to tell the speleological community what had been found in Flint Ridge. Another motive was to make the officials at Mammoth Cave National Park gag on the myth that Mammoth Cave was 150 miles long. To call a twenty-three-mile cave system the longest in the world—right next door—might force the Park Service to make a realistic statement about Mammoth Cave. At that time, Mammoth Cave was the longest known cave in the world, with approximately forty-four miles of mapped passages. Most of that mileage, however, depended on the survey done in 1908 by Max Kaemper. Kaemper's original survey data had been lost, and survived only in the form of a single beautifully drawn map without scale, north arrow, or surface features. In 1955, no one knew how accurate the Kaemper map was (although later it was determined to be remarkably good). The Park Service officials chose not to defend their claim. The ploy succeeded only in annoying those officials, who knew how long Mammoth Cave actually was, but could not say so because they were bound to the official figure of 150 miles that appeared on millions of pieces of National Park Service literature.

The claim of the Flint Ridge cavers produced a storm that raged in newspapers and letters while the diggers moved tons of earth and rock from the valley wall of Three Sisters Hollow on the Floyd Collins' Crystal Cave property. The tunnel inched toward the flowstone terminus of Pohl Avenue.

Park Service officials dropped by to see what the blasting was all about.

"Puttin' in a back entrance to Crystal Cave," Bill said.

Bill was quite aware that it would take years of work if tourist trails and lighting down Pohl Avenue were ever to be installed. Furthermore, Bill and Dr. Pohl never had any intention of commercializing Unknown Cave from the new Austin Entrance. However, the Park Service officials did not know this, and Bill did not bother to tell them. If they wanted to think that the new entrance was being put in to establish a new tourist route in Crystal Cave—thus enhancing the value of that property—that was their privilege.

On weekends, the blasting and rock-moving were accompanied by the cheers of Antioch students and members of the Central Ohio Grotto of the National Speleological Society. During the week, on slack tourist days, Bill and Kanah Cline would go down to work. One day they set off a blast that surprised them. It sounded like a blast-out, much different from the usual impact against the rock wall. They were not far enough in to have blasted into the end of Pohl Avenue if the surveys were at all accurate, and Bill knew that they were.

Inside the tunnel, Bill found four times as much debris as a charge of that size usually generated. Much of it was sand and clay, and the ceiling had blown upward to reveal on one side a bedrock wall, and on the other a slope of loose rocks. The black hole of a cave passage loomed above.

Bill climbed over the pile and up the steep slope, chimneying delicately against the wall of loose rocks. When he was partway up, a rock slumped down onto his arm, breaking it. He climbed back down and out of the tunnel. He was soon back, arm in a sling, to clear a safe passage up to an attic room full of flowstone, stalactites, stalagmites, and columns. The room had been discovered only because when the tunnel was started the best rock for blasting was found to be eight feet higher than originally planned. On the original line, they would have missed the Attic.

The tunnel was continued under the slope leading up to the Attic. When the distance was covered to the end point over Pohl Avenue, the drill was brought in and aimed straight down. After eight feet it broke through into open space. The hole was enlarged, and finally the ceiling was blasted out of the underlying cave passage. They had hit only five feet off the center of Pohl Avenue.

The Flint Ridge Cave System lay at their feet.

To Explore
the Longest Cave

The explorers organize a power base, the Cave Research
Foundation. A new kind of caving is developed. The National Park
Service opens the caves of Mammoth Cave National Park
to exploration and scientific research.

9 Phil Smith and Roger Brucker were certain that the Flint Ridge caves would inevitably be connected with each other, and that the resulting Flint Ridge Cave System then would be connected with Mammoth Cave. This connection effort would require years of sustained, intense work. Connecting depended on a broad base of exploration and mapping by a large number of cavers, many of whom caved only for science or sport and did not particularly care whether the caves were connected or not. And everyone agreed that only a scientific-research organization could gain access to the caves of Mammoth Cave National Park.

This last was easiest to take care of immediately. Phil and the others were already interested in scientific cave research. Realistically, the big connection between the Flint Ridge Cave System and Mammoth Cave was a dream. The finest cave-research organization in North America, however, could soon be a reality. Phil and the others started organizing the Cave Research Foundation in 1955, and incorporated it in 1957.

The first scientific program of the Cave Research Foundation was cartography. Maps are the necessary base of all other cave research. It was also apparent to everyone that cartographic exploration was the best way to connect Great Onyx Cave, Salts Cave, and Colossal Cave in Flint Ridge with the already integrated Unknown Crystal Cave System. A connection of this Flint Ridge Cave System with Mammoth Cave could come later.

Thus, during the winter and spring of 1955–56, Phil and Roger began planning the exploration, mapping, and scientific study that could be undertaken through the Austin Entrance. The result was that a new kind of caving was developed in Flint Ridge. The new cavers were hard, skilled, and as fit as any who had ever entered the cave. But the style of iron-man

caving perfected and carried to its limits by Bill Austin and Jack Lehrberger was to be no more.

There were several reasons for this. One was simply that Phil Smith
forbade it. Another was that after the opening of the Austin Entrance in
May 1956, it was impossible for cavers to be as alone and remote as Bill
and Jack had been. In the past, only Dr. Pohl and Jacque Austin had
known—roughly—where Bill and Jack were exploring. Jack Reccius,
Charlie Fort, and other Louisville cavers had explored and discovered
some of the passages in Salts Cave and Unknown Cave. However, Bill and
Jack took some trips to places no other cavers knew. Had they not returned, it might have taken weeks or years to find them.

Like many Flint Ridge cavers, Bill and Jack caved at the limits of
both remoteness and endurance. The ultimate was a set of loner trips
taken by Jack. He says little about them, except that he was on his own
for as long as twenty-eight hours, and he speaks fondly of one place—
Cama de Juan—where he slept. He was quite alone.

After completion of the Austin Entrance into Unknown/Crystal Cave,
there was to be no more loner caving.

Surveying techniques were standardized, with the four-man party as
ideal. Elaborate controls included sign-out and sign-in sheets and trip
reports. The maps now became familiar to all.

Oh, there was still remoteness. It has always been possible to get way
to hell and gone in the Flint Ridge caves. The difference was that now the
organization kept track.

Bill and Jack grumbled that the good old days were fast drawing to
a close. They were right: Those days were gone forever. Once big-time
caving was organized in Flint Ridge, it was no longer possible to do real,
old-time, far-out, totally self-dependent caving there. And that may be
why Jack Lehrberger eventually gave up caving.

Phil Smith, Roger Brucker, and Roger McClure had begun with very
romantic ideas about cave exploring. They were soon disabused. Bill's
rinky-dink and the C-3 expedition had shown them that much work and
a long-term commitment would be required to explore the caves of Flint
Ridge. They needed more people.

One of their first recruits—Red Watson—had had no previous caving
experience worth mentioning. Neither had stalwarts Dave Jones and Dave
Huber. Phil and the others decided that lack of prior caving experience
could be to the good. Likely prospects who did not have preconceived
ideas and habits could be introduced to hard caving trips that they would
take to be normal. Most of the dropouts from the C-3 expedition and from
the Overlook discovery trip were seasoned cavers. This was just their
problem, Phil decided. They already knew too much about one kind of
caving, and resisted learning new ways in Flint Ridge. The standard joke
at Flint Ridge was about the caver who insisted that he did not need
knee crawlers, and then, with his knees ground to hamburger, went away
never to be seen again.

The new program worked. It never occurred to many of those, like Red, whose introduction to caving was in Flint Ridge, that one could cave other than all-out, or in any other place. Now what was needed was an organization to make these new cavers acceptable to the officials of Mammoth Cave National Park.

Two people inspired Phil Smith to organize the Cave Research Foundation. Jim Dyer had planted the dream of connecting the big caves. He had trained the modern Flint Ridge cavers, and now his personal style pervaded the new organization. Then Phil was advised, encouraged, and intellectually stimulated by Dr. E. Robert Pohl. As a professional geologist, Dr. Pohl gave the organization scientific stability. He stressed the necessity and the value of having experienced cavers work underground in support of scientists.

Dr. Pohl was crucially important in another way. He opened Floyd Collins' Crystal Cave to exploration. Without this land base on Flint Ridge inside Mammoth Cave National Park, the hundreds of scientific publications of the Cave Research Foundation—and the exploration described in this book—might never have been undertaken.

The early opposition of the National Park Service stimulated two further responses that guaranteed the cohesion and the determination of the Flint Ridge cavers. It made secrecy necessary, because although the cave passages under Flint Ridge were usually entered through the Crystal Cave Entrance and the Austin Entrance on private property, the cavers ranged far out under Mammoth Cave National Park land. And then by opposing exploration in Park caves, the Park Service became the ultimate villain. The cavers were held together by sharing secrets, and because the frustrations of internal dissension often could be vented against the Park Service bureaucrats who did not know their cave from a hole in the ground.

Cavers just as cavers had no clout against the Park Service bureaucrats. Cavers had no long traditions of rights in national parks, as did mountain climbers and skiers. Phil had designed a legitimate program of scientific research, through which the cavers would be able to gain access to the caves in Mammoth Cave National Park. And several of the Flint Ridge cavers were training to be scientists. What was needed, however, was instant scientific prestige for the fledgling Cave Research Foundation. Phil found that there were very few established cave scientists in the United States, and that some of these had no interest in being labeled as cave nuts through association with cave explorers.

The problem of how to establish the Cave Research Foundation was solved, not by trying to attract already established scientists, but by helping college students to become cave scientists. Some who started in Flint Ridge are now professors who direct their own students to Cave Research Foundation projects.

Friends and foes alike were not slow in pointing out how fortuitous was the incorporation of the Cave Research Foundation. It allowed the

Flint Ridge cavers to crawl under the prestigious umbrella of science, and thus gain permission to do exactly what they had always wanted to do— explore and map the caves of Mammoth Cave National Park.

Exactly.

Superintendent Perry Brown weighed the new Cave Research Foundation image against the backgrounds of its members. Dr. Pohl and Bill Austin were running a private cave—Crystal Cave—within Mammoth Cave National Park. The C-3 expedition had resulted in publicity that increased the value of Crystal Cave, which the National Park Service was trying to buy. Robert Halmi's *True Magazine* article had stressed danger, and Superintendent Brown was not interested in encouraging dangerous activity in the Park. Joe Lawrence, Jr. and Roger Brucker had written *The Caves Beyond*, in which some of the ineptitude of the cavers had been outlined. Finally, the Superintendent knew or suspected that some of the new scientists had in the past trespassed in the caves of Mammoth Cave National Park. It was hard for anyone to ignore the loud rattling of skeletons in the Cave Research Foundation's closet.

However, Phil Smith's hundreds of hours of negotiation with Park Service officials paid off. In 1959 the National Park Service signed an agreement allowing the Cave Research Foundation to conduct scientific research in Mammoth Cave National Park.

The Foundation and the Service achieved much mutual understanding over the years. One reason for this is worth comment. When Gifford Pinchot formed the U.S. Forest Service, he designed a system of rotating management, so that all Forest Service officials would move every two years or so. In this way he intended to make his personnel loyal to the Forest Service itself, rather than to any local situation or interest. He further instilled a desire for rotation by making almost all promotions dependent on transfer. The National Park Service rotation and promotion plan was based on that of Gifford Pinchot's Forest Service.

From 1954 through 1972, a small central core of cave explorers and scientists had a stable work base. During those eighteen years, six different superintendents and as many chief naturalists passed through Mammoth Cave National Park.

What happens in such situations is that the amount of specialized knowledge about a park that can be amassed and comprehended by the long-term members of an organization like the Cave Research Foundation often exceeds that of the transient management of the park itself. Depending on many factors, such local groups either are resented by Park Service officials, or are viewed by them as valuable resources. Over the years, Cave Research Foundation contributions have been seen as valuable, not just for Mammoth Cave National Park, but for the Park Service as a whole wherever attention is focused on caves.

If one man had remained as Superintendent at Mammoth Cave National Park during those eighteen years, would he have been able to accept the transition of some of the cavers from college kids to university professors conducting major research projects? Or would he have maintained old suspicions and antagonisms? Doubtless he, too, would have grown. Ironically, the Park Service Ranger—Joseph Kulesza—who caught Jack Reccius trespassing in Unknown Cave in 1955 had rotated full circle by 1972. Joseph Kulesza was Superintendent of Mammoth Cave National Park and was encouraging underground exploration and research when the connection between the Flint Ridge Cave System and Mammoth Cave was made.

Jack Reccius examines a stalactite and stalag-
mite that have grown together to form a column,
1974.

Joseph Kulesza, 1951.

Lost

Getting lost is a serious risk. Bob gets
lost and Jacque remains calm.

 Soon after the agreement was signed between the Cave Research Foundation and the National Park Service concerning research in the caves of Mammoth Cave National Park, a chilling incident occurred. In August of 1959, Roger Brucker, Roger McClure, and Micky Storts were surveying in Ralph's River Trail, a passage that trended north from the southeastern end of Pohl Avenue. It had been a fine trip. They walked along surveying in ankle-deep water between light gray limestone walls that glistened beautifully in the light of the carbide lamps. Survey stations were going in rapidly when they heard a caver splashing up behind them at high speed. It was Dave Jones.

"Have you seen Bob Rose?" he asked as soon as he saw the surveyors.

"No."

"Then he's lost," Dave said. Catching his breath, he explained. "Bob said he had a cramp in his leg and wanted to return to Ralph Stone Hall to wait for us until we completed our survey. He obviously needed a rest, so we said sure. When we got back there, he wasn't there."

"Did you look around?"

"Sure. He just isn't there. The others are looking for him. We need more help."

Micky and Roger McClure had already started putting up their surveying gear.

"No note?"

"No. Nothing. He just isn't there. I thought maybe he got turned around and came out this way. He could be anywhere between here and there."

"Maybe he just went on out of the cave," Micky said.

"Maybe," Dave said. "But it isn't like Bob not to leave a note."

By this time they were moving back toward Ralph Stone Hall. At each pit and canyon they played their lights downward, hoping not to find what they were looking for.

They decided to send one man out of the cave to tell the cavers topside that a full-scale search and rescue might be necessary. The situation was explained carefully—Bob's condition when he was last seen, the locale, and the condition of the party remaining.

"I don't like sending another man off alone," Roger Brucker said, as Don Peters set off on the run toward the Austin Entrance. "Why did you let Bob go off alone?"

"He seemed capable of taking care of himself," Dave answered.

Dave was uncomfortably aware that Bob sometimes did need help. Bob had diabetes, but this had never stopped him from caving, or climbing, or canoeing, or doing anything else he wanted to do. He was determined and competent, but he was also somewhat happy-go-lucky. Two or three times he had gone into insulin shock. He might be in shock now. Roger wished he had not forgotten to tell Don to ask for candy bars or, better, sugar to be sent down. However, Jacque Austin would remember.

A couple of hours later Don trotted into a late-night camp. Only Jacque, camp manager for the weekend, was awake.

"Bob Rose is lost," Don panted. "We have to wake everyone up and go look for him."

"Sit down," Jacque said. "Don't talk until I get you a cup of coffee." When she had him seated with a cup in his hand, she said, "Now. Just tell me the whole story from the beginning. Softly, you don't have to shout. I'm sitting right here listening, all ears."

When Don had told his tale, Jacque sat thinking for a moment. Then she said, "All right, you've done all you can do now, Don, and you've done it well. Now what I want you to do is go to bed and get some rest. And don't you say anything about this to *anyone,* or I'll skin you alive. And don't you dare wake anyone up. They've all had hard cave trips today, and they need their sleep."

"But—but—" Don began.

"You'll need your sleep, too," Jacque interrupted him. "If we have to conduct a rescue here, we'll all need all the rest we can get. So finish eating, and then get right to bed. I'll certainly call you if you're needed."

After Don had gone, Jacque poured herself another cup of coffee. She thought over who was in the cave, looking for Bob. She knew them all well, and she knew Bob. And she had known caves and caving since she was a child. She resolved to do nothing until morning. She sat where she was. She was not even particularly worried. Not yet, for goodness' sake. Bob had been lost—if he really were lost—for barely four hours.

Belowground, the group separated into teams of two persons each to explore every lead between Ralph Stone Hall and the end of Ralph's River Trail, if need be.

It was slow work. However, the urgency of the search keyed them up. They were half afraid that they might find Bob in real danger, or injured, or dead. And they were half afraid that they might not find him at all.

"Why do you let someone with diabetes cave?" someone asked.

Roger Brucker grunted, but said little in reply. It had been Phil's decision. "All right," Phil had said, "he does good work, and we can use him. If he wants to take the chance, then it's his choice. Just keep good watch on him. The Park Service would have kittens if we had a diabetic who had an accident in the cave."

Roger had agreed that Bob was a good worker, but now he was beginning to worry that they had gambled too much. However, he still basically approved. Phil's decision was the beginning of a long tradition among the directors of the Cave Research Foundation. When Red Watson was president some years later, he had to decide whether or not to let Burnell Ehman continue caving. Burnell was over fifty, and he had an odd heart condition: Burnell's heart was perfectly strong, but the prognosis was that someday it would just stop—maybe when he was ninety. Some people thought Burnell ought to be barred from caving in Flint Ridge because of this. "Just as well in a cave as anywhere else," Red had said. And then as an afterthought, "Better in a cave than anywhere else, as much as Burnell loves caving."

Bob Rose, also, had been allowed his own choice.

Roger shivered involuntarily as he imagined how they would feel if they did not find Bob. And how would Bob feel, alone? Each caver had enough carbide to provide light for about twenty-four hours. There was also food and water for about that length of time. This meant that soon the searchers would have to leave the cave for new supplies.

Roger poked his head into an obscure lead high above the main route. He was surprised to see a dim light shining from around the corner. Then he heard a shout.

"Who's there?" Roger asked.

"Bob Rose."

"Don't move. I'll be right there," Roger said.

Bob was sitting comfortably.

"We found him," Roger yelled back down the passage, and soon all the searchers were jammed into the small room.

"I thought I knew the way back, you know," Bob said, grinning. His bony face was dominated by big eyeglasses. He was very pleased to have an audience. "I didn't know I was off the route until I reached this little room. I built a cairn, set up my trusty candle, and started looking back the way I had come. The floor was hard stone with no tracks, so I decided to wait it out. I shut off my light to save water and carbide, and yelled and pounded a rock on the wall periodically. Then I decided to put the candle back in the passage entering the room to guide searchers. There wasn't much else to do but eat something and sit and wait for someone to find me, which I did"—he looked around happily—"and which you did."

Back in Ralph Stone Hall, the cavers sat down to eat a meal. However,

Roger Brucker was visualizing what might be happening on the surface. The message they had sent might have started a frenzy of activity. A rescue crew was probably on the way. So he and Roger McClure rushed on out of the Austin Entrance to turn back the rescuers. They hoped Jacque had not phoned Phil.

All was quiet on the surface.

"Sit down and have a cup of coffee," Jacque said. She was amused at the two Rogers' obvious consternation. "I haven't told anyone anything yet," she said.

They sighed with relief.

"Found him, eh?" Jacque said. "I thought you might. Is he all right?"

Roger nodded, and so did Jacque. Down in the cave they had kept busy searching. Her job had been harder.

Bob Rose continued caving for many years.

The Colossal/Salts Link

Explorers trying to connect Colossal Cave with
Unknown/Crystal Cave surprise everyone.

11 Unknown/Crystal was one great cave system. Other caves were tantalizingly close, but so far only one link had been connected in the great chain of Flint Ridge caves. To the northwest, Great Onyx Cave was 600 feet from Crystal Cave. To the southeast in Salts Cave, Indian Avenue was 160 feet from the Lower Crouchway in Unknown Cave, and Upper Salts Avenue was 600 feet from the Overlook in Crystal Cave. Farther to the south, the River Route in Colossal Cave was 600 feet from Argo Junction in Unknown Cave. A connection between Colossal Cave and Salts Cave seemed less imminent, because the closest surveyed points in them were nearly a mile apart. However, some of the long passages in Lower Salts explored by Jack Lehrberger had not been surveyed.

Bill and Jack had failed to find a connection between Unknown Cave and Salts Cave, despite their closeness. The next-best prospect for connection seemed to Phil and Roger to be between Argo Junction in Unknown Cave and the River Route in Colossal Cave. It was thus decided to ask permission to explore in Colossal Cave, to work toward a connection from both sides. The Park Service gave permission immediately. Phil Smith led a number of survey trips down Grand Avenue of Colossal Cave. And how grand it was! It was a different kind of cave from the wilderness out Left of the Trap and in Eyeless Fish Trail in Crystal Cave. It was not the pristine virgin passages of Turner Avenue in Unknown Cave. There a caver felt the awe of being the first person in passages of alien stone not made for man. In Colossal Cave, one walked through large passages on man-made trails back into history.

There was a bit of history in Floyd Collins' Crystal Cave, of course. To find Floyd's rusty bean cans in the Lost Passage gave one a sense of warmth, of continuity, and of understanding with a single man long dead who had also loved to poke down crawlways to see what was there. But now, in Colossal Cave, the surveyors trailed steel tapes along the footsteps of hundreds of predecessors. They stared up at the heights of fabled Colossal Dome, and read on the walls signatures of tourists from long ago. In Colossal Cave there was a constant sense of many old-time explorers who had gone before.

Phil launched one party to search for Colossal River, a stream known from an old map, by hearsay, and by the existence of a smoked message over the entry to a low crawlway that said: "ROY HUNT—TO THE RIVER." Some wet low-level passages were found, but nothing that really resembled a river. This was not surprising, because the old-timers were liberal about what they called a river. Mammoth Cave had a big river with a boat ride on it. Therefore, Colossal Cave also must have a river, even if it were a mere trickle. Beyond that, however, these muddy passages were only 600 feet from Argo Junction in Unknown Cave.

While Phil and others explored in Colossal Cave, Dave Huber, Burnell Ehman, Fred Benington, Micky Storts, and others tried to find a way through or around the breakdown in Argo Junction. They moved boulders. They dug through mud. They jammed themselves into the tiniest of leads. Nothing went.

Roger Brucker led the 1960 August expedition, a week-long affair with thirty or so people on hand for the weekends, and fifteen remaining throughout the week. A promising area was Candlelight River in Unknown/Crystal Cave, named by Micky Storts, who had set up her own candle beside a candlelike stalagmite. It was far from the Austin Entrance, and it tired out explorers without results. Jack Lehrberger then led another attempt to find Colossal River in Colossal Cave. Perhaps it connected with Candlelight River in Unknown/Crystal Cave.

Jack's party carried survey gear, but Roger Brucker was resigned to getting no survey. A survey chain was a tether to a free spirit. However, if anyone could find Colossal River, it would be Jack. True to form, Jack's party brought back an empty survey booklet and the description of a route to a stream of water that might be Colossal River. The important thing was that it was a going lead.

Roger Brucker then led Roger McClure and Spike Werner out to survey Jack's marked passage. It was a low, dry crawl to a canyon that intersected a ten-foot-wide passage with a stream in it. Although water flow was low, it had entrenched itself deeply in the limestone, making a sinuous canyon passage generally wide enough to walk in. The explorers stuck to the top of the canyon, and emerged in a passage twenty feet wide. Here the water dived under a pile of rocks, so the plan of following the water until it emerged as a tributary of Candlelight River in Unknown/Crystal Cave seemed foiled.

The explorers stopped to rest while deciding what to do. They could continue surveying down the wide passage they were in, although it got lower immediately and was terminated by breakdown. They could go back, because they had already surveyed a substantial portion of the passage.

As they talked, Spike looked at the floor, thinking. As deeply as the water had entrenched the passage they had been surveying, surely it would have cut just as deeply under the rock pile. Deliberately, Spike moved a slab of rock. At once his hypothesis was confirmed. He had revealed a hole wide enough to chimney down into. Roger gave Spike fifteen minutes to check the passage. Spike scrambled about 600 feet along the lower level before returning. He had gone far beyond the breakdown that closed off the upper passage. Tired but excited, they left Colossal Cave.

The next morning Roger sent a fresh team in to explore the passage in Colossal Cave now called Lehrberger Avenue. How does one make up such a party? Its members must have knowledge, experience, and push. Roger sent a party of people who had almost everything the years of Flint Ridge experience could provide. First, Spike would be leader. He knew the way, and was a natural caver. Then there was Jack, for experience. He had almost dropped out of active caving, and was soon to marry and move far from Kentucky cave country, but he had been lured back for a fling in Colossal Cave. Finally, for push, there was a long-legged nineteen-year-old biochemistry student from Ohio State University. Dave Deamer was overwhelmed by Flint Ridge caving, and by the people of the Cave Research Foundation. Years later he said it was the turning point of his life. He had not known that such ambitions and experiences existed outside books. He was awed almost to inarticulate deference by Jack. Nevertheless, Dave had been caving for three years, and knew his way about. He would follow Spike and Jack anywhere, and do anything.

The party, then, had knowledge, experience, and push.

Lehrberger Avenue in Colossal Cave turned out to need plenty of push. The explorers spent hours crawling from level to level through multi-tiered canyons and cut-arounds. Roger had predicted the discovery of a great roaring torrent—he always predicted such discoveries—but Spike, Jack, and Dave found nothing of the sort. Finally they crawled through a small—but still very wet—stream on their bellies to emerge in a pit just large enough for all to sit up. They were wet and cold.

Resting in the pit, Spike decided that something must be physically wrong with him. He was having painful cramps. Sitting in the pit, he dreaded the low crawlway and all the narrow canyons on the way back to the entrance of Colossal Cave. Anything was better than taking his aching body back through the contortions of Lehrberger Avenue.

The pit looked difficult to climb, but there was a ledge fifteen or so feet up. The higher-level main passage of Colossal Cave might be reached that way. Spike climbed while Jack and Dave watched. After a few moments a faint yell came down.

"Hey! Walking cave!"

The distance and the echoes sounded like big cave. Jack and Dave quickly climbed up to find Spike looking into a passage the size of a railroad tunnel that led away into the blackness.

"Hot damn!" Jack said.

Everyone forgot Spike's cramps and ran down the muddy passage. It looked virgin, but Jack recalled a similar passage that he had visited several years earlier on a solo trip in Salts Cave.

"We should come to my initials soon," Jack said.

"Great balls of fire!" Spike shouted when they came to Jack's initials—"J.L."—smoked twelve inches high on a rock. Spike mocked the dawning of a great light. "We have connected with Salts Cave!"

Jack confronted Dave with a solemn expression. "Dave," he said seriously, "you haven't been caving long enough to deserve a treat like this. It is a once-in-a-lifetime experience. I hope you appreciate it."

Dave could only nod in reply.

Then they raced northeast, toward the entrance of Salts Cave. With Jack in the lead, there was never a thought of returning to Colossal Cave. Spike yelled for his long-legged companions to slow down. Jack paused at confusing junctions, but he always remembered the way. They quieted down.

"It's easier to go out Salts than back through the crawlways to Colossal," Jack said, "but it'll be long, perhaps five hours. Have you got enough carbide?"

Spike and Dave assured him that they did. What Jack did not say was that the trip might be even longer than five hours, for there was a locked National Park Service gate at the Salts Entrance. No one aboveground had any idea that they might connect Colossal Cave with Salts Cave, so there would be a long trip back to the Colossal Entrance if the Salts Cave Entrance gate were impassable.

"Salts is awfully dry," Jack said. "Conserve your water."

They had enough carbide and water to get to the gate all right. Getting all the way back again to the Colossal Entrance if they could not get through the Salts gate might be touchy. Jack had been on narrow-margin trips before, however, and if they had to use only one carbide lamp on the way back, they would still make it.

Some hours later the weary party reached Indian Avenue—a walking passage eight to fifteen feet high and five feet wide. Ledges covered with red sand jutted out into Indian Avenue. The walls were the color of coffee with cream. Jack Lehrberger and Jack Reccius had found a way into this beautiful passage from the upper levels of Salts Cave some years before. Jack Lehrberger knew it like the back of his hand, and now he began to whistle softly. They were home, but Indian Avenue was long. And after it would be a bit of difficulty with a gate.

The Corkscrew is a narrow succession of spaces between angular breakdown blocks. It is a shortcut that Jack and Bill Austin found connecting upper to lower levels of Salts Cave. The total vertical distance is about twenty feet, but in this distance you must twist, turn back upon yourself, kneel, and adjust. Anyone with legs as long as Jack's or Dave's could get into real difficulties. Phil Smith was once stuck in the Corkscrew about fifteen minutes, although he believes it was longer. It would be easy if your knees bent backward as well as forward.

Jack went up through the Corkscrew first, remarking that he and Bill always left a boulder over the hole at the top to conceal it, and it might be a problem to push that boulder aside. However, the boulder was gone. Then Jack talked the other two up. It went slowly and took several tries. But when they stood up in the huge upper passage of Salts Cave, Spike and Dave forgot their insulted bodies. The immense blackness of Salts Cave swallowed the tiny yellow pools of light from their carbide lamps. Spike and Dave had thought Grand Avenue in Colossal Cave was large. Now they were in a passage 100 or more feet wide and thirty to fifty or more feet high. They could say nothing. They had never been in such a cave. They put one foot in front of the other to follow Jack down the time-worn trail, half a mile to the Salts Entrance. When they came to Mummy Valley and worked their way down 100 feet into that vast abyss and then 100 feet up the far side, Spike could finally stand it no longer.

"What a monster!" he screamed with delight. Jack and Dave laughed while Spike had his full say, and then they plodded on through those dim corridors. Salts Cave is darker than any other cave in Flint Ridge because walls and ceiling are covered with soot left by the torches of prehistoric cavers 3000 years ago.

Jack was satisfied. This expedition was his last in Flint Ridge. Salts Cave was his special love. He had been the first to tread in the footsteps of those ancient cavers on the lower levels, and now Salts Cave had rewarded him again.

"There may be some problem with the gate," Jack said, as they came over a rise and started down into yet another immense, dark room.

"What?" Dave and Spike asked with alarm.

"It may be locked," Jack replied.

"Oh," Spike said. He calculated how much water and carbide he had left. It would be close if they had to go back to the Colossal Entrance. They had better get out the Salts gate.

Dave was even more apprehensive. It had not been long since he had read *Tom Sawyer*, and he recalled with horror the story of Injun Joe, trapped behind a locked door in a cave. Injun Joe had eaten a few bats. They had not seen a bat since leaving the Colossal Entrance, many hours before.

Jack walked slowly up the slope toward the gate.

"It was fixed once so you could get through, but that was some time back," he said. "The Park Service may have found it and repaired it by now. We'll see."

Spike and Dave strained to see daylight showing.

"There's still another big room on beyond the gate," Jack said. "They

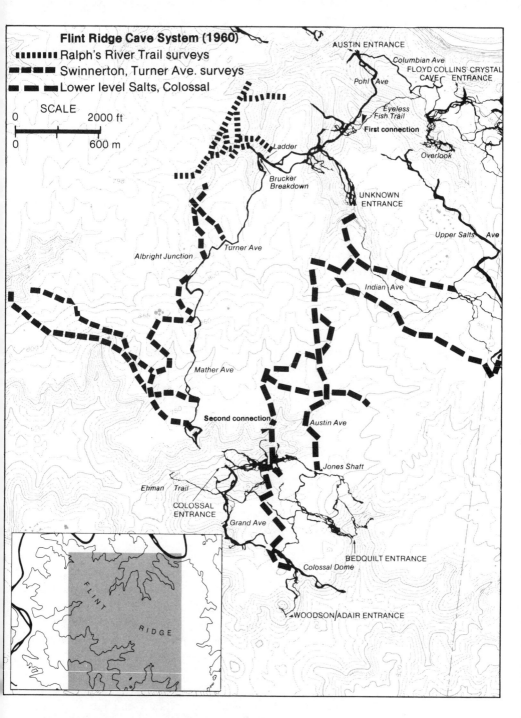

Flint Ridge Cave System (1960)

▮▮▮▮▮▮ Ralph's River Trail surveys
▬▬ ▬▬ Swinnerton, Turner Ave. surveys
▬▬ ▬▬ Lower level Salts, Colossal

SCALE
0 ———————— 2000 ft
0 ———————— 600 m

AUSTIN ENTRANCE
Columbian Ave
FLOYD COLLINS CRYSTAL
CAVE ENTRANCE
Pohl Ave
Eyeless
Fish Trail
First connection
Overlook
Ladder
Brucker
Breakdown
UNKNOWN
ENTRANCE
Upper Salts Ave
Albright Junction
Turner Ave
Indian Ave
Mather Ave
Second connection
Austin Ave
Jones Shaft
Ehman Trail
COLOSSAL
ENTRANCE
Grand Ave
BEDQUILT ENTRANCE
Colossal Dome
WOODSON/ADAIR ENTRANCE

FLINT RIDGE

Map of the Flint Ridge caves in July 1960
includes newly surveyed passages in Ralph's
River Trail, Swinnerton Avenue, and Turner
Avenue areas of Unknown/Crystal Cave. An
explosion of surveying followed the August
1960 discovery of the Colossal/Salts Link.

put the gate here where the breakdown comes almost to the ceiling and the walls are close so they wouldn't have to do much building."

The gate came into view in the light of their carbide lamps. It was about four feet square, set in rough stone walls two feet thick. The rusty bars cast long shadows down the steep slope on the other side.

Dave and Spike watched anxiously as Jack examined the gate. He looked at it a long time.

"Uh-huh," he said at last, unable to restrain a grin. "How would you all get out?"

Spike examined the gate carefully. Then he reached through the bars with a rock in one hand. He pounded sideways on the edge of the large wooden sign that was strapped with steel bands to the bars on the front of the gate. The sign moved a few inches sideways. Then Spike pushed on the back of the sign. Nothing happened. He pushed again. There was a dry creaking, and suddenly the whole sign swung out like a door. The opening through the bars of the gate was a foot high and eighteen inches wide.

"Congratulations!" Jack said. "You see," he went on, "someone cut segments of the bars out behind the sign, leaving an opening large enough to get through, but covered by the sign." The hidden opening extended along the sign's length except for one bar on which the metal straps holding the sign acted as a hinge. All you had to do to enter Salts Cave was to move the sign to the side a little and pull it open.

"Who fixed the gate that way?" Dave asked.

Jack flashed a dazzling Flint Ridge smile, but said nothing as he slid gracefully through the hole in the Salts gate. Dave and Spike dived after him. Jack carefully closed the sign door again, then turned to go down the steep slope into the large vestibule room. They could see welcome sunlight sparkling through the waterfall that sprayed down into the entrance of Salts Cave. Jack never looked back, but before going down that last slope, Spike and Dave did. Their carbide lights played over the sign on the gate.

In large letters it said: "NO ADMITTANCE."

Spike bowed, and declaimed, "Kind sirs, no offense. We were exiting."

Afterward, Dave Deamer beamed with joy. Spike Werner wore his pride modestly, telling the story of the Colossal/Salts Cave connection no more than 365 times during the next year. Jack Lehrberger kept his private counsel. Roger Brucker had glowed with excitement and envy when he heard. It was the second connection he had helped to set up and then just missed.

On the remaining expedition days, surveyors poured through Werner's Colossal/Salts Link to map the lower levels of Salts Cave. There was intense satisfaction in this. The Park Service had not yet given permission to work through the Salts Cave Entrance, but who needed it?

Now there were two long caves in Flint Ridge—Unknown/Crystal Cave and Colossal/Salts Cave—but neither one was yet as long as Mammoth Cave. And Mammoth Cave itself was not now the longest in the world, for its forty-four miles had just been surpassed by Hölloch in Switzerland. Like the Flint Ridge caves, Hölloch was growing rapidly as the result of an active exploration program. However, if Unknown/Crystal and Colossal/Salts were connected, the resulting Flint Ridge Cave System would some day catch even Hölloch. It would be the longest cave in the world.

Futility in Great Onyx Cave

Explorers try to connect
Great Onyx Cave with Unknown/Crystal Cave,
but are stopped dead.

12 The Flint Ridge Cave System would be the longest cave some-day only if Colossal/Salts Cave could be connected with Un-known/Crystal Cave. No one wanted to push the endurance bar-rier Left of the Trap in Crystal, however, and from the Colossal/Salts side nothing obvious headed toward Unknown/Crystal. Upstream Candlelight River in Unknown/Crystal ended in a dome. Out at Argo Junction in Unknown Cave a great pile of breakdown stopped exploration toward Salts Cave. Another breakdown pile below the Unknown Entrance had so far cut off exploration toward Salts. Colossal/Salts and Unknown/Crys-tal grew by leaps and bounds as new exploration continued, but not to-ward each other.

No one was overly distressed. Most of the cavers were primarily in-terested just in the adventure of caving. Even connection fanatics like Roger and Red were satisfied with the general round of cave exploration and discovery. Besides that, there was a new cave in which to play.

Like Crystal Cave, Great Onyx Cave had been purchased by the gov-ernment and made a part of Mammoth Cave National Park. The Park Service had opened all Flint Ridge caves to Cave Research Foundation work, so for the first time Great Onyx Cave, with passages lying only a few hundred feet from Pohl Avenue in Unknown Cave, could be explored. Micky Storts, Fred Benington, Dave Huber, and Burnell Ehman decided to connect Great Onyx Cave with Unknown Cave. There was a story that Edmund Turner had discovered Great Onyx Cave from Salts Cave. There were rumors that Lucy Cox, co-owner and manager of Great Oynx Cave before the Park Service bought it in 1961, had hired cavers from Indiana to explore some long crawlways off the tourist route. It was said that then she had had these passages sealed off with rocks and sand. She had not been interested in having cavers coming underground from, say, Crystal Cave to pop their heads out an opening into Great Onyx Cave.

The best exploration route in Great Oynx Cave—out the drain of the deepest pit—had evidently been dammed to make a pool for a boat ride. Explorers probed, but could find no way out of that pit. The next move was to search for lower levels under the large walking passages. Water always sinks below the floors of large passages to carve narrow tubes and canyons. Often these parallel the main passages, but sometimes they trend diagonally away from the passage that mothered them. If a caver can find a pit this way, he can climb down or up to other tubes or canyons through which big passages on other levels might be reached.

A lower-level passage a few hundred feet long was discovered, but it had no leads, and was so small that Micky Storts—not over five feet tall and weighing about 100 pounds—had to survey it practically alone, shouting data back to Fred Benington, whose forty-four-inch chest prohibited him from enjoying the lesser delights of the Flint Ridge caves.

Because it seemed impossible to go down, Dave Huber started to dig through the sand that filled a side passage. The easy dig was extended thirty or forty feet. At the end the passage is still filled with sand.

The explorers were baffled. It was true that sand and rock had been piled along the walls of the main passages during trail construction. Crawlways could thus have been concealed, whether intentionally or not. Expert cavers walked the trails, watching the wall features. Here and there hunches were foiled by a little digging at the side walls. Nothing appeared.

One day some cavers checked behind a rotting wooden stairway in Great Onyx Cave to find a hidden crawlway stretching a long distance. It had been used as a trash dump for accumulated flashbulbs and candy wrappers.

Denny Burns led a party in to survey the find. There was first a difficult climb, and then hundreds of feet of belly-crawl. On the way out, Denny noticed that the surveyors had crawled in under a massive, unsupported limestone block in the passage. It teetered as they passed beneath it. If it fell, it might crush someone, or trap a party exploring behind it. So Denny and Jack Freeman designed an Erector Set of steel angles to place under the block. Beyond, the passage led to a vertical shaft. With great difficulties, Joe Davidson bolted up to a shadowy alcove that turned out not to be a passage. Great Onyx Cave had won its independence again. There seemed to be no way to connect it with any other cave in Flint Ridge.

The only other possibility was a canyon passage a few hundred feet long. The difficulty was that it was near the top of the cave, so any extension would soon run into hillside breakdown. Any lead out of the canyon would have to go down. Furthermore, this canyon felt deader than any other place in that dead cave. A quarter of an inch of dust covered the large breakdown blocks that impeded travel in every direction.

Red Watson led an exploration party into this side passage. Because it was close to the Great Onyx Entrance, Tom Brucker, now aged ten, was allowed to go along. He was thin and small. He was also wild to explore caves. Red kept watch on him while several other cavers poked out leads. Tom was energetic, and not inexperienced in caves. He pushed up between two boulders.

"It goes big! I'm going to look," he yelled back excitedly.

"All right," Red replied. "But don't go far, don't get out of earshot, and . . ."

Tom had fled. Red shouted. No answer. If this raised a bit of anxiety, it also raised interest. Red poked his head up between the two boulders. There he was confronted by a tiny triangular hole. Nobody larger than Tom could have gotten through it, and Red was not quite convinced at first that Tom had. Red took off his carbide lamp and shoved the light through the hole. Yes, a passage opened up. One could easily walk back there. However, it appeared to be just part of the canyon, closed off by big breakdown blocks. Red yelled again. This time Tom answered, from not far away.

"It goes big."

"That's okay," Red said. "Come back now."

"I'm going to look the other way," Tom replied. "It's *walking* cave."

"How big?" Red asked.

"Big!" Tom shouted.

"Well, be careful," Red said. "No one can get in there to help you, you know."

Tom was gone again. He climbed fifty or sixty feet in and around boulders from one end of the room to the other. "The ceiling is full of big rocks hanging down," he yelled back.

"Come on out," Red shouted. He could hear Tom scrambling. "At least tell me what you're doing."

"There's a little hole here at the end," Tom said. "I'm going to try to get through."

"Tom! Come out of there, *now!*" Red knew that orders would not work, so he added, "We'll find a way in from the other side."

Tom did not reply. After another five minutes that aged Red considerably, Tom consented to leave the room he had discovered. He proudly drew a sketch of it in Red's survey booklet. No one has ever been back. Red swore that he would never take that kid into the cave again. He also decided that there was just no way to connect Great Onyx Cave with any other cave in Flint Ridge.

Others still maintained hope. Pohl Avenue has high-level leads going toward Great Onyx Cave at approximately the same elevation as the main passage in Great Onyx. Roger Brucker for many years continued periodically to get the feeling—in Pohl Avenue, which is *live* cave—that one of those passages would lead to Great Onyx Cave. So he sent himself and others to pound heads against the pinch-downs, the mud and rock fills, and the flowstone terminations of those small crawlways. As yet, the living passages of the Flint Ridge Cave System have not been joined with the dead ones in Great Onyx Cave.

Across the Top of Colossal Dome

Adventure is found looking for a passage
the old-timers thought would
connect Colossal Cave with Mammoth Cave.

13 Spike Werner had come to Flint Ridge as an experienced caver, and on his first trip he had discovered the crawlway that on his next trip took him to the connection between Colossal Cave and Salts Cave. For a while thereafter, Colossal was the only cave for him. The best thing about his new love, he said, was that at last he had found a cave in which he could get lost. Dave Deamer also learned most of Colossal Cave. It seemed appropriate for this pair to do something that had to be done in Colossal Cave.

Around the end of the nineteenth century, the Louisville & Nashville Railroad reportedly offered their Colossal Cave manager, L. W. Hazen, $500 to find a connection between Colossal Cave and Mammoth Cave, 3000 feet to the southwest. Hazen subsequently scratched his name and "MAMMOTH CAVE" on the wall of a passage near the top of Colossal Dome in Colossal Cave. He did not collect the $500, but could he have found a likely lead? Did he scratch "MAMMOTH CAVE" on the wall as a joke? Was he just hoping? In any event, there were leads up there that needed exploring. Maybe Hazen did know a way from Colossal Cave to Mammoth Cave. Before the Cave Research Foundation explorers could look for it, a survey had to be run across the top of Colossal Dome to tie those passages into the map of Colossal Cave.

Colossal Dome is an enormous vertical shaft. Its floor dimensions are forty by 120 feet, and its walls soar upward 130 feet. At the top it is more of a canyon than a typical cylindrical shaft, for the walls come together until in places they are no more than six feet apart. At the very top, there is a narrow ledge of shaly limestone along one wall. The ledge is level in places, but generally slants into the pit. Nowhere is this ledge wider than eighteen inches. There are only two or three feet of headroom, which pinches down to the ledge along the wall.

You can reach the top of Colossal Dome from the tourist route. You climb up an old wooden ladder fifteen feet into a narrow canyon. Then you can scramble easily up to the passage that leads to the ledge of the abyss. The prospect is breathtaking. In times past, tourists were taken across the top of Colossal Dome to walking passages on the other side. The old-timers had built a catwalk across the abyss. Steel pipes were laid across the opening. One end of every pipe was jammed against the far wall, and the other end lay on the ledge. Planks had been placed on the pipes, but now they lay shattered on the floor far below. No one in his right mind would trust that pipe bridge now.

Several cavers had crossed the top of Colossal Dome by crawling on the narrow ledge. They lay on their bellies to inch along, reaching out tentatively and cautiously to press on the wobbly pipes for balance. They reported many leads on the other side. It would be tough to survey across the top of Colossal Dome, however.

If the passage across the top of Colossal Dome had been straight, a single shot would have spanned the void. However, the passage was not straight. Surveyors would have to take several shots, some with the compass held directly over the 130-foot drop. That survey would have to be done by experienced cavers like Spike and Dave. They took along another capable caver, Sandy Irwin. She had not, however, done much climbing.

Dave, over six feet tall, stretched out to reach the pipes to maintain his balance while crossing the widest part of the dome. He was secure, although scared. Dave set the survey points and carried the front end of the steel tape.

Joe Davidson on pipes across Colossal Dome, 1962.

Spike did not feel at all secure. His body was not so sure that it wanted him even to try, but he hugged the narrow ledge, reading the compass held in one trembling hand.

Sandy carried the rear end of the steel tape to chain the distance. As she started to crawl along the ledge, the camera case strapped to her belt caught on the first steel pipe. Sandy inched back and forth to dislodge the camera case.

"Take it off," Spike said.

"I can't, I need both hands to hang on," Sandy said.

"Back up, then," Spike ordered.

Sandy tried to back up, but the case caught again.

"I can't back up, it's too far," she said.

She moved on to the next pipe, where the case caught again. The other surveyors held their breath as she moved forward, catching each pipe on the way.

"Oops," Sandy said.

One of the pipes rolled, teetered, and then peeled off into the void. On the way down it clanged on the walls sepulchrally, like a great bell. After an unbelievably long time, the pipe thudded on the pit floor. As the echoes damped, everyone breathed again. Sandy slowly continued to the other side of Colossal Dome. Without pausing, she crawled on down the passage well away from the dreadful hole. Then she sat back, took the camera off her belt, and stowed it in her shoulder bag. The trip was not over yet. She would have to crawl back across the top of Colossal Dome to get out of Colossal Cave.

They had not been able to survey across the Dome, but they surveyed a passage on the other side. The trip back across the top of Colossal Dome —the cavers unencumbered with survey gear and hanging cameras this time—was uneventful. The job was still to be done, and the episode made Spike very thoughtful. Dave was young—twenty-one—and he had those long legs. But Spike was a thirty-year-old man. He had been frightened on the traverse, but it was no more than the healthy and health-making fright of someone who knows the dangers exactly. He was skilled, the risk (for him) was not high, and he knew he could do the job. Of course there was that other kind of fright, the animal panic that sometimes cannot be quelled, no matter what one does. A bit of it had risen from that bell-ringing pipe sounding the toll of doom. That was over quickly, and perhaps was cleansing. Certainly everyone proceeded calmly after it happened.

They had been belayed, but horizontally by a seated belayer who was not tied in. Had anyone fallen, he or she would have taken the belayer along in a hail of pipes. Why were they not better protected? The answer was chilling. There was not enough rope available just then to belay anyone properly over the top of Colossal Dome.

Why not wait for the protection of a more adequate belay? Once again, there was more rationalization than reasoning in the answer. The distance across the top of Colossal Dome was as far as the distance to the floor, 130 feet. A proper belaying situation would require bolts placed every twenty feet or so along the ledge, for the belay rope to be clipped into. Otherwise, if you fell, you might swing down on as much as 100 feet of rope, and when you hit the wall, you would not ring like a bell.

Spike knew that putting in the bolts was a bother no one wanted to take the time to do. Why rig the pit until they knew that the passages on the other side would go? The whole point of running the survey across the top of Colossal Dome first was to get the leads on the Colossal Cave map, to see where they were. Then it could be determined whether any of them was worth pushing.

Spike and Dave and others, like Barbara MacLeod and Wayne Amsbury, had been willing parties to numerous makeshift climbs around Colossal Dome that set outdoor rock-climbers' hair on end. Like most cavers, Spike believed—and the safety record demonstrated—that most climbs in caves were not as difficult as they seemed. Wet, muddy limestone *is* slippery. But limestone is knobby, and solution leaves sharp edges and points that are firm grips for rubber soles. Furthermore, climbing in caves is often chimneying. A toe, a knee, a back, even one's head jammed against an opposing wall keeps one from falling. In some situations the chimney is so tight that the climber need merely breathe deeply to hang on. Over the top of Colossal Dome, the task was to crawl along a sloping, narrow, crumbling ledge. The old-timers had done it without aid. It was done later by Red Watson, who stubbornly refused to touch any of the pipes. It was a kind of cave climbing pioneered by Bill Austin and Jack Lehrberger, and carried as far as it could go by Bob Keller. People learned to do it.

Knowing this, Spike still felt uncomfortable. What really had gotten him out there on those pipes with a compass playing human fly over the top of Colossal Dome was his desire to be one of the gang. Spike had risked his life not to prove himself, nor to show off in front of others, although there was a need for a bit of each in his make-up. No, overriding all else was his desire to do something with the group that all would appreciate.

Was it worth it? Well, it was all right to risk his own life. That was his choice. But Spike kept coming back to the fact that he had risked the lives of others. He should have spoken up and said that Sandy should not cross the top of Colossal Dome. Had she fallen, Spike could never have forgiven himself. Like many cavers, Spike worried more about danger to those he was leading than he did about danger to himself. It had *not* been worth it. He swore he would not risk the lives of *others* again.

A week later, Bob Keller, George Deike, and Ralph Powell surveyed across the top of Colossal Dome. There were places to explore on the other side, so Joe Davidson took a crew out to place bolts for a proper belay. After all, an old-timer had scratched on the wall back there: "MAMMOTH CAVE."

Jones Pit

Injury is a serious risk.
A brush with death leads to preparations
for emergencies.

14 In 1961, it was still not absolutely clear that Colossal River had been found. There might still be a river in Colossal Cave that would lead to a connection with Unknown/Crystal Cave. And the old stories were that Colossal Cave was connected to Mammoth Cave. If cavers could get down to a low river level, they might be able to follow the water from Colossal Cave to Mammoth Cave.

In one Colossal Cave passage you crawl along, crouch along, and then come to a low room. It is a warm, pleasant room with a sandy floor. One wall is a rock ledge with a belly-crawl leading out from beneath it, and there, many years before, had been smoked the most frustrating message ever written on a cave wall: "ROY HUNT—TO THE RIVER." Roy and his cousin Leo Hunt had been red-hot cave explorers. Maybe there really was a river out there. If you could find it, you could go.

The belly-crawl under the ledge goes for a few hundred feet and then opens into a crouching canyon. After a while the canyon turns into a sandy belly-crawl that opens as a window into the side of a large pit thirty feet in diameter. You can look down about fifty feet and up another thirty. A slender silver ribbon of water twists down. A rocky ledge some three feet wide leads across one side of the pit to a continuation of the canyon beyond. The only hazard is a large boulder, about three feet in diameter, perched about halfway across. However, one can easily straddle it to walk on over. Across the pit the passage goes on. Openings large enough to get into yawn in its floor. Below one of them is a large pit. Explorers wanted to get down into all these openings, in hopes of finding a drain that might lead to Roy Hunt's river on a lower level.

One August afternoon in 1961, Dave Jones, Bill Hosken, and Red Watson went to Colossal Cave to try to find Roy Hunt's river. Dave and Red were old Flint Ridge hands, and though Bill was new to Flint Ridge, he had already proved himself. No matter how low down and muddy, if a passage went, Bill would push it.

The walk from the road was a mile and a half through dense woods and underbrush full of ticks and chiggers and spider webs. The three cavers carried denim jackets, and Red even had a wool sweater, in eternal optimism that he might get soaked in a cold underground river. It was miserable in the woods, and the clinging haze of heat left the cavers drenched in sweat. When they reached the Colossal Entrance, they hardly paused to light their lamps before descending into the cool breath of that wonderful Colossal Cave.

The crawl out to Roy Hunt's ledge was familiar, and they moved along quickly without talk. They slid on their bellies in the cool sand under that ledge and admired the water-worn sinuosities of smooth limestone as they inched along. They felt the coolness of the cave finally penetrate bodies that had been overheated only an hour or two before. It was a comfort and a pleasure.

It can be a special joy. Caving is tactile in a way that no other contact with the inanimate can be. There is no other sport where one crawls through mud and slides through sand. One is *in* a cave, but not as a swimmer is in the water. In the cave one is clasped in solid, ever changing walls of stone that provide variegated patterns of visual and tactual delight. Caving can be almost totally sensual.

In three hours, the voluptuaries had reached the first big pit. They stood up and paused to look across at what appeared to be a completely inaccessible lead on one sheer wall of the pit. Well, that was not their job that day. Someone could bolt a way around there sometime. Getting to the bottom of the pit was a possible goal, but there was no obvious way down the walls in view. For now, the explorers had other prospects in mind, much farther along. They were to cross a ledge to a passage continuing on the other side of the pit where perhaps a hole in the floor would lead to a chimney down. They were intent on finding an easy way down to lower levels where a river—possibly the fabled Colossal River— might indeed lead to a connection between Colossal Cave and, well, not Mammoth Cave, but possibly with Unknown/Crystal Cave.

Bill was having trouble with his carbide lamp, but after three hours of burning it probably needed nothing more than a change of carbide. He sat down in the canyon, took off his pack, and began laying out materials for changing his carbide.

"Might as well go on across," Dave said. He and Red had crossed the pit several times. It was new to Bill, however, so Red paused.

"Go on," Bill said. "I see where to go."

Dave then started walking the ten feet across the three-foot-wide ledge. Red stood in the opening of the canyon, watching. He would wait until Dave reached the passage on the other side of the pit.

Dave walked along the ledge with unconcern. Halfway along, the boulder three feet in diameter rested on the ledge. Rather than stepping behind or over it, Dave stepped up to place his outer, left foot on the boulder's slanting top. The boulder was slippery with mud, and Dave's foot slid down it, toward the pit.

Now slowly, incredibly slowly, Dave's body tipped over toward the pit. He said nothing. Red, frozen, watching, said nothing. Time slowed until motion was almost nonexistent. Dave tipped slowly to the left, and then his body was free of the ledge, but still above it. Still slowly, slowly . . . slowly . . . slowly, in the light of Red's lamp, Dave's body turned completely over in the air and then, still turning, sank without sound into blackness. Many minutes later—it seemed—there was a distant, heavy, solid, quiet thud.

At the sound, all that lost time sprang back into being like a watch spring wound too tight that has finally let go.

"Dave!" Red screamed. "Dave! Dave! Dave!"

Bill had not seen the fall, but he had heard Dave hit the bottom. Bill's light was out. The tip of his lamp was jammed. He could do nothing but work on it, because he could not help until he had light.

Red has no memory of what he did until he was halfway down to the bottom of the pit. Bill saw him run across the ledge so rapidly that he was terrified that Red would fall, too, and he would be left without light, alone, and uncertain that he knew his way out.

In the passage on the other side of the pit was a narrow crack in the floor. No one had been down it, but Red knew it was there. Without thought he had raced to lower his body into it. He found himself in a tight chimney that belled out into the pit below. As he slid and climbed down, he shouted over and over again, "Dave, I'm coming! Dave, I'm coming down! Dave!"

There was no sound from the bottom of the pit. Red knew that it was at least forty and probably fifty feet from the ledge to the floor. Dave had hit with a heavy sound. He weighed around 190 pounds.

"He's dead." The thought was involuntary. Red shouted, hoping desperately for a reply. No one could have survived such a fall.

It seemed to Red that he was descending the pit faster than Dave had. Red's mind was racing. He was building the psychological strength to stay with the corpse while Bill went for help. He did not consider that Bill might not know the way. He just assumed that Bill would go and he would stay.

Red knew that there was no good reason for staying, perhaps as long as half a day, in a dark cave with a corpse. Nevertheless, as he slid and scrambled and yelled his way to the bottom of the pit, he nerved himself to sit with that dead body. Dave was his friend.

Just as Red paused at the top of a final ten-foot drop that was covered with mud and offered no obvious way down, he heard a groan. He yelled again, incredulous: "Dave, are you all right?" Then he let go to slide down the wall. He could not stop at the top of a thirty-degree slope of hard mud, and slid down another ten feet to the edge of a deep pool of water.

There Dave lay, his head pointing down the slope, almost in the water. He was twitching and he muttered that he was all right. Just to his left a large boulder with a heavy pointed finger rose four feet. To his right, several smaller boulders flanked him. Red saw at a glance that Dave had landed extended on his side on the steep mud slope, six feet farther up, among other boulders, and had slid down to his present position.

So far, Dave's luck had been phenomenal. Later measurements showed that the fall was forty-four feet. Dave had tumbled end over end to land with his body stretched out on the steep mud slope. The slide had cushioned his fall. He had missed the rocks and boulders on the slope. And his slide had stopped just before his head reached the pool. So far so good.

Now Dave was conscious. He wanted to roll over, to sit up, to turn around.

"Stay put!" Red was now yelling in anger.

Dave stayed still.

"How do you feel?" Red asked.

"Okay, I guess," Dave answered.

"Does that hurt?" Red was feeling Dave's arms, ribs, and legs. Dave did not complain of tenderness anywhere.

"I don't think anything is broken," Dave said. "My head hurts, my ear."

Dave's hard hat had smashed down over the top of his left ear, which was cut badly.

"Look straight at me," Red said. He looked into Dave's eyes. Both pupils seemed the same size, so presumably there was no bad concussion. Dave could move all his fingers, and claimed that he could move his toes.

"I just feel pretty beat," Dave said.

"You were damned lucky," Red said.

"Can I sit up now?" Dave asked.

"Yes, let's see you sit up." Red could not believe it. The corpse was alive, and evidently not even badly hurt. It sat up.

Now Bill was at the top of the ten-foot pitch. The flame of his lamp blazed out three inches, and his eyes reflected reddish in the deep, hollow eyesockets of his skull. Bill's bony face was the spookiest among the Flint Ridge cavers. The pupils of his eyes seemed lodged permanently almost all the way up under his eyelids, so that, straight on, his eyeballs were almost completely white. Not many people could stare him down. Now he was an apparition.

Bill unrolled a twenty-foot piece of hemp rope. Dave protested that he did not need it, and that it would impede his climbing, both of which were probably true, but he tied in anyway. Bill belayed him as he climbed up the wall ten feet to the chimney, with Red right behind offering support. No one spoke of the fact that Dave would have to cross the ledge again. Pausing in the passage at the top of the climb, Dave deliberately untied the rope and handed it to Bill, who coiled it and put it in his pack.

Then Dave walked calmly across the ledge, straddling the boulder this time. He did not look down. Red and Bill quickly followed.

Dave led out of Colossal Cave. They joked about the fall. The distance to the entrance was over two hours, and then there was the long, hot walk through the woods to the car. Dave kept saying that there would be no need to go to a doctor.

"I feel fine."

Red and Bill kept their counsel. They exchanged a glance to confirm their plan. It had been bad enough already. They worried now about Dave going into shock. It had been, after all, *his* glorious adventure.

It was Dave's car. He wanted to drive, but did not protest when Red took the keys and slid behind the wheel. Dave muttered, but sat still when Red turned not back toward camp, but toward Cave City.

"I'm all right, you know. I don't need a doctor," Dave said after a few minutes.

"Fine, fine," Red said. "We'll go check, anyhow."

The fifteen-mile drive over back roads was quiet. Was Dave's breath getting shorter? They did not talk.

The doctor stitched the cut over Dave's ear, and peered into his eyes. (This latter gave Red some satisfaction.) The doctor then startled Red and Bill by asking if *they* needed anything. Sleeping pills, maybe? No, no.

On the drive back to camp they talked and laughed. Dave said he knew he was very lucky. That night Bill had nightmares. Red did not sleep much. Dave slept quite soundly. It figured. Dave had driven all the way from New York to Kentucky and had entered Colossal Cave with only a couple of hours' sleep. Falling into the pit was like falling asleep. It had not worried Dave when he slipped. A tune was running through his head as he fell, one that he had heard on the radio over and over again on the drive from New York. He had not been frightened at all. He remembered nothing more until Red was beside him asking if he were all right. Even then he felt that he was fine, just fine. He needed sleep. He fell asleep.

It was a hard lesson. Not only could you get lost, you could get killed in Flint Ridge. Dave's fall demonstrated the need for formal emergency preparations including first-aid kits, caches of food and supplies, and practice rescues. Dave participated heavily in these new activities. They would not forget the lesson nor the place.

They named it Jones Pit.

The Unknown/Salts Link

The connection between Unknown/Crystal Cave and Colossal/Salts Cave
integrates the Flint Ridge Cave System. It is still not
the longest cave in the world, but it is the third-longest.

15 During the year after going on the Colossal/Salts connection
trip, Dave Deamer did a lot of caving. He also helped plot the
surveys, and while making the map he was struck—as others
had been—with the fact that the Unknown Entrance pit was only 160
feet from the end of Indian Avenue in Salts Cave. Dave asked Roger
Brucker, the August 1961 expedition leader, if he could explore the Unknown Entrance pit. Roger knew of a crawlway out that way, so he said,
"Sure, go ahead."

Dave intended to poke around and move rocks in the huge breakdown
pile in the Unknown Entrance pit. This interested Spike Werner and Bob
Keller. Bob had done an enormous amount of exploration in Salts Cave,
and very much wanted to make a connection. Judy Powell, the vivacious
fifteen-year-old daughter of caver Ralph Powell, had been agitating for a
trip. The area was easy to reach through the Austin Entrance, so they
took Judy along. They did not have much hope, but they enjoyed being in
Unknown Cave.

Their thoughts turned to the Indians. Dave and Spike had been in the
lower-level Indian Avenue in Salts Cave, into which they were going to
try to find a way from Unknown Cave. Everyone who had been through
Indian Avenue was impressed by the evidence that prehistoric explorers
had gone into Salts Cave as far as two miles from the Salts Cave Entrance.

"What were they looking for?" Dave asked. "Epsom salts and mirabilite" was one obvious answer. These minerals melt in your mouth with
a salty taste, and they are laxatives.

"But why would the Indians go so deep into the cave just to mine
salts?"

"Well," Spike declaimed, "their main purpose was not to mine salts
at all. Who could be that constipated? No, it was this way: Every August

it got so hot and muggy on the surface that everybody got together for their annual August cave expedition, to explore Salts Cave."

The prehistoric cavers had been in almost every passage modern explorers had discovered in Salts Cave. The ancients must simply have enjoyed cave exploration. Spike and the others happily accepted that conclusion as they reached the breakdown pile below the Unknown Entrance. Their four carbide lamps did not begin to light the blackness of that huge room. Bob and Spike climbed to the top of the pile to look for leads. They did not expect to find anything because Bill Austin and Jack Lehrberger had checked it thoroughly. Eventually Bob and Spike went into the crawlway Roger had told them about. They were gone a long time.

Dave and Judy searched the base of the breakdown pile, but after fifteen minutes it seemed hopeless. They looked in an adjacent pit, but found no passable drain out of it. Partway up its breakdown pile, they sat down together to wait.

Dave immediately felt a cold draft against his back. He and Judy began to move the rubble. The breeze increased with every loose rock that they pulled out.

It was a weird scene in the elongated room. Down one wall water fell into a pool and gurgled away among the rocks. Two tiny lights bobbed up and down on the breakdown pile. Stone echoed on stone as boulders rolled down the slope. Dave and Judy babbled with excitement as they worked. Finally, only one large rock remained. They grabbed hold of it together and heaved back. It crashed down the slope behind them, and their lights went out. For a moment, they had seen an open hole, then the wind had rushed out to extinguish the flames of their lamps.

Quickly relighting, they yelled for Bob and Spike. There was no answer. Holding up a hand to protect his flame, Dave crawled into the hole, with Judy close behind. After thirty feet they reached a fork. Dave went left and Judy right. Dave's lead soon pinched out, and he had to back up. At the junction he yelled for Judy, but there was no answer. He followed her lead. Fifty feet along he found her sitting in a walking passage, her feet dangling into a deep canyon with running water at the bottom. She smiled up at him. It was the prettiest sight he had ever seen.

Judy's eyes twinkled at Dave. Obviously she had been prepared to wait there, however long it took, until he came along. Such discoveries are not made often, and not by everybody. Judy and Dave were very pleased.

Dave did not need his knowledge of the maps to realize that this was virgin cave. Dragging a reluctant Judy back through the crawlway, he squeezed out into the pit at the Unknown Entrance to fetch Bob and Spike. They had returned from a futile attempt to force a too tight bellycrawl along the wall behind the breakdown pile. Their mood changed quickly when they heard the news. The four rushed back to explore the new passage. All had visions of another long trip out through the sign gate of Salts Cave.

The wet-walled walking cave soon turned into a high, very narrow canyon with passage at several levels. It was not comfortable to travel on any of these levels because sharp blades of limestone projected from the walls. The explorers sweated through the canyon for thirty minutes, each one at a different level. As usual, Bob was climbing along the most difficult route. They emerged more or less at once. They had reached the familiar pit at the end of Indian Avenue in Salts Cave.

They had connected Unknown Cave to Salts Cave. They *could* go out the Salts Entrance, just as Dave and Spike had done with Jack Lehrberger on the Colossal/Salts connection trip the previous year.

However, no one had expected to make this connection, and the trip to the breakdown pile in Unknown Cave from the Austin Entrance was so short and easy that everyone had brought a minimum of carbide and water. After discussion, during which Judy's heated arguments for proceeding out the Salts Entrance were mostly ignored, they decided to return through Unknown Cave to the Austin Entrance.

First, however, they did a little sightseeing, looking at 3000-year-old footprints and cane-torch fragments. Before starting back, Spike copied in his notebook from the wall of Indian Avenue: "J.L. F.P A.R.M. WELCH." Then Spike put a small piece of charred cane from an ancient Indian torch into his pack, and turned to follow the others back through the connection from Salts Cave to Unknown Cave.

They had entered Unknown Cave in the afternoon, but it was night when they returned to camp. Into the glow of the Spelee Hut they marched two by two. Roger and Micky Storts were plotting a survey. Fred Benington was telling them about experiments he was conducting—as a pharmaceutical chemist—with hallucinogenic drugs. "That was one cat you wouldn't want to meet in a dark alley," he said.

The explorers stood silently. Spike solemnly extended a closed fist to Roger, who automatically held out his hand. When Roger saw the cane fragment, his face broke into a broad smile. Fred snatched up the cane and held it close to his good eye.

"They did it, by God! They did it!" Fred yelled.

Now there was one giant cave in Flint Ridge. The Unknown/Crystal/Colossal/Salts complex was truly an integrated Flint Ridge Cave System. The first connection had been made in September 1955, between Unknown Cave and Crystal Cave. The second was made in August 1960, between Colossal Cave and Salts Cave. Now, in August 1961, Unknown/Crystal Cave was connected to Colossal/Salts Cave to form what the explorers thought was surely the longest cave in the world. But again they were wrong. Although the Flint Ridge Cave System was about thirty miles long, both Mammoth Cave and Hölloch in Switzerland were still longer. At least the Flint Ridge Cave System would soon be the second-longest cave in the world, for exploration in Mammoth Cave was at a standstill, and there were leads and unsurveyed passages everywhere in Flint Ridge.

All the old-timers, everyone from E.-A. Martel and Hovey to Hazen,

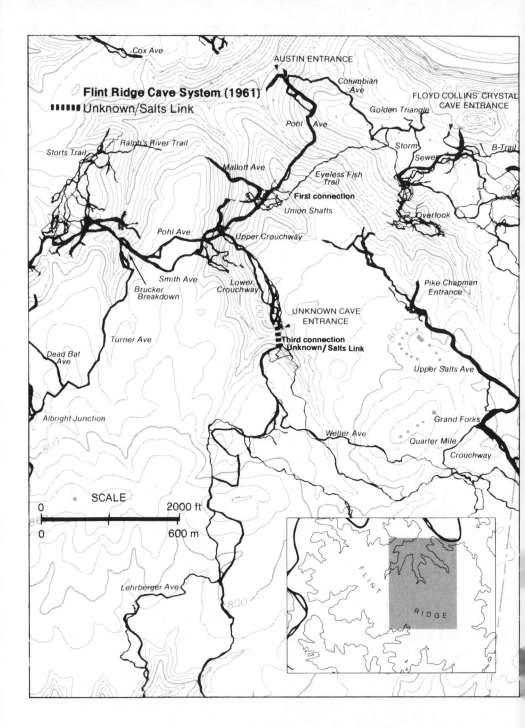

Map of the Flint Ridge Cave System resulting
from the discovery of the Unknown/Crystal–
Colossal/Salts Link in August 1961.

Edmund Turner, and Floyd Collins, and all the old guides, right down to Ellis Jones, "knew" the caves in Flint Ridge were connected. Now it had been demonstrated. Now—putting aside that pesky Great Onyx Cave, which resisted all attempts to connect it with the other Flint Ridge caves —there remained one thing to do. Three thousand feet to the southwest across Houchins Valley lay the next connection target.

Mammoth Cave, here we come!

Why Does a Good Caver Quit Caving?

Six major explorers drop out,
but others take their places.

16 The connection between Unknown/Crystal Cave and Colossal/Salts Cave increased the fame of Flint Ridge caving. Many new young cavers applied to join the adventure. However, six major explorers quit caving about this time: Bill Austin, Jack Lehrberger, Phil Smith, Spike Werner, Dave Deamer, and Bob Keller. Oh, they continued to cave now and then, but they spearheaded no more trips of discovery.

Why *does* a good caver quit caving? Caving is frustrating, wearing, and dangerous. A caver might quit because of annoyance, age, or fear. And some cavers get bored.

The question of why cavers quit arises in part because there is no obvious terminating goal in caving. Big caves seem to go on forever. Thus, for many cavers, there is no well-defined accomplishment on which they can close a career.

The goal of connection is stressed in this book. Yet, Bill Austin and Jack Lehrberger did not stop caving after that first connection trip between Unknown Cave and Crystal Cave. And, however appropriate it was that Jack's great connection trip between Colossal Cave and Salts Cave came near the end of his caving career, his blaze of glory was not planned. It was not exactly fortuitous, either. A dozen other Flint Ridge cavers of top caliber did not reach some objective and then stop. They caved awhile, and then moved on.

Spike Werner and Dave Deamer had two big connections to their credit. Could they do it yet a third time? Great Onyx Cave and Mammoth Cave still remained to be linked into the system. Alas, it was not to be. Dave did not quit, exactly. After he got his Ph.D. in biochemistry at Ohio State University, he took a post-doctoral fellowship and then a job at the University of California. He did not return to the Kentucky caves.

Spike also went on to get a Ph.D., in speech pathology. He *did* quit

Flint Ridge caving, and he knows exactly why. During the mid-1960s, Cave Research Foundation directors got involved in environmental politics. The National Park Service was exerting pressure on them to prove the Foundation scientifically, and there were local enemies in Kentucky. Part of the Foundation response was a more authoritarian internal regime. Some of the caution seemed to people like Spike to be rather silly. Some of it was.

Spike disturbed some of the Cave Research Foundation directors. He and Fred Benington used to do a burlesque routine, with Fred in the role of Mr. Bones, and Spike in swimming trunks flourishing a towel. They came on a mile a minute with a wildly obscene dialogue, mostly in verse. What if the Park Superintendent came over while this was going on?

The final straw was Spike's riding a trail bike down through the woods to the Austin Entrance. This shocked the conservationists and raised visions of motorcycle gangs. Joe Davidson callously chewed out Spike in front of the whole camp. Spike left in a huff. He was a conservationist himself, like most cavers, but he could see no harm in riding a trail bike down the old road to the Austin Entrance.

Spike was one of the most colorful of the Flint Ridge cavers, a man who in caving determination, mannerisms, and looks bears a strong resemblance to Baron Jacques Sautereau de Chaffe, a French explorer of the Pierre St. Martin, the deepest cave in the world. It was a real sign of change when Spike left. The institutionalization was reflected in the fact that the connection between Colossal Cave and Salts Cave that was originally called the Werner Link appears on Cave Research Foundation maps merey as the Colossal/Salts Link.

But that was okay with Spike. He was one of the many who thought that nothing in the caves ought to be named after anyone who was not dead. Phil Smith argued against that. "If we don't name some of these passages after ourselves," he said, "they'll be named after National Park Service directors and Secretaries of the Interior." That settled the matter.

Why do cavers quit? Jack Lehrberger went off to climb castles in Turkey, to explore hill villages in India, and then became Senior Lecturer in Linguistics at the University of Otago in New Zealand. Bill Austin explored crevasses in the Antarctic, took a jet boat up the Grand Canyon, and now manages Mammoth Onyx Cave in the Kentucky cave country. Phil Smith navigated traverses across the Greenland Ice Sheet and the Ross Ice Shelf in Antarctica, became Deputy Director of Polar Programs for the National Science Foundation, and then Special Assistant to the Director of the National Science Foundation. Bob Keller took up automobile racing and white-water kayaking and became a research chemist.

Red Watson abandoned caving periodically, went on expeditions with his archaeologist wife, Patty Jo, roamed the mountains of Iran and Turkey, studied a glacier in the Yukon and landslides in New Mexico, and became a professor of philosophy at Washington University in St. Louis, Missouri.

Roger Brucker?

"Roger," Red said one day while working on this book, "you are the only Flint Ridge caver who during all these years has lived on the same corner in that same small town of Yellow Springs, Ohio, working in the same advertising agency. I realize that now that you are president of the agency your work has changed. Anyway, you're the only one of us who has year in and year out kept on caving, caving for the last twenty years, kept on taking hard trips, kept on being enthusiastic."

"Yep," said Roger.

Under Houchins Valley

A 3000-foot-wide valley stands in the way of connecting the
Flint Ridge Cave System with Mammoth Cave.
Then the explorers find a muddy
lead. It goes to a complex of passages under Houchins Valley.

17 Flint Ridge Cave System passages now extended as one big cave through Flint Ridge on four main offset levels, like four subway systems under Times Square or Piccadilly with offshoots to the suburbs. Overland, the distance between Unknown Entrance and Crystal Entrance is less than a mile. The grand tour underground from entrance to entrance is nearly five miles long, and involves five north-south traverses under Flint Ridge.

Despite all these magnificent miles of passages, there seemed to be no way out of them, no way to get out from beneath the sandstone caprock canopy that covers Flint Ridge. Even under Flint Ridge, which is over four miles long, passages were confined to a transverse band two miles wide. The big rectangle was filling with newfound cave, but in all those immense avenues no lead would *go*. Go where? Out of Flint Ridge, out under Houchins Valley, to Mammoth Cave, of course.

The most promising lead off the great Flint Ridge avenues was a 6000-foot sand-floored passage: Swinnerton Avenue, running northwest parallel to the south flank of Flint Ridge from a point near Argo Junction. It was the walking lead that Roger Brucker had asked about on an early trip into Unknown Cave. Swinnerton Avenue is walking and crouching cave that would be spectacular in most other caves, but in the Flint Ridge Cave System it was just a side passage. However, one day in 1959, Bill Austin said, "You fellows ought to look into a lead Jack and I noticed off Swinnerton Avenue."

"What's it like?" Roger wanted to know.

"Well, you worm your way through a muddy tube, and then it's a crawlway," Bill replied.

Bill had firm theories about caving: You get out of caving what you are willing to put into it. He understood the value of sharing knowledge

as well as anyone. But when it came to sheer *caving*—the skill of exploring caves—Bill might tell you where to go, or lead you there, but then you
had to learn on your own. The thing to do when you could, of course, was
to watch how Bill did it.

The cavers knew Bill Austin. They raced to that small hole off Swinnerton Avenue. They found Gravel Avenue, a long passage with many
leads. Roger used the maps to find the Duck-under, a shortcut that bypassed the muddy tube. During the next two years, they explored and
surveyed Gravel Avenue, Lower Gravel Avenue, and, inevitably, Lower
Lower Gravel Avenue. Joe Davidson cut his caving teeth on all that gravel.
Jake Elberfeld found his Breathing Trail leading south from Gravel Avenue to Shower Shaft. And by moving some rocks in Shower Shaft, Roger
Brucker uncovered a south-trending drain that led to the discovery of
Candlelight River in the Unknown/Crystal Cave complex. This was the
route to all further exploration out toward Mammoth Cave.

The water in Candlelight River flows north from Houchins Valley,
which separates Mammoth Cave Ridge from Flint Ridge. Maybe the headwaters of Candlelight River are under Mammoth Cave Ridge, the explorers thought. You would have to follow the water upstream under
Houchins Valley to see. No one had ever been in cave passages under
Houchins Valley. There were surely passages there, but the question was
whether they were large enough for cavers to crawl through. *Could* cavers
crawl under a deep, wide karst valley like Houchins Valley, from one of
those great karst ridges on the Mammoth Cave Plateau to the other?
Could you go from Flint Ridge under Houchins Valley to Mammoth Cave
Ridge? No one anywhere had passed under such a wide karst valley. It
was one of the greatest unknowns, and one of the greatest challenges, in
world speleology.

So the explorers pushed upstream in Candlelight River. It went a long
way, but finally ended in a shaft—Candle Shaft—which it drained. The
surveys showed Candle Shaft to be under the south flank of Flint Ridge.
The explorers had not found their way out under Houchins Valley at all,
and there were no side leads. Maybe it would be impossible to get under
the valley.

Then they went downstream in Candlelight River. You drop into
Candlelight River from the Shower Shaft drain through a chimney where
someone has smoked on the wall the sign: "BEST WAY DOWN." From Best
Way Down you crawl downstream in water a foot deep. It is clean and
refreshing for the sweaty caver. Then you chimney along the top of a
canyon with water flowing inaccessibly ten feet below, beneath a series
of sharp protruding ledges called the Turbine Blades. Beyond the Turbine
Blades area, a tributary comes in from the right. It is a fine smooth-walled
canyon in which you can walk with your shoulders turned sideways. The
explorers named it Bretz River in honor of a famous cave geologist, just
as they had Swinnerton Avenue. Bretz River was heading south, toward
Houchins Valley.

*The Duck-under connects Swinnerton Avenue
with Gravel Avenue. Its discovery bypassed a
muddy tube and a long crawl.*

Again the survey teams poured in. Bretz River went for half a mile, but it turned east and ended in shafts and breakdown under (but with no open connection to) Argo Junction. The explorers were still under Flint Ridge, but this time there were side leads.

Bill Hosken explored long distances upstream in some of these side leads—toward Mammoth Cave—and found that they, too, had further side leads. It was tight, yet those giants John Bridge, Fred Dickey, and Walter Lipton went out there. They are the sort you send through a crawl-way first to demonstrate that anyone can do it, and to sop up some of the excess mud and water.

Bill Hosken was doing graduate work in mathematics at Ohio State University in the spring of 1964, and he had an office just around the corner from Joe Davidson's office in the Department of Mechanical Engineering. Soon, Bill had Joe primed for a far-out trip up Candlelight River.

Most intriguing to Joe was a narrow, muddy walking passage that had never been looked into, or at least not very far. It was downstream in Candlelight River, between the water crawl and the Turbine Blades.

"You didn't explore it?" Joe asked.

"Nah." Bill shuddered in mock horror. "It was too *muddy,* and besides, you could *walk* in it. We were following the water."

Bill spun out his yarn. He had also been out a couple of low leads off Bretz River for 1000 or so feet each. One got very small, but there was a possible lead at the end. The other went on. And there were side passages that might go. It was going cave, right toward Houchins Valley.

Joe had trouble finding an experienced party. Bill was swamped with work. Joe recruited Red Watson, who was eager to go out toward Houchins Valley, and Bob Hough, a green caver who did not realize what he was getting into. But Joe could not find anyone who had been out there who would go again.

Finally, Joe decided to go without a guide. Some years before, Joe had been out to the beginning of Candlelight River. Red had not even been out Swinnerton Avenue, and Bob had not even been in the Austin Entrance. But they would find the way.

Joe remembered some rocks at an important junction. At Shower Shaft before Candlelight River, he remembered the tiny hole that no one in his right mind would crawl into, given the wide open passage leading out of the pit. The hole was the way.

"Come on, this is right," Joe said. "I remember this slippery canyon."

"Why didn't you remember it before we got soaked from this damned waterfall?" Red complained.

Then they found the welcomed Best Way Down and clambered down into Candlelight River to follow the water through the crawlway. They took Bill's walking lead, which loomed to the left, but after 300 feet it pinched down to a crawl. It did not go south, and Red was a fanatic about going south toward Mammoth Cave. "Leave it for another day," he said.

They returned to Candlelight River to go downstream again. Past the Turbine Blades, the intersection with Bretz River drew a blank in Joe's memory. Without hesitation they turned to follow the water downstream. Unfortunately, the drain for water from Bretz River and Candlelight River begins as a crawl through a big pool that requires complete body immersion. Joe got rather damp before realizing that this was not the way they had intended to take, and was chagrined when the others refused to come through the pool to see the fine passage he had found beyond.

"It's heading *north,*" Red said with disgust.

Six hundred feet upstream in Bretz River, the explorers finally confronted the lead that was said to pinch out but might have a side passage. The lead branched to the south, heading straight for Mammoth Cave.

"Go on until it ends, then we'll look around," Joe said.

Red led on, and on, and on. Obviously they had long passed the point where the passage was supposed to pinch down to impenetrability.

"Who said this passage pinched out?" Red asked.

"I think they were right," Bob groaned. "Let's stop a minute." It was Bob's first trip to the extremities of the Flint Ridge Cave System.

"How do you feel?" Joe asked with concern.

"A bit tired, cold," Bob replied.

"Someone must have thought the passage ended at that tight spot where I moved a few cobbles," Red said, ignoring Bob's comments.

"Yes, that was a chest compressor for me," Joe replied. All three lay on their bellies in the low, narrow passage: Red, Bob, and then Joe. Joe looked absently at the soles of Bob's boots.

"I guess I feel pretty bad," Bob said.

Joe and Red felt each other's disappointment. They had been in the cave eight hours, and were prepared to stay in the Candlelight River area for another eight hours before turning back, if anything interesting turned up. Neither felt the least bit tired. Hard crawling would soon warm up the chill from lying still on the wet passage floor. However, they had been well trained by years of Flint Ridge caving. The dangers of having a sick or incapacitated caver deep in the Flint Ridge Cave System are too great to take chances. If you say you have had it, the party wraps up and goes home. Over the years, half the parties out Candlelight River way turned back because of sickness or chills, or because someone did not feel right about being there. In such situations, the weakest man on the party directs the order of the day. Within reason, that is. A strong young caver like Bob might be encouraged some.

"Let me look on ahead a bit," Red said.

"Sure," Joe said. "I could use a rest. I'll stay with Bob."

"Say twenty minutes," Red said, crawling on.

"Fine." Joe moved forward so he and Bob could talk. "How far do you think we've come since we left that last junction?" he asked idly.

"A mile?" Bob guessed.

"I'd say that's pretty close."

"Actually, it seemed longer."

"It always does," Joe replied. "But wait'll we go back. It'll seem a lot shorter." They lay quietly a moment. "Think you can lead out?" Joe asked.

"Boy," Bob laughed uneasily, "I don't know."

"It'll be a lot easier than you think," Joe said.

"Well, this is only my second trip in Flint Ridge," Bob said.

"Is that right?" Joe said, his face splitting in a toothy smile. "And we may really find something. Did you know that Red's first trip was the Overlook discovery trip?"

"Was it?" Bob asked with interest.

"Yes," Joe said, laughing. "And you think this crawlway is tight, let me tell you about *my* first trip in Flint Ridge." Joe launched into his tale, and soon Bob was chuckling with him.

Then there was a muffled shout from Red. "Come on up here."

"Where are you?" Joe shouted.

"Right here," Red said, surprisingly close.

Joe could not see him anywhere. Up ahead was the same tight crawlway, so where could he have gone?

"Where?"

"Here," Red said. "I'm standing up."

Joe felt the rock ceiling against his back and the rock floor against his chest. It was a classic belly-crawl.

"You're crazy," Joe said.

"No, no," Red said. "See my feet?" Suddenly, just down the crawlway, two rocks leaped into focus for Joe. They were moving up and down. They were feet.

Joe crawled on toward the feet. His face almost scraped them before he could see where Red was standing. He was in a vertical slot as narrow as the belly-crawl.

"What's up there?" Joe asked.

"Big cave," Red replied.

"I'm not sure I can get up," Joe said, as Red's feet disappeared after his body. "How did you get up? Must have been desperation," Joe muttered to himself. The crawlway pinched down just ahead, and unless there were a way to turn around, it was a long way to back out. Joe rolled on his side, stuck his head, an arm, and a shoulder up into the crack. It was a hard bend for his legs, but finally he was upright.

Red had squirmed up into a low, flat room a couple of hundred feet long with several holes leading out of it. He watched with amusement Joe's difficulties in jacking his long bones up into the vertical crack.

"If Joe can do it, anyone can," he said to Bob, whose incredulous face had appeared below.

Red and Joe were so excited that Bob decided they had forgotten that he was tired. Or at least no one mentioned it again. Bob dutifully shoved himself up through the crack. Anything was better than staying down in that belly-crawl.

"Well?" Red said to Joe, glancing at Bob.

Joe followed Red's glance, sighed, and said, "I suppose."

Joe started taking out the survey gear. They surveyed down through the crack and back out the belly-crawl. Now they would see where they were. However, surveying on your belly is difficult, and after fifty stations they agreed that it was time to start out toward the Austin Entrance. The survey was left hanging—that is, they had not yet connected it to the main survey grid. Leaving a hanging survey is a cardinal sin. Red Watson had committed it.

"I don't care if Red *is* president of CRF," Fred Dickey said the next day. "He's a madman, and ought to go out and pick up his own hanging survey."

But by then Red was off politicking with Park Service officials. Fred, John Bridge, and Denny Burns marched indignantly into the cave. Miles along, Fred lost their only pencil. Survey notes cannot be taken without a pencil. They turned back.

"We can't leave a hanging survey out there," Joe shouted when he heard their story back in camp. "You'll have to go back."

"Are you coming?" John asked. Denny was not returning because he had an upset stomach, and a third man was needed.

"Yes," Joe said, "and everybody take three pencils."

That made ten pencils for the three-man party that went back to Candlelight River, because Fred found one in Turner Avenue that looked suspiciously like the one he had lost earlier.

They surveyed fifty more stations, and the passage was on the map. Further, Joe had established himself as an iron man, because taking two such hard trips two days in a row was unprecedented at that time.

Fred was six foot three, and weighed over 200 pounds. His pleasure in the Candlelight River area was never intense. However, he had joined the cavers just in time to see Dick Sims eased out. Dick was not a strong caver, but weaker ones have been useful. However, Dick's love of amateur photography hindered the work of nearly every party he was on. That did him in. Fred saw what happened, and did some things he did not enjoy to show what a devoted Flint Ridge caver he was.

These two trips began the saga of exploration beyond Candlelight River. Red took Stan Sides and Dennis Drum to the Bretz River tributary and surveyed the upper level beyond the crack. To the southeast it ended hopelessly in a beautiful orange- and cream-colored flowstone plug only 100 feet from Ehman Trail in Colossal Cave. Connecting caves can be frustrating work. After determined exploration for many years, these two passages were still ten miles apart by way of the shortest passable route through the Flint Ridge Cave System. And still Houchins Valley appeared to be an impassable barrier between Flint Ridge and Mammoth Cave Ridge.

John Bridge, Fred Dickey, and Kim Heller returned to survey the walking lead that Red and Joe had ignored in Candlelight River between the water crawl and the Turbine Blades. It turned out to be a cut-around back to Best Way Down, but there was a side lead off it to the south. This soon pinched down to a belly-crawl. It is seventeen feet from John's finger-tips to Fred's toe tips when the two are stretched out end to end. In girth, they are as impressive. When asked, each would say, "Oh, 190, 195 pounds, something like that." This is what all modest men say who actually weigh upward to 230 pounds.

"Fred," John said around the huge curved-stem pipe that he managed to smoke even in belly-crawls, "we're too big to be surveying in this passage."

They continued for 700 feet before managing to turn around and flipper their way back to the larger passages.

The next victim was Denny Burns. At least he was small. He was a graduate student in forest entomology at Ohio State University and had only recently become a Flint Ridge caver. He had, however, read *The Caves Beyond* fifteen times by actual count during an unadventurous Ohio childhood. His thesis professor was Joe Davidson's father.

"Interested in caves, eh?" Professor Davidson said one day. "You ought to get to know my son, Joe. He's a caver."

Denny then discovered that marvelous, cluttered place on the Ohio State University campus where for so many years so many Flint Ridge eggs were warmed and some were hatched, Joe Davidson's office in the Department of Mechanical Engineering.

Denny soon took over the cartography program from Roger Brucker. Denny also soon found himself walking down spacious Turner Avenue toward Swinnerton Avenue to continue the big men's survey in the belly-crawl out beyond Candlelight River. He asked himself apprehensively how he had gotten into this situation. It seemed fairly obvious. Each member of his party was five foot eight or under, and not one weighed over 150 pounds. This passage called for men like Denny, his younger brother Dick, Jake Elberfeld, and Bill Morrow. Four hours of fast caving from the Austin Entrance, and the agony began.

Denny took notes in the survey book. In a mud crawl, this calls for sacrifice. The book must be kept dry at all costs, so the body is not. Denny lay on one side with his elbows in the mud, the book held high. But not too high. The mud-covered ceiling was only twelve to eighteen inches from the floor. The floor mud was like molasses six inches deep, relieved here and there by pools of clear water.

Then there is the matter of moving along in a muddy crawlway. After each sighting, the note taker can put the book in a plastic bag, put the bag in a pocket, and put gloves on to keep his hands clean, which is essential for keeping book. Then he slurps, slides, and soaks on through. For the next reading, he takes gloves off and book out. Most cavers, however, prefer to hold the book in hands or teeth and crawl or drag along using elbows for traction with hands waving free, rather than bother with gloves. Denny used his elbows, but it was excruciating because of sharp chert lying under the soft mud in the passage the party was surveying. Denny named the belly-crawl, Agony Avenue.

Here and there in Agony Avenue a novel method of locomotion is possible. You set your right shoulder in the mud, clasp your hands across your belly, and shove with your feet. This is fine for ten or fifteen feet, but then the plow effect starts dumping muck down your neck. You have to rise up over the ridge you have plowed, and then you can press on.

Denny's party pushed on until they came to a pool of water that stretched from wall to wall. They had surveyed a respectable 670 feet in four hours. They were cold and miserable. The easiest way to read a Brunton compass is to position your head over it. In low passages, there is often no head room. In low passages half filled with water, the job is maddening. You must not get the compass wet, for if it gets waterlogged, it will not work.

This passage had to be surveyed, so Denny's party slid gingerly into the water—no matter how you do it, you get wet—and continued. After three stations, to their great relief, the compass needle stuck. They could do no more surveying. Someone else could come back to continue the survey of Agony Avenue. Denny and Jake collected the survey equipment, Dick decided he had never been so miserable in his life, and Bill pushed a little farther down the tube through the water.

It is enjoyable to be on a surveying party, Denny thought. There you are, with tape, compass, and book. The data of direction and distance are shouted out. The book keeper draws neat sketches. Where are we? Right here! Right on this page. It is a delight to know *exactly* where you are. And never mind that your little book shows only a few hundred feet of passage which may be two or three miles from the entrance along routes that no member of your party knows completely. Surveying is often comforting as well. As wet and cold as he was, Denny took great pleasure in the making of the map of Agony Avenue.

Denny, Dick, and Jake struggled to clean the tape and compass before retreating from Agony Avenue. Suddenly, a wave almost large enough to flood the passage came bearing down on them. It was Bill Morrow's return.

"There are dry passages ahead," Bill said.

"Oh?" Denny said. "How big?"

"Well, three to five feet wide and four to six feet high."

The others strapped up their bags and turned around. Enough was enough. Bill had probably found a little ceiling pocket where he could stand up.

"No," Denny said. "Wait a minute."

Denny had been working with the maps, and knew that they were under the edge of Houchins Valley. He was almost certain that Bill had made an important find.

"Lead on," Denny said to Bill.

Calling his brother, Dick, who was on his first (and last) trip into Flint Ridge, Denny oozed into the water to get completely wet at last. It could, after all, be the breakout. Dick and Jake reluctantly followed.

After 300 feet of crawling, they came into walking cave running east and west. Denny restrained his enthusiasm, and sent Jake and Bill down the passage. Meanwhile, he got a meal under way for Dick, who was badly chilled. Jake and Bill returned breathless. They had gone west for about 1400 feet, finding several leads including a canyon heading southwest, toward Mammoth Cave. Denny then went alone to the east, where he came to a big T-junction within 200 feet, with a strong breeze beckoning him onward. Denny smiled through his fatigue. He had no doubt about what the map would show when their survey was plotted.

They had penetrated beneath Houchins Valley. It had at last been demonstrated that cavers could get under that deep karst valley. They had walked under the axis of Houchins Valley, and they had found leads on the other side. The eventual connection with Mammoth Cave on the other side of Houchins Valley was surely now just a matter of time, and of work. But they had done enough for one day. Everyone in Denny's party was very cold by then, so they started back to the Austin Entrance. It was to be the last trip in Flint Ridge for three of the four party members. After that Agony Avenue trip, only Denny returned for more caving.

They got back to camp at five in the morning. Denny shook Roger awake. "Roger, Roger, we've broken out under Houchins Valley, found a lot of walking cave."

"That's nice," Roger groaned, falling back to sleep.

Denny thought Roger had not comprehended. He went to bed, annoyed. However, late that morning when Denny read the party assignments, he saw that Roger was to lead Claude Rust, Alan Hill, and Spike Werner out to survey in Agony Avenue. It was a very strong party. Roger had gotten the message.

Roger's party surveyed 2148 feet under Houchins Valley. Much of the survey was in a beautiful clean limestone river passage, some walking and some hands and knees crawling. It runs underneath the axis of Houchins Valley, so they named it Houchins River. There were leads everywhere, and a strong wind bore to the south, toward Mammoth Cave.

Red Watson rushed out to Houchins Valley to survey some of the south-trending passages. He was the eternal optimist. If a passage pinched to four inches wide and eight inches high, he would write in the notebook: "Good lead, possibly goes, needs checking out by a small person." Then he would back out. Whoever later read the notes needed to know that Red was five feet eight, weighed under 150 pounds, and had a reputation for pushing tight leads. The smaller caver would have to be five feet and 90 pounds.

One such caver was Micky Storts. She had universal talents as caver, cook, photographer, draftsman, baby sitter, and automobile mechanic. Best of all, she was small, thin, tough, and fearless. Once she crammed her body into a crack, and then her muffled voice came back, "All right, I've gotten in here just like you wanted and it doesn't go. Now do you have any suggestions as to how I can get out?"

Red and Joe started taking parties out to survey under Houchins Valley. They took along a couple of young cavers from Missouri, Stan Sides and Dennis Drum. Red, Joe, and Stan explored while talking, talking, talking. In 1967, Joe took over from Red as president of the Cave Research Foundation, and in 1972, Stan took over from Joe. Dennis became a member of the Board of Directors of the Cave Research Foundation. It is a voluntary organization. Those who do the most work run it.

Houchins River was surveyed for nearly a mile upstream to the southeast, but always it stayed under the axis of Houchins Valley. Downstream, Houchins River water had once flowed through Agony Avenue, but now it escaped through a small hole under the south wall of the passage. Did it go to Mammoth Cave? The new watercourse had to be lower than Candlelight River, and Houchins River might be it. However, Houchins River was very small and wet, and larger leads to the south beckoned at a higher level.

A winding, muddy hands and knees crawl continued to the southwest at ceiling level where you drop into Houchins River from Agony Avenue. John Bridge set 1200 feet of N-survey in it, and named it N-Trail. It continued, getting smaller and smaller. A side lead near the end with some belly-crawling went for several hundred feet generally south until exploration was stopped by a large slab that had slumped from the ceiling. Joe Davidson put an A-survey in it. The area was only 2000 feet from Mammoth Cave, closer than anyone had ever been from a Flint Ridge entrance. Surely the connecting of the Flint Ridge Cave System with Mammoth Cave would be just a matter of time.

Under Mammoth Cave Ridge

The hell of far-out Flint Ridge caving is close to the
heaven of Mammoth Cave Ridge. A breakdown pile at Q-87 closes the
passage only 800 feet from Albert's Domes in Mammoth Cave.

18 John Bridge and Fred Dickey could not stay away from those
small passages out under Houchins Valley. Those crawlways,
after all, had the best potential of any known so far for leading
to a connection between the Flint Ridge Cave System and Mammoth Cave.
They took Scooter Hildebolt there on his second trip into Flint Ridge.
Scooter was a contrast to their height and bulk, and he was terribly
earnest. He admired and trusted these big, tough Flint Ridge cavers. John
had told him that they might connect with Mammoth Cave on this trip.

Alan Hill, who is famous for having made the scientific discovery
that a steel tape dragged in Candlelight River during a thunderstorm can
shock you, told John that the best energy food was Eagle Brand sweetened
condensed milk. John decided that he, Scooter, and Fred should each take
two cans.

They went out to continue the survey at the end of N-Trail which
trended south toward Mammoth Cave from Houchins River. The passage
ahead was tortuous, and they knew the surveying would be as difficult as
any in Flint Ridge.

"I don't particularly want to," Fred said.

"Yes, but now that we're here," John said, "we might as well."

They began a Q-survey.

They had made the mistake of bringing a 100-foot tape. In the twisting
muddy passage it sometimes took all three of them to pull it forward.
They discussed breaking it in half, but Expedition Leader Joe Davidson
would not like that. Then, after twenty stations, the tape broke at the
50.1-foot mark.

John was having a lot of trouble with his lamp, and at Q-34 asked
Fred if he wanted to start out.

"Might as well," Fred said thoughtfully, concealing his eagerness. "Coming, Scooter?"

When they got back to Q-1, John, Fred, and Scooter settled down for a big meal before starting on out. Fred watched John down his can of Eagle Brand sweetened condensed milk.

"You had a funny expression on your face when you were drinking it," Fred said.

"It tastes like a mixture of honey, maple syrup, and heavy cream," John said reflectively.

Scooter took out his can, punched holes in it, took a swig, and immediately threw up. John and Scooter buried the rest of his Eagle Brand condensed milk. Scooter and Fred were indignant when they discovered that John had brought only one can. They would have to carry their own unopened cans out of the cave.

They retreated through the mud and cold and water, and finally got through Agony Avenue on their way back toward the Austin Entrance. Scooter was not feeling well. Fred and John did not appear to Scooter to be concerned about his condition. They did not want to show that they had begun to worry about him.

Climbing out of Candlelight River at Best Way Down, Scooter could hold out no longer.

"How far is it now?" came the eternal question.

It was miles; but the worst was behind, so John answered, "Half an hour."

Two hours later they stopped for a rest, and all fell asleep. When they woke up Scooter asked again, "How far?"

"Maybe half an hour," John replied.

Much more than an hour of crawling and trudging passed. Scooter had believed in these two implicitly. Now they were torturing him.

Time passed. Scooter could not remember *any* of the landmarks. For all he could tell, they might be going into, rather than out of, the cave.

At the ladder down into Pohl Avenue, less than a mile from the Austin Entrance, Scooter blew up.

"Damn it! I'm not going another inch unless you tell me how far it is to the entrance." He was tired, hungry, sick, and angry. He sank to the cold mud floor of the passage.

"Don't you remember the ladder?" Fred asked.

"I guess so," Scooter replied.

"We're about forty-five minutes from the entrance," John said, taking his pipe out of his mouth to stress his sincerity.

Scooter was skeptical, but got up to follow. They plodded along until they were about 600 feet from the Austin Entrance. Here Scooter declared his independence.

"I'm stopping. You guys lied to me, and I'm hungry. I'm going to eat a meal here."

"But the entrance is only five minutes away," Fred said in desperation. "In twenty minutes we can be up the hill and in camp and have a good hot meal."

John took his pipe out of his mouth again and started to say something, but he saw that it was of no use. He sat down, opened his pack, and set out a meal. He would have no trouble eating again when they got out. If Scooter spoiled his own appetite, that was his business. Fred shrugged, sat down, and ate a full cave meal himself.

None of them said anything when they reached the Austin Entrance a few minutes later as predicted. Scooter was full of pain and hurt pride, but nothing would stop him. He is still a regular Flint Ridge caver, but Eagle Brand sweetened condensed milk has never again appeared on the cave menu.

The Q-survey headed directly across Houchins Valley toward Mammoth Cave Ridge.

John Bridge knew the way. Roger Brucker was eager to continue to survey toward Mammoth Cave Ridge from Q-34, the last point reached. Roger decided to send supplies out to Shower Shaft in the Candlelight River area. Then a lightly burdened assault party could go on out to continue the Q-survey. This plan had not worked Left of the Trap, but Roger optimistically thought it was worth another try.

John led Stan Sides and Tom Conlan out with supplies to set up the dump at Shower Shaft. Tom got very ill. He did not return to Flint Ridge after that trip. The trip had tired John, too, and he was reluctant to lead the assault party the next day. However, Roger took him aside and told him seriously that he had concealed a dime above the toilet door in Snowball Dining Room of Mammoth Cave.

"If you connect on this next trip, use the dime to telephone us," Roger told John, "and we'll come and pick you up."

It was the old Flint Ridge Con. John knew it well, but he could not resist Roger's enthusiasm.

"All right," he said, "I'll go."

For John's party there were old-timer Roger McClure and newcomer Art Palmer. Art and his future wife, Peg Vogel, were doing an incredible job mapping Blue Spring Cave, nineteen miles long, full of cold chin-deep rivers, in Indiana. They could afford only one second-hand wet suit, so they alternated. Peg wore the top and Art the bottom on one trip, and then on the next they swapped ends. Like Walter and Barbara Lipton, who had also mapped a large cave of their own before coming to Flint Ridge, Art and Peg were attracted by the possibility of learning and mapping new cave passages without end. Art was ideal for the mud and water beyond Candlelight River.

The day of the assault on the Q-survey dawned. At the supply dump at Shower Shaft, John said, "We'll never make it."

"Why?" Art asked, his high spirits damped for the first time.

"No Eagle Brand condensed milk," John replied.

Through Agony Avenue, Art kept almost entirely dry by doing a perpetual push-up through several thousand feet of crawlway. John was worried about Art's wearing himself out, but reflected that he must know what he was doing.

John smoked his pipe. "We're well under Houchins Valley now," he said.

They felt fresh and happy when they reached Q-1. Roger McClure's non-stop joke-telling marathon had entered its eighth hour by the time they reached Q-34; they were glad to get to work so Roger would shut up.

At Q-34 begins the most impressive mud in the Flint Ridge Cave System. Q-survey mud resembles printer's ink. It covers everything, including the ceiling. This is a sign of recent complete flooding. The conscious mind quickly represses this information. After being churned by a passing caver, the mud assumes a character that Julia Child would admire. A finger dipped into it and lifted produces delightfully firm peaks, like egg whites perfectly whipped. Such mud *must* be good enough to eat. It is not.

Where the mud is deep, it is like axle grease. One caver stood crotch deep in this mud for five minutes taking a difficult compass sighting. His companions had a most difficult time extracting him.

Mud is a problem, but the primary difficulty with the Q-survey passage is that it is filled with breakdown blocks. The surveyors had to crawl among, around, under, and over many friable slabs of rock. At one point a large block closed the passage almost completely.

"I'm afraid this is it," John called back to the others.

"Let me look," Art said. The block was composed of thin sheets. Art reached to his side for the heavy-duty putty knife he always carried in caves. Although he had convinced few others, not even Peg, he swore that the putty knife was the universal caving tool. He used it to open cans and to eat the contents. Now he attacked the boulder with his putty knife. He pried and stabbed, and soon the block was reduced to rubble. The explorers went on.

Exfoliating slabs of rock hung from the walls. The slabs dropped at a touch—one on Art's foot, reminding him vividly of how Floyd Collins had been trapped.

The surveying was not easy. Some shots were less than three feet long. Art now led because of his expertise with the putty knife. Several times they seemed stopped, but Art pried a way. Then the passage character began to change. The limestone was now massive. There were little balls of popcorn on the walls. Popcorn is hard on hands and knees, and it rips clothes to shreds.

They moved on until there was no more mud, not even on the floor. The passage was dry and there were gypsum encrustations on the walls and ceiling.

John, Art, and Roger McClure looked at one another, afraid to say what they knew: Gypsum forms only under the sandstone caprock.

"We're under Mammoth Cave Ridge," Art said finally, in awe.

John and Roger nodded. John fumbled with his pipe. They were the first people ever to cross completely under Houchins Valley, to move underground from one to another of the great karst ridges of the Mammoth Cave Plateau, the first to go from Flint Ridge under that deep, wide karst valley that separates it from Mammoth Cave Ridge. Surely they would go on, would come to cross-passages and pits, to leads that would take them up into Mammoth Cave. One *could* cross under those wide karst valleys from ridge to ridge. They had done it. For a moment they dreamed that they had already connected the Flint Ridge Cave System with Mammoth Cave. But they knew that was not true, not yet. They continued the Q-survey.

The passage was now a fine, open, clean crawlway with solid walls and ceilings and no more breakdown blocks. John had taken the lead again. Suddenly he yelled: "You'll never believe this!"

Art and Roger McClure crawled quickly up to where John was stopped. The canyon passage there was bell-shaped, four feet wide at the base, closing up into a fissure a few inches wide up which they could peer six or eight feet. Just ahead the passage narrowed to a foot and a half wide, and was blocked by a floor-to-ceiling choke of sandstone cobbles and boulders. The reddish-brown rocks made a stark contrast against the white limestone walls. The pile was wet, and looked mobile.

"I don't believe it," Art said. "It proves we're under the sandstone caprock of Mammoth Cave Ridge, all right. But we're *at least* 150 feet, and probably considerably more than that, below the sandstone. So how did all these sandstone rocks get down here? It can't be the valley wall— we're well back from that, surely. It must be breakdown in the bottom of a pit."

"If it's a pit, we can climb up," Roger McClure said.

"Dig!" Art said.

They tore at the pile, moving several hundred pounds of rock. They had to be at the bottom of an enormous vertical shaft that stretched all the way to the Mammoth Cave Ridge caprock above.

If we get in, there will surely be an intersection with another passage, and that passage will have to be in Mammoth Cave. All thought it, but no one said it.

They opened a belly-crawl around one side of the pile, but it looked too dangerous to crawl into. The passage behind them was beginning to fill up with rocks they had moved back from the pile.

"We'll have to start swallowing them soon," John said.

The rocks came down as fast as they pulled them out, with ominous rumbling in the distance. However, after four hours they had to stop. A boulder nearly four feet long was settling down, almost filling the opening. To remove it would take more than bare hands and a putty knife.

They set their final station, Q-87, right at the boulder choke, vowed to return with a long bar, and sadly crawled back down the Q-survey, out from under and away from Mammoth Cave Ridge, back toward the Austin Entrance.

They looked for leads they might have missed on the way in. Perhaps they could find a side passage around the breakdown pile. But there were no leads until they were far back down the Q-survey, once again out from beneath Mammoth Cave Ridge, out under Houchins Valley, whipping up peaks of mud meringue.

They were hungry, so they stopped at the first place where they could halfway stand up. The opposite wall was six inches from their noses, and it took some care to get spoons from the cans into their mouths.

"Isn't this great?" John asked.

They remained cheerful until they reached the supply dump at Shower Shaft. Wool blankets, food, carbide, and foam mattresses sat incongruously on the wet, rocky floor.

"God, how depressing," Roger McClure said.

"I suppose it would be best if we got a little rest," John said.

They had been in the cave about twenty-two hours so far. They sat eating candy bars.

"We're not supposed to get the blankets muddy," Roger said.

John strapped up his pack and stood up. "I can't take my clothes off to take a nap," he said. "I could never face putting them back on again."

They went on toward the Austin Entrance.

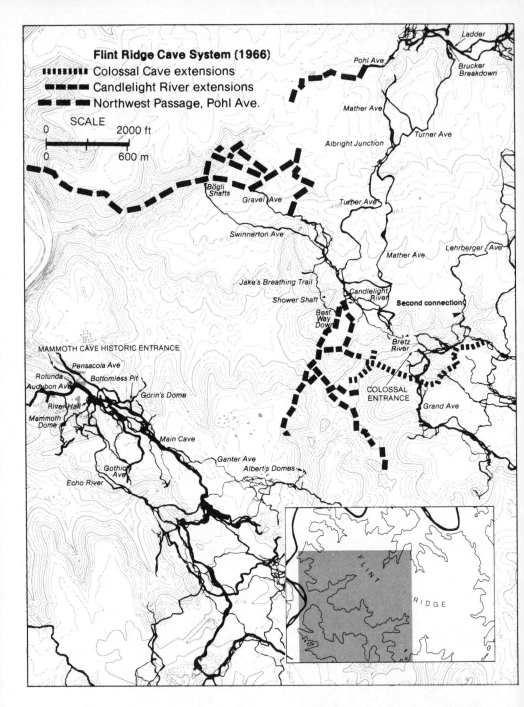

Flint Ridge Cave System (1966)
- ||||||||| Colossal Cave extensions
- ▬ ▬ ▬ Candlelight River extensions
- ▬▬ ▬▬ Northwest Passage, Pohl Ave.

SCALE

| 0 | | 2000 ft |
| 0 | | 600 m |

Ladder

Pohl Ave

Brucker Breakdown

Mather Ave

Turner Ave

Albright Junction

Bögli Shafts

Gravel Ave

Turner Ave

Swinnerton Ave

Lehrberger Ave

Mather Ave

Jake's Breathing Trail

Shower Shaft

Candlelight River

Second connection

Best Way Down

Bretz River

MAMMOTH CAVE HISTORIC ENTRANCE

Pensacola Ave

Rotunda

Bottomless Pit

Audubon Ave

Gorin's Dome

COLOSSAL ENTRANCE

River Hall

Mammoth Dome

Grand Ave

Main Cave

Ganter Ave

Gothic Ave

Albert's Domes

Echo River

FLINT RIDGE

Map of the Flint Ridge Cave System showing the extension of passages from Colossal Cave toward Houchins Valley in 1963, and from Candlelight River in 1964. Discovery of the Tight Tube in 1966 put Houchins River and Q-87 within easier reach. Other 1966 discoveries include the Northwest Passage and a Pohl Avenue extension.

Then came those pleasant but long crawlways and walking avenues. It is two miles from the Candlelight River area along Jake's Breathing Trail, Gravel Avenue, Swinnerton Avenue, Turner Avenue, down Brucker Breakdown into Pohl Avenue, and then another mile on to the Austin Entrance. John practiced turning off his mind as he plodded through Turner Avenue. It was difficult for Art and Roger McClure to do this because the soles of John's boots had come loose. That did not bother John, but Art's and Roger's wills were weakened with fatigue, and their bodies kept maddening time to the flap, flap, flap.

After having been in the cave for nearly twenty-seven hours, they arrived near the Austin Entrance just as Roger Brucker entered with a party. Listening to their story, he grabbed Art's foot and addressed it:

"You were there!"

Under Mammoth Cave Ridge!

"We didn't use your base camp," John said disgruntledly. "It was a waste of energy to carry in that stuff."

"You might have needed it," Roger Brucker said.

John knew that was right, but he puffed on his pipe without comment. You have only so much energy—physical and psychological—for such trips. John would not supply another camp himself.

John, Art, and Roger McClure went out through the Austin Entrance and up that long hill over the edge of the escarpment onto the top of Flint Ridge.

Art had never seen such excitement about someone else's trip. Everyone *really* cared, and was as pleased about what they had done as they were themselves. Like Scooter and many others before him, Art found himself inextricably bound to Flint Ridge caving on his first trip.

When the Q-survey was later made into a map, it was found that Q-87 in the Flint Ridge Cave System was within 800 feet of Albert's Domes in Mammoth Cave under Mammoth Cave Ridge. If that rock pile at Q-87 could be penetrated, it seemed likely that a connection between the Flint Ridge Cave System and Mammoth Cave would soon be made.

Q-87

Red thinks about caving on the
first big attempt to connect with Mammoth Cave.
That is where your friends are.

19 The news spread quietly that John Bridge's party had been under Mammoth Cave Ridge, having crossed beneath Houchins Valley from the Austin Entrance of the Flint Ridge Cave System. Joe Davidson listened attentively to the description of the breakdown choke at Q-87. He thought he could force his way through, but he knew it would be difficult to get a party together for a trip out to Q-87. There was no question about its being the hardest trip in the Flint Ridge Cave System. And those passages beyond Candlelight River had already provided many cavers with their final cave adventure.

Joe found two good cavers to go with him to Q-87. Dave Roebuck had a reputation as a daredevil. However, those who watched him lead a scuba-diving team into Pike Spring of Crystal Cave saw that, although he was very daring, he also paid careful attention to safety details. If Joe wanted someone to dig at a dangerous breakdown pile at the extreme end of the Flint Ridge Cave System, fine. Pete Barrett was ready, too. He viewed the trip almost as a lark. Pete was a New Zealander doing graduate work at Ohio State University. He had heard about Flint Ridge, and decided that it was for him. Like several other cavers who drifted in from New Zealand, Australia, England, and Canada, Pete made a number of valuable contributions.

This was Joe's introduction to the Q-survey. The last 1000 feet were as tough as any passage he had been in. He was ready to work for a long time, but after they reached the boulder choke at Q-87 and dug for two hours, they had to quit. There were several three-foot boulders that they probably could not remove without killing themselves. They needed a long pipe so they could poke at the boulders safely from a distance.

The danger was exhilarating. One thing about it—the Q-87 pile was moving. If they could keep pulling rocks out, eventually they might get through. And on the other side they might find Mammoth Cave.

After looking long and hard at the belly-crawl that led around the base of the pile, they decided not to enter it. So far, that dangerous way was just too dangerous. Eighteen hours after entering the cave—a fast, efficient, and work-filled trip—they exited. However, Joe was dissatisfied.

What was needed was proper digging equipment. The best would be a long pipe that could be jammed up into the pile so that the rocks would tumble down. A long bar was also needed to lever boulders out. It would be dangerous. First-aid supplies ought to be added to the old cache at Shower Shaft. Walter Lipton reluctantly, but dutifully, led a party out to deposit the emergency gear. It was a way to acquaint more people with the route in case they were needed for rescue. As they left, Denny Burns said, "The emergency kit out there should be a pint of Jack Daniel's and a .45 pistol."

Nine months after Joe's first visit, his second assault on Q-87 was assembled. There were eight Candlelight River veterans. Fred Dickey, Bob Fries, Bill Hosken, and Terry Preston entered the Austin Entrance an hour before the second party. Each carried a four-foot chunk of steel pipe, threaded to screw together easily. They would carry the pipes to S-21, half a mile from Q-1, and then go on to survey in a relatively easy passage. Joe Davidson, John Bridge, Art Palmer, and Red Watson entered the cave an hour later than the first party. They would pick up the pipes at S-21, carry them on out the tortuous way to Q-87, and use them to dig at the rock pile.

Who was left behind to mind the store? Every one of the eight was of party-leader caliber. Having as many as eight top leaders was unusual on any expedition. Obviously, everything in the book was being thrown at Q-87. In fact, the prospects for connecting the Flint Ridge Cave System with Mammoth Cave on this expedition were so great that there remained people topside—Roger Brucker, Barbara and Walter Lipton, and several others—who could immediately provide further support if necessary.

Red was eager to go to Q-87. But he was also apprehensive, because he did not like breakdown piles. Trying to calm a nervous stomach before going into the cave, Red decided to record this Q-87 trip by writing down as many impressions as he could on the way. The resulting thirty-three-page manuscript was in some part responsible for no one's going to Q-87 again for the next six years.

Those were the days when Flint Ridge cavers slept late. It was always dark inside the cave, so what matter when one entered? Pity the mountain climbers who had to rise before dawn. Cavers could dress slowly.

The cavers casually entered the Austin Entrance around noon. There had been no meeting beforehand. Most Flint Ridge operations were coordinated with little obvious leadership. "Don't lose those walking sticks," was the only instruction Joe had given to the pipe party when John Bridge and the others had set out.

This might really be *the* trip. Q-87 was within 800 feet of Mammoth Cave. Yet there was no fanfare. Roger Brucker was preparing to take a short exploration trip around the Overlook in Crystal Cave. He would be the main man in case of rescue. He hardly glanced at the departing cavers.

What would the French have made of it? Red loved to read French caving books. Every French expedition was a confrontation of man with himself, with the cave, and with the cosmos. Had the French done all the work that so far had been done in Flint Ridge, there would be a dozen books about it. The Americans had written only one book, *The Caves Beyond,* and it covered only the one-week National Speleological Society C-3 expedition in Crystal Cave in Flint Ridge in 1954. For Q-87 there were no trumpets, no priests, no announcements to the press, let alone to the cosmos, that man was once again setting out to push beyond the limits of the known. Americans are so matter-of-fact, so unemotional, so dull. So Red wrote in his notebook.

They hurried from the Austin Entrance a mile along Pohl Avenue, climbed eighty feet up Brucker Breakdown, walked 7000 feet down Turner Avenue and out Swinnerton Avenue 2000 feet to the Duck-under down into Gravel Avenue. It was another two miles of crawling and crouching from the Duck-under to Q-87.

In Turner Avenue, Red had been walking far behind the others with a very dim light. As the others picked their way on ahead, their lights illuminated an ever changing view of passage size and configuration for Red to enjoy. The smoothly arched ceiling reached down to sharp-angled wall contacts with the jagged floor. Of all the senses the greatest is touch, but the most intoxicating is sight.

Beyond the Duck-under, they moved steadily along the crawlways. It would have annoyed them to have to stop to rest. However, there was no hurry. Everyone automatically conserved strength. A minimum trip out to Q-87 for two hours' work had taken Joe's previous party eighteen hours. Joe intended to work at the Q-87 breakdown pile four to six hours this time.

The cave is hard on cloth. About half the explorers wore Levi's pants and jackets. This was the uniform Bill Austin and Jack Lehrberger favored. Others wore coveralls. Coveralls give no ventilation at the waist, and pull down at the shoulders when pants legs get caught on rocks in crawlways. Levi's allow dirt and sand to get in at the waist. Which is worse? The argument will never end.

When they reached Shower Shaft, 3000 feet from the Duck-under, they saw that nine months of moisture had nearly destroyed the emergency cache. Plastic bags had broken open to spill wool blankets out in sodden, moldy masses. Food cans had rusted through, and the carbide was ruined. However, the first-aid supplies were usable. The odor of the

*An obscure drain in Shower Shaft at the end
of Jake's Breathing Trail leads to Candlelight
River. Note the explorer crouched in the drain.*

dump was offensive, so they hurried on through a tiny crawlway that had a clean smell of clay and wet stone.

As you move along familiar passages, sometimes you realize that you do not remember what the passage looks like for some distance back. It is like suddenly noticing when driving a car that you cannot recall having seen the road for some time. Yet, there you are, far from where you last focused on the route. In the cave, wide-open side passages have been missed this way by somnolent explorers.

Orientation is tricky, too. A sense of direction is sometimes a handicap underground, because it is so easy to get disoriented. It is better simply to have a feel for the cave, a sense of where one is.

Sometimes you follow a virgin passage until it joins a traveled passage. If you recognize where you are, and if you are disoriented, then directions may reverse and reorientation come in a great, almost physical whirl. The experience can be so striking that you remember it vividly forever.

Joe's Q-87 party took an easier route through the recently discovered Tight Tube to avoid the mud of Agony Avenue. The Tight Tube has turns too sharp for four-foot pipe lengths, so Bridge's pipe party was doomed to wallow through Agony Avenue.

John Bridge had been through the Tight Tube before, but now he was having trouble. His six-foot-four-inch frame filled the belly-crawl with no room at all to spare. He lay with his arms stretched out in front of him.

"I just don't seem to have it today," he said. "I can never make this right-angle turn."

"I suggest you turn around and back through," Red said. He had gone on ahead.

Behind John was Joe, who said, "You might roll over on your back and go through with your stomach facing the *outside* of the curve."

John sighed at this cave humor, and managed to shove through, a fraction of an inch at a time. "Here is my torn body," he said on emerging. He filled his pipe and started smoking up a storm. "I really don't feel well," he said.

"A little sick at the stomach?" Joe asked relentlessly.

"Perhaps a slight constriction of the chest?" Red joined in.

"A prickly sensation along the back of the nasal passage?" Joe asked.

"It affects me the same way," Red said.

"Okay, okay," John said. He knocked out his pipe and put it away.

Art Palmer, thin and a foot shorter than John, slid easily through the Tight Tube at the rear of the party. John peered over to watch Art come through, then leaned back and sighed again.

"Who went through the Tight Tube first?" Red asked.

"Tom Hall," John said.

"Who's Tom Hall?" Red asked.

"I don't know," John replied. "He never came back. But he was a hell of a caver to have pushed around that right-angle turn."

Smoke penetrates long distances through cave passages. The members of the Q-87 party could smell Bill Hosken's cigar several hundred feet before they caught up with the pipe party at the rendezvous junction. Then eight people were strung along a passage six or eight feet wide, but not quite high enough to stand up in. There was too much noise. A pipe was wedged across the passage near the ceiling for a chin-up contest. The cavers had to bend their knees to keep their feet from hitting the floor to compete. No one could beat Art's one-armed chin-ups.

Eight men were too many. Only on two- or three-man parties do you begin really to feel the loneliness and remoteness of Flint Ridge.

"I can always tell when I'm beyond Candlelight River," Bill said, "because I lose control of my face muscles."

"I wish you'd lose that cigar," Red said.

"I need the flame to dry off," Bill replied. "How come you guys aren't wet and muddy like real cavers?"

The four who had gone through the 1450 feet of Agony Avenue were soaked.

"I've had it," John said.

The bantering stopped.

"Do you want to go out, John?" Joe asked.

"No, no, I'm not that bad. I just don't feel up to going out the Q-survey." It had taken previous parties four hours to cover the 4000 feet of crawling left to get to Q-87.

Joe looked around. Everyone was sober. Bill and Fred could certainly go on to Q-87, although Fred was a very big man. Fred said later that carrying one of the pipes out to Q-1 was one of the several hundred difficult things he was willing to do to keep from going to Q-87. Fred was horrified when John's illness made him a candidate after all.

"I'll go if necessary," Bill said without enthusiasm.

"I've just," Terry Preston said, blushing, "well, I've just gotten married, and I don't think I'm up to a twenty-four-hour trip."

Joe turned to the youngest man present, Bob Fries.

"What worries me," Bob said, "is what if that passage beyond Q-87 goes someplace?"

"That's been worrying us for months," Joe replied dryly.

While this was going on, the oldest man present, Red Watson, reflected. Sitting on the cold mud did not help, for the body stiffens. For a fleeting moment Red thought of not going on himself. Why not go out with John? What am I doing here? Why? The trip was not half over.

How old is John? Red thought. Not over thirty, anyway. Earlier that day on the surface, Fred Benington had asked Red how old he was. Red told him.

"Thirty-five, eh? I'm fifty-five," Fred said, and turned away.

Now it was decided that Bob would replace John. The Q-87 team took the four steel pipes and started off toward Q-87. After half an hour into the Q-survey, Bob said, "Damn, I have the tape for the other party."

Everyone stopped.

"They sure can't do any surveying without it," Joe said.

"Shall I take it back to them?" Bob asked.

"It would be an hour round trip," Red said, "and they might not be there."

"Forget the tape," Joe said firmly. "Our goal is Q-87. They'll just have to abort their survey." He turned to lead on.

They made the 3500-foot crawl from S-21 to Q-1 and on to Q-87 in the record time of three hours. Rocks had fallen down until the Q-87 rock pile looked almost as it had when first discovered. Mud and water oozed through the sandstone cobbles and boulders.

Only one man could dig at the pile at a time. The others moved rocks back down the passage. Art worked first, then Joe. The pile brought out a manic glee in Joe, who began heaving rocks back down the passage wildly.

Sometimes there would be a rumble that could be felt through the walls of the passage. What *was* that noise? The sound seemed far away, but maybe rocks were rolling down a slope in a big pit, just on the other side of the pile. Maybe Mammoth Cave was just on the other side of that pile.

"We're okay here, this passage is in solid bedrock," Joe said.

"Yeh, and once it was solid bedrock up where that breakdown pile is coming through the ceiling, too," Red said.

Part of the Q-87 pipe party in Agony Avenue:
Red Watson (writing memoirs), Bob Fries,
and Terry Preston, 1966.

Red took his turn at the breakdown face, but was not enthusiastic. He removed a few rocks until a big boulder settled down. He was not optimistic about getting the boulder out, so he turned it over to Bob. Soon Bob came back down the passage to sit beside Red.

"Are you chicken, too?" Red asked.

"Yeh," Bob replied. "I want out of their way if they come roaring back down the passage."

"Me, too," Red said.

Bob meant Joe and Art, who were still manic about the rock pile. Red meant the rocks. He fancied he heard them rumbling directly overhead, although he was twenty feet down the passage away from the hole in the ceiling at Q-87. Bob and Red continued to move rocks down the passage to make way for the ones Joe and Art were removing from the Q-87 pile.

Bob shouted up to Joe, "This is a bad place to have an accident. I won't have one if you won't."

There was another rumble, then Joe shouted back, "Agreed."

Art tied his geology pick onto the end of a pipe and brought down a few more stones. However, after two hours and the removal of 1000 pounds of rock, the settling pile stabilized. Even Joe could move no more stones. Joe and Art crawled back to join Bob and Red.

"What if we had opened it up?" Bob asked. "Would it be safe to go through?"

"Probably not," Joe said. Everyone knew that they would have gone through anyway, to see if Mammoth Cave were on the other side.

"We'll probably mark *finis* to this one," Joe said gloomily. "This is the third attempt at it."

They opened cans and ate a meal. In all those eighty-seven Q-stations, there would be no better place. While they sat there, they heard a distant rumble.

"The rock pile must have shifted again," Red said.

"Should we go look?" Bob asked.

Joe looked up, startled, from his canned chicken. "Nah," he said finally. "That rumble was way up above." He went back to his food.

Red was secretly delighted. No one *really* knew. Something might have opened up. Someone would go back someday to see. It is the combination of uncertainty and the unknown that is the lure.

Red led the way out the Q-survey. It was new for him, but not much of a challenge because there was no place to go wrong.

At one place a fallen block balanced across the passage. On the way in, Joe had stopped a moment just after going under the rock, leaving Red lying on his belly under it. On the way out, Red slid under the block in one quick movement. Another stretch is just high enough to stand up in. However, it is barely a foot wide, and at about knee level there is a projecting ledge. To slide along, you must lift each leg until the ankle is level with the ledge. Where it is impossible for you to raise your legs, you can feel the serrated edge of the ledge cut into your calf muscles as you force yourself along.

They went on without rest for five hours until they reached the pleasant little room just below the stairway-like climb up to the Duck-under into Swinnerton Avenue. From there it is a mere three-mile walk and crawl out of the cave, interrupted by the climb down Brucker Break-down. They relaxed in the cozy, warm room and had another meal. The tension was gone. Shoulders and hips began to protest those hours and miles of contortions. Hamburger knees began to hurt. Eyes drooped. Finally they cranked up creaky muscles and bones for the long haul to the Austin Entrance.

Far-out caving starts where the passages are unknown, whether just inside an entrance, or at the end of the Q-survey. Joe's party had been as far from an entrance as you could get in the Flint Ridge Cave System, about four and a half miles, and now in familiar passages, three miles from the Austin Entrance, the cave trip was as good as over.

After that Q-87 trip, John wrote Red a letter:

> I don't understand what made me feel the way I did. I had prepared myself physically and psychologically for the trip. I have heard vague, sometimes embarrassed descriptions of it happening to others who should on any occasion be unquestioned as to their performance in the cave.

Red replied that anyone was susceptible to such a letdown, and went on:

> I don't hesitate to say that out at Q-87, on the first occasion of that long rumbling high in the pile above our heads, I had a pang of fear in my stomach so intense for a moment that I thought I would be sick.

John had done the right thing. Everyone remembered one caver who had told no one he was having stomach cramps until he was almost incapacitated, far back in the Flint Ridge Cave System. Another simply became exhausted, and argued violently that he should just be left in the cave to die. His fellow party members had to drag him out.

After the Q-87 trip Red tried to derive from his notes some sense of what caving is all about. He decided that cavers are fortunate because they are often aware of some of the best moments of their lives. Like climbers, surfers, long-distance runners, and a few others, cavers often know that they are doing exactly what they want to do while doing it.

But Red could not rise to the French model. What understanding does one gain from such a trip? He and Joe could look at one another afterward and grin. There was nothing mystical or cosmic about it. Some good friends had done a difficult and neat thing together. Caving at the limits is fun. It demands exotic skills that few people have. It brings out the joy of life in an activity where the possibility of death is clear and clean. It takes you where few people or no one has ever been before, and in the midst of life it makes you aware of where everyone must eventually go.

Red tried again to distill the essence of caving. He asked the question: What is out there? He answered:

There is one simple thing to say: I want to connect with Mammoth Cave. Beyond that, that is where you live, out there. That is where your friends are.

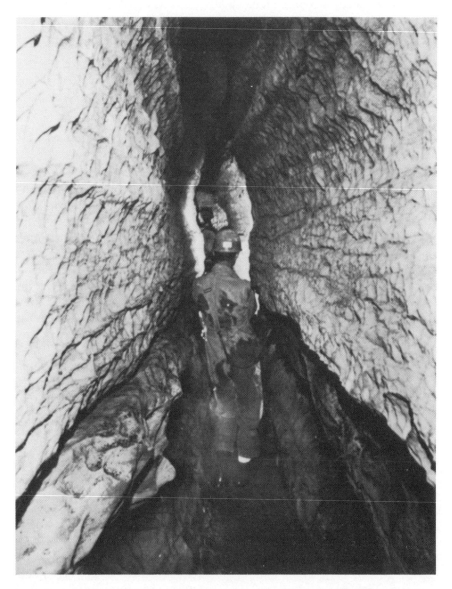

Bretz River.

Fatigue and Politics

A threat to the future protection of the longest cave halts efforts to connect the Flint Ridge Cave System with Mammoth Cave, so Joe finds some cave in a blank spot in Flint Ridge. Bureaucrats say there is no cave in Joppa Ridge, so Gordon discovers the Joppa Ridge Cave System and a new connection dream emerges.

20 And now comes an embarrassment. After so many years of *claiming* to have the longest cave in the world, and so many years of striving to connect caves to integrate the longest cave in the world, the Flint Ridge cavers got so wrapped up in exploring the Northwest Passage, crawlways and rivers in Colossal Cave, new routes off Pohl Avenue, and the far reaches of the cave under Houchins Valley, that no one marked the occasion on which the Flint Ridge Cave System exceeded Hölloch in length and actually became the longest cave in the world. Over a year later, someone noticed that in March 1969 the Flint Ridge Cave System finally did surpass Hölloch in length. At that time the Flint Ridge Cave System had about sixty-five miles of surveyed passages to Hölloch's sixty-four and a half. Mammoth Cave had forty-four miles. The momentum of discovery was so great in Flint Ridge that after that date Hölloch never again had a chance. In the process, Mammoth Cave had dropped from second to third place. And though the cavers knew that Mammoth had the potential of being once more the longest cave, they were determined to connect with it before it ever reached that status again. They were going to make Mammoth Cave a *part* of the Flint Ridge Cave System.

To connect the Flint Ridge Cave System with Mammoth Cave would be to climb the Everest of world speleology, and *that* was the real goal. However, a new question arose: What if the Flint Ridge Cave System *were* connected with Mammoth Cave?

It dawned on the cavers that there might be political implications. Cave Research Foundation directors had been arguing that in Mammoth Cave National Park, the Flint Ridge Cave System—the longest in the world—should be designated as *underground wilderness*, under provisions of the Wilderness Act passed by Congress in 1964. However, during one discussion, Park Service officials made a debater's point. In Mammoth Cave there were trails, lighting, toilets, and even an underground restaurant, the Snowball Dining Room. If the Flint Ridge Cave System and Mammoth Cave were connected, as cavers claimed but had not yet shown, then it was not wilderness.

This should probably have been laughed off, but Cave Research Foundation directors took it seriously. It seemed possible that bureaucratic heads in Washington might nod solemnly over such a point and deny wilderness designation to the Flint Ridge Cave System. For a while, then, Foundation policy was to curtail attempts to connect the Flint Ridge Cave System with Mammoth Cave. No one would turn back if he found himself in a virgin lead going toward Mammoth Cave from the Flint Ridge side, but exploration beyond Candlelight River dwindled.

(In fact, for whatever reasons, and after ten years of immense and heartbreaking efforts by conservationists, the National Park Service denied wilderness status to any surface or underground area in Mammoth Cave National Park. This is a decision cavers everywhere are working to have reversed, for the longest cave in the world should be protected under the provisions of the Wilderness Act.)

It is difficult to reconstruct exactly how some of the cavers allowed themselves to be diverted from attempts to connect the Flint Ridge Cave System with Mammoth Cave. There were many other things to do in the Flint Ridge Cave System, of course. Many people were involved in major programs of hydrological research directed by Will White, biological research directed by Tom Poulson, and archaeological research directed by Patty Jo Watson. But had the Candlelight River area been easy to get to, the slowdown of connection work probably would not have occurred. Fatigue came to the support of politics. Many good cavers had been flushed down the Candlelight River drain, never to return to Flint Ridge caving. And the old regulars had gone too many times out under Houchins Valley.

One of the last trips beyond Candlelight River for several years was itself partially political. A new Park Naturalist, Alan Mebane, had come to Mammoth Cave National Park, and he was truly interested in caves. He was athletic and obviously would make a natural caver.

By October 1966, upstream Houchins River had been pushed very hard. The miserable passage continued, but Roger Brucker decided that the way to connect with Mammoth Cave was *downstream* Houchins River from the Flint Ridge side. The water flowed in from the Mammoth Cave side, but it flowed out under the south wall of the passage and might lead back to Mammoth Cave Ridge. On such a trip, Alan Mebane could see what Flint Ridge was *really* like. If the connection were made, it would also be of some political value to have had a conservationist Park Service official along. And everyone liked Alan, anyway.

Red Watson and Stan Sides completed Roger's party. They all talked politics until they came to the Tight Tube. Roger climbed up to show the way, and then Alan followed. He chimneyed up the wall, stuck his head into the Tight Tube, put his knee against the edge of it, hunched up, and twisted. Something went pop.

Alan lay in the tube sweating. Then feeling came back.

"I've twisted my knee."

"Badly?" Red asked from behind him.

"I can't tell, lying here," Alan replied, "but I think maybe I did."

"Can you go on through, or do you want to go out?" Red asked.

"What's up ahead?"

"After about forty feet there's a sharp right-angle bend, and then about twenty feet farther on there's a small room with a dry, sandy floor. You could sit up or lie down there."

Alan knew he could do neither of those things back in the narrow canyon, so he crawled on. After a short rest, Alan said he thought there was no point in just turning around and going back. Everyone wanted to get some surveying done.

They surveyed 250 feet down the drain in a small stream of water, stopping where the water disappeared through a crack in the floor. Various cavers have since tried to penetrate the Houchins River drain, but your body plugs the small passage, causing the water to rise behind you at an alarming rate. The water probably does go to Mammoth Cave, but no one has tried to follow it twice.

On the way out, Alan favored his tender knee, but he moved along all right. The cavers were wet anyway, so on the way back they avoided the Tight Tube and took the easier, but muddy, Loose Tube.

The problem was Roger. He was seriously worried about his knees. The long crawlway back to Swinnerton Avenue caused him a great deal of pain. However, the only way out is to keep on crawling.

Finally, they reached the Duck-under up into Swinnerton Avenue. There Roger took off his new fishnet long drawers to discover that the bulky cords had sawed into the flesh of his knees. The fishnet was supposed to hold outer clothes away from the skin to form air pockets which would increase warmth. Roger's fishnet had ground sand and gravel into his skin. It was the first and last time Roger wore fishnet bottoms. He had a miraculous recovery.

Alan said little about his knee. They crouched and walked along the 2000 feet of Swinnerton Avenue to the junction with Turner Avenue. There they stopped for a meal.

Stan sat down beside Alan. Stan was a medical student. Having seen to Roger's cure, he now insisted that Alan let him examine his knee.

"Well, look," Alan said, pulling up his pants leg, "I can walk on it, but I can't straighten my leg out."

Stan dug his thumbs into the swollen joint, making Alan wince. Stan looked up at Alan. "You've surely got a torn cartilage," he said.

"That's what I figured," Alan replied, pulling his pants leg back down.

Alan took himself on out the Austin Entrance. His knee had to be operated on. Red, president of the Cave Research Foundation at the time, was greatly embarrassed. The rumor went out that the days of the rinky-dink were not over in Flint Ridge. "The CRF president broke the Park Naturalist's leg, ho ho."

Red tried to reassure Alan by remarking that he had had a cartilage removed from one of his own knees years back and now it was good as new. Alan's knee would recover just as well. That was true, of course, but it was feeble to say it.

Alan replied that he would not have missed the trip for anything. That, also, was probably true, but the cost seemed high.

Alan's injury cooled interest in Candlelight River for a while. Joe Davidson had found another passage, anyway, into which to channel talent, or to drain it off. Through the Duck-under off Swinnerton Avenue, down into Gravel Avenue, to Lower Gravel Avenue, Joe led an exploration party out a mile toward the northwest corner of Flint Ridge. Joe was searching for a northwest passage to a blank place on the map. This Flint Ridge passage was going away from Mammoth Cave, but not everyone can be a connection fanatic, and Joe often claimed that he was not. He was just looking for new cave.

Joe took Scooter Hildebolt, and a New Zealand caver, Pete Anderton, and an English caver, Mike Goodchild, out there to survey a miserable little belly-crawl in mud and water. Scooter was keeping book, and kept clucking about the difficulty of it. Pete and Mike also voiced skepticism about the passage.

"You never know in Flint Ridge," Joe said brightly.

"We are blessed," Scooter explained to the other two, "with CRF's most nauseatingly cheerful party leader."

"Well, it really *might* go," Joe said.

The others groaned.

They ended the survey at a pit which was too difficult to climb down into. Pete went around the pit through a side passage while Mike traversed across it. They emerged into a large virgin walking passage. Scooter and Joe quickly followed.

"See!" Joe said. "I told you so. This is really *it!*"

The passage was twelve feet high and six to eight feet wide. They quickly ran 1000 feet in one direction, and then turned back. In the other direction, they followed a canyon six feet wide and twenty-five feet high. A meandering stream between mud banks made travel difficult, so after only a mile of walking, they turned back.

The passage trended to the northwest. They had found Joe's Northwest Passage. Joe and Scooter were ecstatic. It was what Pete and Mike had traveled halfway around the world to experience (although they had not known it before they started).

"It was a real Austin-Lehrberger-type trip," Joe said later to everyone who would listen.

Scooter could not get over the trip. He wrote in his trip report that his greatest dream had come true. He had always wanted to stick his head through a hole into virgin cave where he could walk until he got tired. "Just like Austin and Lehrberger." He had half worried whether the Austin-and-Lehrberger legend were actually true. Now he knew it was. Flint Ridge was a place where you could find virgin cave and walk until you were tired.

All the old veterans went out to survey and to explore the Northwest Passage, dragging along ever more new recruits for inevitable erosion and attrition. There was no thought of connection. They were going northwest away from Mammoth Cave. The lure was two square miles of blank space on the map of Flint Ridge. Red even managed to get lost out there for an hour or so. This greatly impressed Gordon Smith, who was taking his first trip into Flint Ridge with Red's party.

Then the Mammoth Cave National Park Preliminary Master Plan was unofficially revealed. Among the proposals was one for a large parking lot south of Mammoth Cave, across Doyel Valley, on Joppa Ridge, the third major karst ridge in the park. Park Service officials said it was a good place for a parking lot, because there was no large cave system in Joppa Ridge.

"So there isn't big cave under Joppa Ridge, eh?"

To doubt that there was a large cave system under Joppa Ridge was to fly in the face of all knowledge of the area. Joppa Ridge had the same geological and hydrological conditions as Flint Ridge and Mammoth Cave Ridge. However, if there were a Joppa Ridge Cave System comparable to those in the other two ridges, it would have to be shown to the bureaucrats, in the ground. So if politics had something to do with holding up the connection between the Flint Ridge Cave System and Mammoth Cave, politics also threw explorers into Joppa Ridge.

Prior to these machinations, however, Gordon Smith and Judy Edmonds spent many weekends tramping around the margins of Joppa Ridge. The underlying limestone is exposed where the valley walls intersect the standstone caprock, and here sinkholes, blowholes, or pits might be entrances to a big cave system.

One November weekend in 1968, Gordon and Judy found a pit along the northwest edge of Joppa Ridge. The drain was a crawlway. About 100 feet along it, over a long, narrow crack in the floor was scratched on the wall: "T. E. LEE 1876." T. E. Lee had been a real cave explorer, so if the crawlway led to a big cave, presumably it would have been known. No such cave was known. The name was elaborately drawn with antique letters. T. E. Lee must have sat there thinking about that crack for a long time. It was a canyon in some places as much as ten inches wide. Was T. E. Lee as thin as Gordon?

Gordon lowered himself down, probing the crack with his toes. It went. By feeling blindly, Gordon established that the canyon opened out below. He hunched down and managed to stretch out horizontally, six feet below the top of the crack. Getting back would be interesting. He moved forward easily on his side for a body length before the crack closed in again. There he worked himself upright to descend another body length. Again he stretched out horizontally to inch along sideways for twenty feet. There the canyon opened up to a foot and a half wide and he could crawl. Soon he looked down into a thirty-foot pit. Beyond was the blackness of a much larger pit. He could hear the music of a water-fall.

The Joppa Ridge Cave System!

Gordon's straight dark hair escaped from under his hard hat to hang over his sweat-drenched forehead into his eyes. His crooked smile became even more crooked. Gordon talked out of the side of his mouth when he was being conspiratorial or confidential. People would probably have to get around behind his back to hear him when he started talking about this discovery. He soon solved the difficult problem of getting back up out of the crack.

Now Denny Burns took charge. He led the survey into Lee Cave. The drain of the smaller pit was plugged, but a wide, sandy crawl led from high on the wall of the larger pit beyond. It looked great, but after a few hundred feet, banks of mud closed the passage almost completely. The explorers pushed on over them to find themselves in a major trunk passage sixty to 100 feet wide and thirty feet high. White crystals of a rare double salt on the walls and ceilings were in such fragile balance that the movement of the cavers through the passage caused crystals to fall like snow. A thick blanket of crystals covered the floor, showing that they had been drifting down for a long time. But more exciting than the beauty was the feeling of big cave, unexplored leads, and discoveries ahead.

And Lee Cave was not, after all, virgin. Although T. E. Lee had not made it down there, the floor of the big trunk passage was strewn with cane-torch fragments left by prehistoric cavers 4000 years before. This passage—Marshall Avenue—continued for a mile and a half.

Soon more than seven miles of Lee Cave were mapped. Now no one could deny that there is a big wilderness cave system under Joppa Ridge, just as there is under Flint Ridge and Mammoth Cave Ridge. The explorers promised that, after the Flint Ridge Cave System was connected with Mammoth Cave, they would then connect with the Joppa Ridge Cave System. There were many years of unfinished connection work ahead.

One connection was soon seen to. After their success in locating Lee Cave, ridge walkers Gordon Smith and Judy Edmonds got married.

From the Mammoth Cave Side

Joe authorizes one party to resume the effort to connect
the Flint Ridge Cave System with Mammoth Cave.
Red sends ten parties.

21 In theory, the Cave Research Foundation had permission to work in Mammoth Cave itself after the 1959 agreement with the National Park Service to conduct research within Mammoth Cave National Park. Will White and George Deike had already gone many places in Mammoth Cave to make geological observations. However, it was more than ten years after that agreement before exploration and mapping began in Mammoth Cave. One reason was that after years of struggle to gain permission just to work in Flint Ridge, the cavers did not want to push their luck.

Albert's Domes in Mammoth Cave were known from Kaemper's map of Mammoth Cave to be only 800 feet from Q-87 in the Flint Ridge Cave System. One could walk underground from the Historic Entrance of Mammoth Cave almost all the way to Albert's Domes. Of course, even to suggest trying to connect with the Flint Ridge Cave System from the Mammoth Cave side was to rub the fur the wrong way on a fair number of Flint Ridge cavers. After twenty years of efforts from the Flint Ridge side, it was thought just *not right* to try to connect, or even to take the chance of connecting, from the Mammoth Cave side.

However, another aspirant—Tom Barr—rose to prod the conservatives. Tom was a good caver the Flint Ridge cavers had known forever. He had been on the National Speleological Society C-3 expedition in Crystal Cave in 1954. He organized the Institute of Speleology at the University of Kentucky. Tom was pretty much of a loner, with round, beady eyes in round horn-rimmed glasses, a pipe, and great caving strength and agility in a deceptively round body. Tom was doing biological work in Mammoth Cave. His field was beetles. Beetles could be anywhere. Tom went everywhere in Mammoth Cave. He was looking out all the passages that went north from Mammoth Cave toward Flint Ridge. He had even been in

Albert's Domes. He was a friend. But what if *Tom Barr* found the connection between the Flint Ridge Cave System and Mammoth Cave *from the Mammoth Cave side?*

The Cave Research Foundation had been growing steadily. The turnover was always great, and many new people had come in. Red Watson, away for a year at the Stanford Center for Advanced Study, returned to Flint Ridge to run a week-long August expedition in 1968.

Red had run many expeditions in the past, and, like Roger Brucker, he was exhilarated by the role of expedition leader. Red arose early to cook breakfast (something he did not do at home), and pity the caver who wanted anything so simple as mere scrambled eggs. Red always assigned himself to one of the hardest and grubbiest trips, and then bullied everyone else to take trips just as hard and grubby. He schemed great schemes and sent parties to the boondocks. The expedition was his body, the cave parties his limbs, and the cavers his senses. The organism was sound and healthy. Red cooked, caved, bossed, and slept well.

But times and people change. When Red returned to direct that expedition, he found sixty people—when in the past half that number had been thought a lot—and he knew practically none of them. There were too many to cook breakfast eggs for. Red had to get Gordon Smith's help in making up parties because he knew the capabilities of only a few of the cavers on hand. The survey program was designed by Denny Burns, and Red did not know some of the areas where difficult caving was to be done. The first day he had twelve parties in Flint Ridge, and none of them included himself. He remembered sympathetically how Joe Lawrence had been overwhelmed at the beginning of the C-3 expedition in 1954 by just about the same number of people Red now had on hand. Bill Austin and Jack Lehrberger had said that the good old days were over. Red decided that they sure as hell were.

Red could not sleep. After a while he sat up in his sleeping bag and said, "This is ridiculous. I'm *worrying* about those damned cavers."

He had never worried about them before. He had always either had confidence in them, or adopted the hard-boiled view that they had taken their necks out there, they could get them back. But now he had sixty people he barely knew, and a dozen parties underground. By the end of the week he had come to know some of the cavers better, had gone caving, and had begun to relax. Nevertheless, after the expedition he wrote an ultimatum. He would direct no more expeditions. When an expedition leader begins to sit up all night waiting for his parties to come in, and cannot get into the cave himself when he wants to, it is time to quit leading expeditions.

Red also made some choice comments about the Cave Research Foundation getting too big. The reverse side of that, said Joe Davidson—who had taken over the presidency of the Cave Research Foundation from Red—was that old Red was too small. He had failed to grow with the organization.

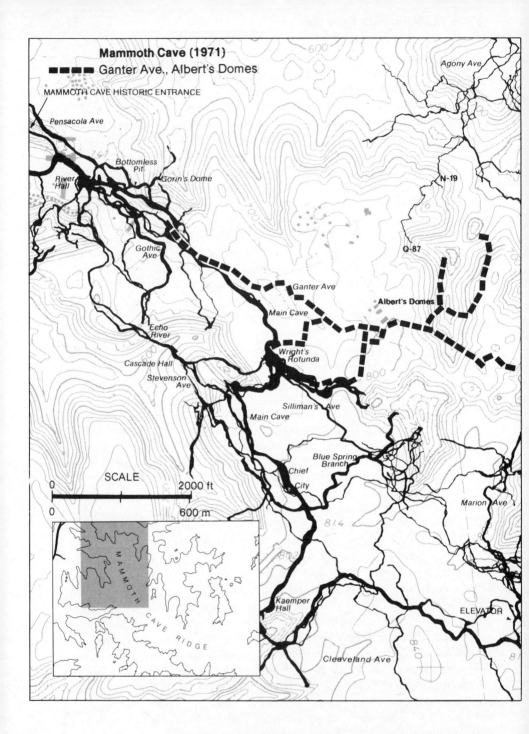

Map of the Flint Ridge Cave System and Mammoth Cave showing the Ganter Avenue and Albert's Domes areas. Red Watson's two-day sweep in May 1969 initiated these surveys, and by 1971 the distance between Q-87 and a passage near Albert's Domes had been reduced to 400 feet.

Despite all that, there was need for someone to lead the 1969 Memorial Day expedition. Joe phoned Red and pleaded with him to run it.

"There won't be a lot of people there," Joe said, "and you can do anything you want."

There was a long pause, and then Red said, "Anything?"

Instantly on the defensive, Joe asked, "What do you want to do?"

"Mammoth Cave," Red said.

Joe paused for a long time. "All right," he said finally. "One party. You can send in one party. I suppose we have to start surveying Mammoth Cave sometime. One party. We have to be very careful. We have to do it *right*. The Park Service . . ."

Red held the phone back while Joe told Red what Red in the past had told Joe about the delicate political relationships with the National Park Service.

More than forty people were on hand for the Memorial Day expedition of 1969. Red did not cook breakfast. It was again an expedition with many people Red did not know, so Joe had assigned two Ohio State University graduate students, Jack Freeman and Bill Bishop, to help. When they told Red this, he gave them the Flint Ridge smile, grinning hugely, his face wrinkling until his eyes almost disappeared. This was a bad sign. They smiled back uneasily. They had been warned that Red was a willful bastard. It appeared that he had already figured out that Joe had sent them down to hold the reins. Furthermore, Denny Burns had prepared the finest summary of what could be done in Flint Ridge that had ever been available. Red glanced at Denny's Flint Ridge lead list and then theatrically tossed it aside. Bill and Jack watched with discomfort.

"Albert's Domes," Red muttered, and unrolled the Mammoth Cave map. He started making up parties for Mammoth Cave—*four* parties—to survey in the Albert's Domes area of Mammoth Cave, 800 feet from Q-87 of the Flint Ridge Cave System.

"But Joe said one party," Jack began.

"Forget Joe!" Red yelled instantly and as loudly as he could.

After a while, Jack tried again. "You know, Red, there *are* a lot of political implications in this. Joe says we can't take any chances."

Red continued looking at the map of Mammoth Cave without comment.

"Joe *is* president of CRF," Jack said.

"Big deal," Red said.

That was obviously the wrong tack to take with a past president of the Cave Research Foundation.

"Where *is* Joe, anyway?" Red asked, putting a lot of concern into his voice.

They all knew that Joe was in Columbus, Ohio. Red was here in Kentucky, standing on top of the Flint Ridge Cave System.

Jack gave up. He joined Bill—who was already eager to go to Mammoth Cave—in planning the first big attempt to connect the Flint Ridge

Cave System with Mammoth Cave from the Mammoth Cave side. Excitement was rising.

Red made a little speech to the assembled members of the expedition. "There are many political implications in this," he said. "We can't take any chances."

Gordon Smith suggested that a triangulation could be gotten on the first effort by surveying from the U.S. Geological Survey bench mark in the Wooden Bowl Room along Ganter Avenue and from the bench mark in Wright's Rotunda along Ranshaw Avenue in Mammoth Cave. Two parties were assigned for each survey.

The Park Naturalist said that Mammoth Cave could be entered through the Historic Entrance between the 9:00 and 11:00 morning tourist trips. At 9:30 a.m., Red got the keys from Ray France, a Mammoth Cave guide everyone had known for a long time because he seemed always to have weekend duty. None of the Park Service personnel appeared to realize what a high-tension, historic occasion this was. This was all the more reason to be careful.

The four parties assembled in the Mammoth Cave parking lot with their leaders, Red, Bill, Jack, and John Bridge. Sixteen cavers tried to march unobtrusively through the Mammoth Cave National Park Visitor Center arcade and down to the Historic Entrance to Mammoth Cave. No one talked. Red thought he was running the show like a drill sergeant. Carol Hill, however, whispered, "He's just like a mother hen." Everyone was nervous.

This was just not right! Any moment some Park Service Ranger was sure to say, "All right, boys, over this way." Such thoughts were silly, of course. Red was remembering the past, when Flint Ridge cavers had had to trespass to explore in Park caves. But this was the present, and the first wave of future exploration in Mammoth Cave.

Red carried the keys ostentatiously in one hand, the letter of permission from the Park Naturalist in the other. Two Mammoth Cave guides walked past without even glancing at the cavers' hard hats and knee crawlers.

Silence was maintained on the walk through Mammoth Cave out to the Wooden Bowl Room. There, watches were synchronized. The four parties were to meet again in eleven hours at the Wooden Bowl Room. The surveying was beginning at 10:00 a.m. It would be over at 9:00 p.m. Everyone would be back in camp by 10:00 p.m. for a good night's sleep so the work could go on again in Mammoth Cave the next day. Flint Ridge cavers' usual inclination would have been to blast out for eighteen to twenty-two hours of work on the first trip. However, people got tired on such long trips, and care was of the essence. Red decided that two days of work on shorter trips would be better.

Red was having the time of his life. He designed everything for his own pleasure. For his party members he had chosen Carol Hill, Roger

Brucker, and Burnell Ehman, all old friends. The survey proceeded without incident. The four parties regrouped in the Wooden Bowl Room within ten minutes of the set time, and walked back to the Historic Entrance to Mammoth Cave very pleased with themselves. Now, with what they had learned this day, a serious push to connect Mammoth Cave with the Flint Ridge Cave System might be possible the next day.

The next day, Red sent six parties into Mammoth Cave. One of them was most appropriately led by Harold Meloy, the dean of Mammoth Cave historians. Everything went perfectly, except that no connection route was found. After all the parties had regrouped to leave Mammoth Cave and to end the expedition, Red remained in the Wooden Bowl Room alone for a few minutes.

It was true that Red had not connected Mammoth Cave with the Flint Ridge System. But the Flint Ridge cavers did now possess Mammoth Cave. In two days, more than three miles had been surveyed: all the previously known passages and some new ones in the Albert's Domes sector of Mammoth Cave lying only 800 feet from Q-87 in the Flint Ridge Cave System. On the first expedition into Mammoth Cave, nearly one-tenth of the known passages in Mammoth Cave had been mapped. It was worth celebration.

When everyone was out of sight, Red yelled, "Whoop!" It sounded awful. Well, Red had never whooped and hollered in excitement. He did not even sing in the cave as some people did, although he talked incessantly. Somehow he had thought he should yell now. However, the glow was enough. More appropriate for celebration was the can of cranberry sauce he had opened for a gala meal during the surveying, and the ceremonial unwrapping of a candy bar in metal foil that had been emergency ration in the bottom of his cave pack for many years. The candy bar was inedible.

Back in camp, Bill and Jack got nervous again as they prepared to go back to Columbus.

"What are we going to tell Joe?" Jack asked.

Bill groaned, "One party, Joe says, and Red sends in ten."

Red gave them a long speech. "If I'd sent in only one party, we'd have had to contend with an elite corps of Mammoth Cave cavers. Now almost everyone on the expedition has been in Mammoth Cave, so it isn't exactly an exclusive club. And no one can go on any more about how hard it will be to work in Mammoth Cave. That was a myth. It's no big deal."

Red could not contain his delight. "Tell Joe they're all alike. Mammoth Cave's just like any other hole in the ground."

Nothing is perfect. One of the surveying parties left an all-time classic note at a junction in Mammoth Cave: "We went on past this point."

They certainly had. All at once a hundred imaginary difficulties—and one or two real ones—had been hurdled. Now it was full speed ahead in Mammoth Cave. There was still some nostalgia about connecting the

Flint Ridge Cave System with Mammoth Cave from the Flint Ridge side, but now serious attempts would be made from the Mammoth Cave side, too.

Some of the cavers complained about Red's authoritarian methods. Jack and Bill explained that Red was under a lot of pressure—there could be no mistakes on that first expedition into Mammoth Cave. Still, Red's direction of that expedition is one explanation of how one of the sweetest, most tractable, reasonable, and relaxed little cavers you could ever meet got something of a reputation for being bossy.

Joe was furious, but he adjusted. Soon he was bragging quietly about how the Cave Research Foundation had surveyed the whole north side of Mammoth Cave. However, there were no obvious leads toward Houchins Valley or toward Q-87 in the Flint Ridge Cave System. It was not going to be easy to connect the Flint Ridge Cave System with Mammoth Cave from the Mammoth Cave side, either.

The Pressure Builds

There are two styles of connection fanatic.
The romantics line up against the engineers. John Wilcox
enters the running.

22 Roger Brucker and Red Watson had long wanted to connect the Flint Ridge Cave System with Mammoth Cave. Each ran several survey lines out under Houchins Valley from the Flint Ridge side, only to dribble off in mud and sand. Neither carried through a long-term plan of attack—although each had developed plans of how it should be done—and neither one was very efficient. They were romantics, chasing virgin leads, but always abandoning the tough ones for newer sirens that always turned up.

Joe Davidson and John Bridge were professors of mechanical engineering who might be expected to be better organized than an advertising executive (Roger) and a philosophy professor (Red). Joe explored under Houchins Valley toward Mammoth Cave from the Flint Ridge side, but he said that for caving problems and adventure, other directions were as interesting. He proved this by discovering the huge Northwest Passage in Flint Ridge. John certainly did want to connect with Mammoth Cave, and he spearheaded the Q-87 work, which was the closest approach to Mammoth Cave from the Flint Ridge side so far. John's size, however, was a definite handicap in the small passages under Houchins Valley.

Work beyond Candlelight River in Flint Ridge had never been especially well organized. It depended entirely on individuals who got hot enough to carry along a party or two of sometimes quite reluctant volunteers. It was still thought that Q-87 was the way to go, out there under Mammoth Cave Ridge, 800 feet from Albert's Domes in Mammoth Cave. But it was a tough way to go, still the hardest cave trip in the Flint Ridge Cave System.

From the Mammoth Cave side it was easier to get within the vicinity of a possible connection than it was from the Flint Ridge side. Red's grand entry into Mammoth Cave with ten survey parties had been the most organized assault yet, but, typically, once the opening move was made,

Red left Mammoth Cave to others. The difficult exploration out around Albert's Domes in Mammoth Cave was taken over by Bill and Sarah Bishop. But everyone seemed to hope for—at least they talked about—a lucky break from the Flint Ridge side. Someone, the daydream went, would one day poke into a neglected crawlway out under Houchins Valley from the Flint Ridge side, crawl along a few hundred feet, and walk—yes, *walk*—right into Mammoth Cave. It could happen.

Of course, even the romantics had a nightmare as well. It carried the cold breath of mud and water a foot deep in a passage eighteen inches high that snaked along for a mile under Mammoth Cave Ridge to end in a siphon, almost but not quite connecting with one of the underground rivers in Mammoth Cave. Those passages under Houchins Valley were very close to base level. And if they were not filled to their ceilings with water, many of them were doubtless too small for cavers to pass through. Oh, no doubt a connection existed somewhere between the Flint Ridge Cave System and Mammoth Cave. But there was plenty of doubt about whether the connecting passage could be found, and about whether cavers could get through it even if it were found. It would not count as a connection unless the cavers went through it from one cave to the other. It would not be easy. Even the romantics knew that. And now another engineer entered the running.

John Wilcox had been a friend and rival of sorts of Joe Davidson since childhood. John and Joe were interested in the same things and did them well. Each earned his Ph.D. in mechanical engineering at Ohio State University. Joe stayed on the faculty there, while John took a research job at Battelle Memorial Institute in Columbus. When Joe became involved with Flint Ridge caving, it did not occur to him to bring John into it. John was involved in collecting gas engines. John Bridge, however, decided that John Wilcox would make a good caver.

John Bridge was close to Jim Dyer—who had stopped caving, but still talked it a lot—and Bridge turned Dyer's Flint Ridge con onto Wilcox. Joe agreed to give Wilcox a trial. Because Wilcox still showed little enthusiasm, Joe took him not to Flint Ridge, but to Carter Caves, which are closer to Ohio than Flint Ridge is. Wilcox was not impressed with Carter Caves, and Joe was not impressed with Wilcox's caving. Bridge, Denny Burns, and Bill Bishop, however, continued to insist that Wilcox had great potential. They thought that someone with his abilities and determination could make a tremendous contribution to the Cave Research Foundation. It did not matter that he was not a caver. None of them had been cavers before hitting Flint Ridge.

Joe reluctantly agreed to take Wilcox to Flint Ridge. Wilcox was captured immediately by the huge, complex cave system. He soon showed a rare aptitude for exploration. A lot of people talk about exploration, but few cave explorers ever actually push into virgin passages more than a few hundred feet beyond known limits. There are some, however, who, if they find a way, will push miles into the unknown. John was one of the far-out cavers.

The next step in Wilcox's progress was his interest in the cartography program of the Cave Research Foundation. He began helping Denny Burns, and when Denny got his Ph.D. in forest entomology and was ready to move from Columbus, Wilcox was his candidate for taking over the cartography program. Again, Joe said that Wilcox just was not interested enough. He might be a firmly committed Flint Ridge caver now, but John Wilcox was still the nation's foremost collector of gas engines. Wilcox would not have time to direct the cartography program. Denny, however, with typical Cave Research Foundation stubbornness, made Wilcox Chief Cartographer despite CRF President Joe Davidson's objections.

Wilcox moved into Denny's old town house in Columbus, Ohio, where the map factory was left intact.

"He's moving away from his parents' home into your old apartment?" Joe asked incredulously.

"Yes," Denny replied smugly.

"All right. You've convinced me. I was wrong about his interest in caving and I was wrong about his interest in the mapping. Hot damn! With Wilcox's efficiency and energy, the cartography program will really roll!"

John Wilcox took Red's place on the Cave Research Foundation Board of Directors. As soon as he became comfortable in the role of Director of Exploration and Cartography, he decided to make the big connection between the Flint Ridge Cave System and Mammoth Cave, from the Flint Ridge side.

In January 1972, Wilcox asked Joe if there were any reason why a serious attempt to make the connection should be held off any longer. Joe did not answer immediately. He knew how determined Wilcox was about anything he undertook. If Roger or Red or John Bridge had asked, Joe might have lit up with his toothiest smile and said, "Sure, try it!" In sheer caving skills, those three were as good as anybody, and if they did hit the right combination, they would go. But they would be asking to shotgun it, as usual, and it would still be a matter of chance, even if based on shrewd guesses.

John Wilcox was another kettle of fish. If Wilcox were turned loose, Joe knew that it would not be a matter of chance. If the connection could be made, John Wilcox would make it. Joe thought about the fact that Joe Kulesza, the current Superintendent of Mammoth Cave National Park, was an old-timer who had been at the Park as a Ranger when all this work had begun, almost twenty years before. At the Foundation banquet the year before, Superintendent Kulesza had reminded some of the old cavers that he had told them back then that the only way to get into the caves of Mammoth Cave National Park was to start a scientific-research organization. Well, the notion was in the air then, what with the scare of Russia's sputnik, and the resultant stress on science. Ranger Kulesza, as well as Dr. Pohl, Jim Dyer, and that fledgling science administrator Phil Smith, could see the lay of the land. Superintendent Kulesza really liked the caves, loved Mammoth Cave National Park, and was genuinely interested in underground exploration. The Cave Research Foundation was firmly established as the most prolific and prestigious cave-research organization in North America. And the whole speleological world was watching for the connection between the Flint Ridge Cave System and Mammoth Cave. Furthermore, it seemed that the publicity of a connection between the two caves at this time might help in the conservationists' battle to get wilderness designation under the Wilderness Act for the longest cave in the world in Mammoth Cave National Park. It seemed to be a good time to make some cave history.

"All right," Joe Davidson said finally to John Wilcox, "go ahead and connect them."

Q-87 Again

John decides to force the Q-87 breakdown pile that may block a way from the Flint Ridge Cave System to Mammoth Cave. His crew learns about engineering the hard way. They do not get through.

23 John Wilcox immediately set about planning the connection between the Flint Ridge Cave System and Mammoth Cave. He expected the project to take a long time, perhaps years. First he examined the maps again, to verify what he already knew. The only passage in the Flint Ridge Cave System that passed under the axis of Houchins Valley toward Mammoth Cave was the walking extension of Agony Avenue beyond Candlelight River that led to Houchins River and N-Trail. And the only passage that got fairly under Mammoth Cave Ridge was the Q-survey leading to Q-87 off N-Trail. Also off N-Trail the branching A-survey led a few hundred feet south. Any new passages reached by a breakthrough from N-Trail would probably be more closely related to Mammoth Cave than to the caves of Flint Ridge.

John Bridge had been talking about the ends of the Q and A surveys for years. Wilcox listened closely and took Bridge seriously. Bridge was always enthusiastic about everything in Flint Ridge, but his observations had been reliable in the past. The two engineers discussed the possibilities. Bridge had a good memory for the cave, so could answer detailed questions about the breakdown plug at Q-87. Q-87 was *under* Mammoth Cave Ridge, and any breakthrough there would probably connect with Mammoth Cave. Three attempts to break through had failed, but Bridge still believed that it would go. Wilcox agreed. For one thing, he himself had not tried it.

John Wilcox zeroed in on Q-87. He studied the trip reports systematically, and talked to people who had been there. Of the three parties that had gone all the way, two had not spent much time at the end. Art Palmer was not satisfied that everything possible had been done there. Joe Davidson said that had he thought there was no more hope, he would not have left the digging tools out there to litter the passage. The Q-survey was

1305 feet long, and there had not been much time or strength for pushing side leads. Red Watson thought some of them might go.

John decided to follow the old route. First he would go to Q-87 and push hard at the breakdown pile. He would take several trips to Q-87 until he was satisfied that it would or would not go. On the way back each time, he would check a few side leads on the Q-survey, beginning with those closest to Mammoth Cave. If no Q-survey leads went, then he would push the A-survey toward Mammoth Cave. If the A-survey did not go, he would explore systematically all the other leads in the Houchins River area. It was an ambitious, long-term program. John knew that it would take an enormous amount of commitment on his part if he were to be able to lead others in such objectively hard and dangerous caving.

John had no illusions about the ease of fielding parties to Q-87. He needed three other strong cavers who were experienced enough to avoid the trip-aborting blunders that had made many of the deep cave trips unproductive. And good cavers had done silly things, such as going in with shoes that were ready to fall apart, and that did fall apart. The best equipment would be required, and the cavers themselves had to be in top physical and mental condition.

Above all, John sought companions who would be optimistic despite hardship, discouragement, and fatigue. They had to keep up the façade of optimism, even when they felt that the situation was hopeless. Such cavers are generally party leaders. Consequently, if John went to Q-87 during a regular expedition, he would strip it of party leadership. Furthermore, a Q-87 trip put a caver out of action for at least forty-eight hours. The first twenty-four would be spent in the cave, the second, recuperating. That would be it for a two-day weekend expedition. John thus considered running the Q-87 trips on non-expedition weekends, but then there would be no surface support in case of emergency.

However, in thinking it over, John decided that it did not matter where the rescue teams were. The important things were how fast they could get under way and what condition they were in. A back-up team at the end of a long-distance phone call probably could reach Flint Ridge just as fast as—if not faster than—and in better condition than a team drawn from cavers already underground in Flint Ridge. It would take many hours to gather together cavers who were exploring in different parts of the cave, and most of them would be tired. Members of a rescue team would be easier to reach in their own homes than in the cave.

Finally, the explorers must be psyched up well in advance for the trip. Mental conditioning is as important as physical conditioning for far-out caving. John knew that he would need a second team in case of dropouts. Very good cavers have been known to get ill just before a hard trip. John started the old con game.

It was harder than he had anticipated. Many cavers had planned their activities around expedition weekends, so he had to revert to expeditions after all. He and Art Palmer, the leader of the 1972 March expedition, decided that a Q-87 trip would disrupt the long-planned purpose of that expedition. Steve Wells, the leader of the April expedition, was willing, but he and John concluded that there was not a Q-87 party in camp. Instead, John led a party into Bransford Avenue in Mammoth Cave to survey an amazing total of 4667 feet. But it did not get him any closer to making the connection.

John decided that it would have to be the 1972 Memorial Day expedition. Gordon Smith, the expedition leader, was as eager as anyone to connect with Mammoth Cave.

John made a list of fifteen people he knew were capable of operating at Q-87. His criteria were quite firm. Although he was thirty-five himself, he wanted no one—neither new cavers nor old-timers—who might have set ideas about Flint Ridge caving, Q-87, and what is and is not possible. He wanted young cavers who were still impressionable enough to do some things without realizing that they had formerly been considered to be impossible. He needed strong cavers, but stability, competence, and eagerness were as important as strength. His list thus included some cavers who were quite new to Flint Ridge caving, and some who were not exceptionally strong.

He made a mental note that he would have to curb his own tendency to pessimistic understatement. Even sardonic jokes could discourage the kids John intended to pipe into the cave.

John was well aware of the very real dangers of cave exploration. Another reason for choosing young cavers, then, was that they were most likely to be unattached like John, himself a bachelor. Having no families, and being newly out on their own, they would give their all without thought of consequences. John did not intend to be careless, but he wanted to minimize the consequences of accident. He was considerably more concerned about danger to his friends than to himself. And what if he organized a project that cost the life of someone's husband or wife or father or mother?

However, when he got down to it, it was not possible to fill the list with the unattached. He put cavers with children at the bottom of the list.

Letters went out weeks ahead of time, first to only a few because John needed only three cavers besides himself. When he heard that someone could not go, he wrote to the next person on the list. As refusals mounted up, John began to worry that the younger cavers did not have the enthusiasm that the old-timers had shown. This was an opportunity possibly never to be repeated to go to the bitter end of the world's longest cave. John knew that he and Joe Davidson, when they were twenty, fifteen years ago, would have dropped absolutely anything else they were doing. They would have gone.

It had never been easy to get people to go out to Q-87. John resisted thinking of himself as an old-timer—for one thing, he had been a caver for only a few years—but he began to sympathize with the old-timers more than he had before. He thought longingly of weekends working on gas engines. He could do that alone.

Letter writing did not work. John started telephoning. It took twelve calls to set up that first trip. These kids had no comprehension of what he was offering them. He was most frustrated by college students protesting that they had to study that weekend. That was just more of the baloney of academia. John longed to say to them, "So drop your grade a notch in that course if you must. Will you remember and tell your grandchildren about your good grades?" But he did not. He let them make their own decisions.

Gary Eller accepted at once, but that had been expected. Gary was twenty-five, but already had a Ph.D. in chemistry. He was doing research at the Georgia Institute of Technology. In all respects he was ideal. From the top of the list to near the bottom, however, John drew blanks. He wrote in his journal:

> I have a theory about the younger generation. I think they are less able to lose themselves in projects of adventure than we are. They have grown up without exposure to major war or major economic disaster. They have not had a demonstration of how helpless man can be to control the destiny of his civilization. And so they have bravely shouldered the cares of the world instead of looking for lesser projects. A better way? I don't know. But I feel for them; it is tearing them apart.

There were only two names left. John worked the telephone dial. Pat Crowther answered. She and her husband, Will, were packing to go to Flint Ridge for the Memorial Day weekend. John felt a chill. It was already so late. If Pat and Will said no, he would have to scrub the Q-87 trip. He fought down the pessimism that came from five months of trying to put together just one party to go to Q-87. Emotionlessly, he asked his question.

"Sure, we'll do that!" Pat answered at once, without even consulting Will.

John now eagerly outlined his plans. He was extremely pleased. Pat and Will had been on the bottom of his list only because they had two young daughters. In caving ability, they were tops. John still felt slightly guilty about asking them. He was taking along both parents. But of all the people on his list, Pat and Will would know best what they were getting into. They were beyond the young impressionable age, being almost thirty, and they were quite experienced. When Will got on the other extension and sounded eager, too, John began to feel very good about asking them. He believed that individuals should be allowed to choose their own risks, commensurate with their abilities.

Late on the night of May 26, 1972, John arrived at Flint Ridge. His car was piled high with expedition maps and gear. He had a Q-87 party, but there were no back-ups. If someone did not show up, or got sick, there would be no trip to Q-87. John went to sleep, apprehensive that something would go wrong.

Pat and Will had heard of Q-87 as the place where the Flint Ridge Cave System ran into a pile of sandstone breakdown under Mammoth Cave Ridge. People used to go out there to knock rocks down with a long pipe, but no one had been there for years. John had promised them a twenty-four-hour trip, and Pat wondered how well she would stand up under the longest trip in the Flint Ridge Cave System.

Pat and Will had gone on outings together in undergraduate days at MIT. Will had taught Pat to climb vertical rock walls in the Shawangunk Mountains in New York, and to ski New Hampshire slopes. On their honeymoon they explored Breathing Cave, a long, confusing labyrinth in Virginia. They had been recruited for Flint Ridge caving by Walter and Barbara Lipton, who promised them the cave trip of a lifetime. The immense Flint Ridge Cave System caught their fancy. Also, for years Will had been looking for people who could follow him in a cave, and at Flint Ridge he found a whole troop of people who could lead him. After one day at Flint Ridge caving, Will and Pat had made their commitment.

Soon Will and Pat began to share direction of the cartography program with John Wilcox, and their home became a map factory. In one corner of their living room stood a computer terminal, provided by Will's employer, Bolt Beranek and Newman Inc. Will is a computer scientist, and Pat is a programmer. Will had permission to use the computer to process the Cave Research Foundation cartography data, so a new era of map construction was begun through the generosity of Bolt Beranek and Newman Inc.

Now Will and Pat were ready to take the big trip to Q-87. Sandy and Laura, aged four and two, were caught up in the excitement as they kissed their parents good-by. Pat and Will flew to Louisville, where Bob Eggers met them at the airport in his bright blue station wagon. Bob was short and solid, with thinning hair and thick glasses. At forty-five, he was the oldest active Flint Ridge caver. He brought his small daughter, Caroline, to the cave, where she had a mad time with Red Watson's daughter, Anna, and other cavers' children, roaming the woods, exploring small hillside cave shelters, and searching for copperheads. Bob was a caver of a kind that had done an enormous amount of work in the longest cave. He probably would not be on anyone's supercaver list, but he was deceptively strong, and everyone wanted him to keep coming back. He was the kind of friend on whom Pat and Will had come to depend. They laughed with satisfaction as Bob backed out the in-gate of the Louisville airport's hopelessly jammed parking lot. Then they hummed 100 miles south down I-65 toward Flint Ridge.

When they arrived, the Flint Ridge Field Station was a midnight blaze of lights. Cavers were carrying equipment from the old Spelee Hut back in the woods to the Austin House up front. This marked the end of an era in Flint Ridge caving. For sixteen years, the crude sixteen-by-twenty-foot Spelee Hut built by members of the Central Ohio Grotto of the National Speleological Society in 1956, had served as kitchen, operations center, and bunkhouse. Now it was to be used for storage. Operations had been moved to the old Crystal Cave Ticket Office, which also served as a laboratory. Floyd Collins' Home and the old Guide House had long since become bunkhouses full of double-decker beds. Now the Austin House would be kitchen, dining hall, and resident researchers' quarters.

The next morning, for the first time in six years, a party was going to Q-87.

John roused Pat, Will, and Gary at 6:00 a.m. No one else would be up for two hours. Pat cooked breakfast while John packed the last of the shoulder bags with pipe fittings, wrenches, a heavy hammer, and a trowel. They reached the Austin Entrance at 8:00 a.m. The steel door clanged behind them, and as they entered the cool breath of the cave, they were caught up in the joy and the rhythm of caving. John relaxed— for the first time in weeks, it seemed. There was actually a Q-87 party underground in Flint Ridge. Nothing could stop them now.

They sped quickly over the mud floor and teetering rocks of Pohl Avenue, up Brucker Breakdown to the bone-dry gypsum sand of Turner Avenue. Down that long walk past Old Granddad, they turned right into

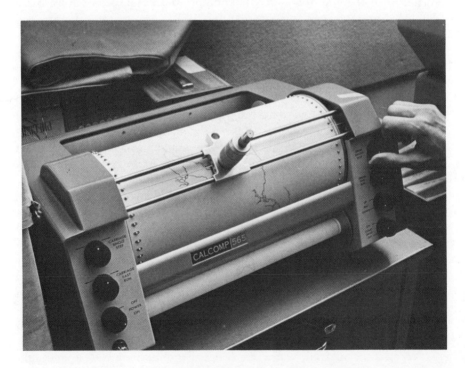

Computer-driven plotter drawing a cave map.

Swinnerton Avenue, where boulder piles broke the monotony of walking. From the Duck-under out of Swinnerton into the crawlways to Best Way Down, and finally knee deep in Candlelight River was all new cave to Pat. She was totally confused by the maze, and there was still a long way to go.

John had decided to push some leads on the Q-survey on the way out to Q-87. He expected no breakthrough on this first trip, but he did want it to be productive. He was determined to demonstrate that the Q-survey was a passage just like many others, and not some terror-filled nightmare. If it could be shown that the superman cavers of half a dozen years ago were simply good, solid cavers who had traversed long and difficult—but not impossibly difficult—crawlways, then he could get more recruits.

The first target was Q-17. John was astonished by the lead. It was ten feet wide, crawling height, and a small stream trickled from it. The previous explorers had stopped only thirty feet in at a mud bank that rose to within six inches of the ceiling. Gary and John dug at the high spot, while Pat and Will checked some tubes and crawlways in the ceiling.

In ten minutes, Gary had dug out enough mud to squeeze through. In another five minutes the passage was large enough for John. This was really good. The old-timers had not pushed this passage *at all.* Gary was already making scornful remarks about them. This was just what John had hoped for.

Beyond the mud bank, the passage split into three parts. The stream snaked through a canyon with limestone walls very close together. It could not be entered even after vigorous pounding on jutting ledges. At the middle level was a belly-crawl, and in the ceiling was a tight crack. By this time Pat and Will had joined John and Gary in the crawlway.

"I think the ceiling crack is too tight," John said.

This was a challenge to Pat. She scooted to the front of the line of prone, muddy figures. Forcing her head and thin body into the crack, she could see around a right-angle bend to where the passage sloped down and continued. A mud bank along one wall prevented her from going on.

Pat backed out to reach for the garden trowel that was handed up to her. She squeezed back in and began to dig. It was very awkward. In the tight passage she could bend her wrist, but not her elbow.

Pat dug until she thought she could wriggle through the opening. The passage seemed to widen a few inches up ahead. Perhaps she could turn around there to get out. She pushed herself up into the broadest corner of the crack, slid her chest around the bend, then turned her legs sideways to work her knees around. She was lying on her back on the sloping mud bank. She would have to go ahead two or three more feet before she could turn her head to see where she was going. She pushed forward and slid down the mud bank. Her shoulder jammed painfully. In no way would she be able to move ahead from this position. What had looked like a wider spot ahead was an illusion. Pat was on her back, head down, and could not turn around. She would have to back up and out.

Pat tried to move backward, but it was uphill and tight. She was too cramped to push with her arms. Pat had explored many tight passages, and was not often bothered by claustrophobia. Now, however, panic began to well up. She was panting, dizzy with hyperventilation.

STOP!

Pat told herself to calm down. Relax. She repeated to herself the old caving adage: Anything you can get into, you can get out of again.

(It is not true.)

When Pat was calm, she concentrated on remembering *exactly* what she had done to get into her present position. With great effort she squirmed back until her feet were at the right-angle bend. The light of her carbide lamp was almost out.

"John, are you there? Can you see my feet?"

John could not understand the question, and Pat could not understand what he was saying. Still, it helped to hear someone out there, just a few feet away. She inched upward, threading her legs back around the corner, her knees bending awkwardly. She was almost out. She dragged her hard hat in the mud behind her, clutching it with a slippery hand. The lamp was still in place on its bracket. She almost dropped it into the crack, which brought her sharply to attention. There would be no going back after it. Her slim waist curled around the corner. Hard hat in hand, she slipped out of the crack to where the others were waiting.

Pat sat on the mud bank with her feet in the stream, shaking. The Q-survey was already a nightmare, and it was still a long way on to Q-87. Perhaps the old stories of this being the most formidable part of the cave were true. She watched as John scraped mud away from the opening of the passage that continued at middle level. He could not squeeze himself over a rock ledge on its floor.

"Pat?" John asked.

No! Pat thought. I will never do anything like that again. She took her time changing carbide. The new light was a comfort. She thought, Well, maybe I should take a look.

It was a flat crawl, with rock floor and ceiling. Pat slipped past the tight place that had stopped John, but three feet farther on mud nearly filled the passage. It looked like an easy dig to an opening eight feet ahead where a small stream flowed into a tiny side passage. Pat started to back out to get the trowel. She missed the widest spot and jammed tight.

Once again came the involuntary spasm of claustrophobia. Pat's eyes darted over the wet mud walls a few inches from her face. She felt faint, and her heartbeat and breathing raced. But her fear of being stuck here was irrational. In the ceiling crack, an assessment of the situation had led to the conclusion that she really was in trouble. It was foolish now to panic in this wide crawlway. She moved forward again, and then backed out. It would be an easy dig, but she was through with that sort of thing for the day. What was it, five or six hours to this point? And then, say, twenty minutes of pushing. Pat was exhausted: This was far-out caving.

They still had a long way to go on that old route to Q-87. Pat followed along. At Q-33 they stopped again. Gary and John had found a lead to the left not noted by earlier explorers. It was a small elliptical tube. John got stuck again, so Gary tried it. He was just slightly larger than Pat.

Gary slid on his side through mud of cake-frosting consistency. As the passage constricted, he peeled off his sweat shirt, his knee crawlers, and then his shirt. At a final squeeze, he removed his gloves to get a better grip on the muddy rock. Ahead he heard water dripping. Perhaps there was a pit up there, but he could go no farther. As he backed out, sharp projections of rock pushed out through the mud to shred his remaining clothes. Gary's body glistened with the wet mud. John used a putty knife to scrape the worst of it from his bare back and stomach.

"Maybe Pat could get through," Gary said.

Pat was not tempted.

Gary had left his gloves at the end of the crawlway. He was still alert and in excellent condition, but he had lost his gloves. As for going back in to get them, no.

John silently opened his pack and took out a spare pair of gloves. Gary started to protest, but then took them. John was right—gloves could make the difference this far underground.

From Q-33 onward, the party crawled and squirmed through the progressively smaller Q-survey passage. Occasionally, someone would put on his pack, only to find that it had to be taken off again to get around the next tight corner. There was a brief delay when a heavy rock slumped off the wall onto Will's legs. Interesting, Will thought. Pat propped up the rock slab so Will could wiggle out from under it. It remained perched diagonally between the walls, a problem for future travelers.

At Q-85, the passage widened into a room big enough for them to crouch in. Rocks were packed along the walls. Rusty pipes lay on the floor.

Fifteen feet beyond the small room the canyon bent slightly to the left. Here they saw the famous pile of sandstone breakdown at Q-87, the bitter end of the Flint Ridge Cave System. It had stopped John Bridge, Art Palmer, Joe Davidson, and Red Watson. Would it stop them, too?

Before disturbing the unstable pile, each caver crept up to have a good look. John had hoped to find a chamber hollowed out of the pile by previous digging. The rocks might even have shifted to open a way right into Mammoth Cave. When Joe Davidson had left on the last trip, there *had* been a rumble that the party had not investigated after they left the face. During the six-year interval there had been plenty of time for slumping. However, John was disappointed to see a vertical face of football-sized rocks completely blocking the end of a canyon eight feet high by one and a half feet wide. He decided that the large rock Joe's party had feared had now slumped to rest almost on the floor. It looked as though many rocks would fall if the right ones were removed.

Only one could work at a time, so Will and Pat were dispatched to check the ceiling several hundred feet back from Q-87 for possible leads. They looked carefully, but could find no way up that was not blocked by thick stone ledges. Nothing but dynamite would break through those ledges, and of course dynamite was not used. The Flint Ridge Cave System was in a national park, and these cavers were conservationists. Besides, dynamite was not fair. If a passage could not be forced by digging out sand and mud, or by removing loose rocks, then it did not go.

Pat and Will returned to a chilling sight at Q-87. John was standing upright with his back lying against the loose pile of rocks, his head thrust up into a wastebasket-sized opening above him. Loosely wedged rocks hung directly down into his upturned face.

Do or die.

There was no way to climb up. John came down and unbuckled his pack. He took out several short lengths of pipe and fittings, and a couple of pipe wrenches. The pipes were screwed together to make a hook like the top of a candy cane, which was fitted to a long pipe Joe had left on the floor six years previously.

They took turns with the hook. Pat viewed the digging as a task in which she was to snag big rocks to pull them down on her head. She did not like it at all. Further, Pat was frustrated because she was not strong enough to shift the rocks much. She grimly took her turn, anyway.

They excavated a chamber in the breakdown above and ahead of the top of the canyon in which they were crouching. They could see into an opening along the wall at that level, but it was too narrow to get into even if anyone had been so foolish as to try.

At floor level along the right side they reopened the crawlway that all those previous Q-87 parties had seen, but had shunned as too dangerous. While doing this, they thought they heard distant rocks rolling down the backside of the pile. It was an eerie, spooky sound, but it also gave hope. If they could crawl around the base of the pile, maybe they could climb up, or on, to Mammoth Cave.

The others watched without comment as John got down on his belly on the floor in front of the pile. He slid into the crawlway, his right shoulder jammed against the solid rock wall, his left shoulder grazing the loose wall of rocks. Slowly he squirmed along the base of the rock pile, and then he carefully backed out again. It did not go.

They attacked the pile again. Thrust, jam, twist, pull. Big, noisy rocks shifted up above. The last group had dug for only two hours. John's party moved rocks at the face for four hours in an attempt to break through into passages that might lead to Mammoth Cave.

It was comfortable to sit back down the passage at Q-85 while someone else was digging, but it was warm only when two or three of the cavers huddled together. Will and Pat cuddled. They also warmed themselves by moving rocks on down the passage. They noticed intermittent drafts. It was breathing cave, and this was surely a sign of big—but elusive—cave close by. Was it Mammoth Cave?

Now several large rocks wedged down across the canyon at Q-87. A few more desultory pokes with the pipe left deep white scars on the brown rocks, but did not move them. The pipe jammed between the rocks with a squishy, sucking sound. The outlook was bad.

Gloomily, everyone recharged his lamp, and then all turned around for the trip back through Flint Ridge to the Austin Entrance. They passed by the Q-survey leads without stopping, all the way to Q-1, and then along the N-survey to the Houchins River passage. They looked into the mouth of the dark crawlway that was the A-survey. This was a place where they could all stretch out. As they lay there, they speculated about the A-survey, but were too tired to explore it. They ate a little, fell asleep for thirty minutes or so, and then woke up to go on. The agony was familiar to them, and their minds drifted away from their aching bodies.

Five or six hours later they walked up that long hill above the Austin Entrance, toward the Flint Ridge camp. They were four scarecrows in mud-covered rags. It was another day. Everyone at breakfast greeted the returning cavers. Red Watson was the only person in camp who had been to Q-87 before. He just grinned at them. They shook their heads wearily, and tried painfully to smile back.

The Brucker Strategy

A romantic fling at connecting the Flint Ridge Cave System with Mammoth Cave leads to the reality of a cold Flint Ridge river. It sends the explorers back to try again from the Mammoth Cave side. They fail to connect.

24 During the 1972 April expedition, only weeks before John Wilcox's first Q-87 trip, the old-timers Roger Brucker, Red Watson, and Denny Burns went out to Ganter Avenue in Mammoth Cave to check some leads. They knew that John wanted people to go to Q-87, but it was almost a five-mile trip one way, and, besides, they had not been invited. It was much easier to get close to Flint Ridge cave passages from the Mammoth Cave side than to Mammoth Cave passages from the Flint Ridge side, anyway. However, additional exploration of the Albert's Domes area of Mammoth Cave failed to uncover any routes to lower levels that might go to Q-87 in Flint Ridge, now only 400 feet away. The water at the bottom of Albert's Domes gurgled through rocks too tightly jammed to move, although Bill Bishop, who had been down there, said it ought to be given another try. Tom Brucker, now eighteen years old, had probed a passage leading off from high in Albert's Domes, but it showed no promising leads.

The leads along Ganter Avenue in Mammoth Cave, however, might go toward Flint Ridge. The three old men—Denny was actually ten years younger than either Roger or Red, but he had been around a long time—walked happily along Ganter Avenue toward the first lead, caving as they had always done. They hoped that one of the small crawlways would turn out to be an abandoned drain from a vertical shaft located under the Mammoth Cave Ridge wall of Houchins Valley. Roger pictured himself creeping to the edge of a drop into a pit where he would toss in rocks to gauge the depth. There would be canyon passages around the pit, one of which would provide a chimney to the bottom. There a trickle of water would go through a bedding-plane drain just large enough to squeeze into. Downstream the drain would become larger. It would run into a stream system, and one of the tributaries of that system would come from Flint Ridge. . . .

Roger glanced up sheepishly at Red and Denny as they moved along the old familiar Ganter Avenue, still in Mammoth Cave, well back under Mammoth Cave Ridge, a long, long way from Flint Ridge. Well, they shared the same dream.

Roger had reviewed the trip logs from previous Ganter Avenue trips. He also had copied off station numbers from the master lead list. They explored the leads one by one, finding that some just cut around back into Ganter Avenue, and that others were too tight to penetrate. One lead looked very promising. Like all the Ganter Avenue passages, it was high in the cave, but it went the right direction, toward Flint Ridge. Neither Roger, who was twenty pounds over his peak caving days, nor Denny, also a bit heavier than he had once been, could get through a tight squeeze at the beginning of the passage. Red, who bored everyone by ostentatiously keeping his weight down, pushed into the crawlway. It was lined with gypsum crust, and the squeeze was tight, for the passage was barely bigger than his body. Red stretched his arms out in front of him and wiggled along. This was the kind of caving he had always enjoyed. Roger and Denny waited at the mouth of the passage, watching Red's feet disappear around a corner. The attenuated sound of scraping and scuffing came back.

"I'm trying to move a rock blocking the passage," Red's muffled voice floated out. "Then I think I can go on to where it opens up a little."

"We'll look around at some other leads down the passage," Roger said.

"Okay," Red agreed.

Roger and Denny returned in about forty-five minutes to the passage that Red had entered. After a few moments they heard him returning. He described with pleasure how he had pounded the blocking boulder into gravel. He had crawled beyond, only to find the passage nearly filled with sand. He managed to squeeze through, only to find another sand fill beyond. The passage was too narrow to turn around in, so he dug his way through the second pile of sand into a small room. The continuation of the passage was almost plugged with a third sand pile. With great contortions, Red turned around in the little room to prove to himself that he could. He then turned back around to dig a while at the forward sand plug. He could see that the passage continued ahead, but it was very narrow. He pushed into it, but decided that a series of three tight places between him and his larger party members was enough. So he backed up, turned around again, and crawled out. Resting a moment, he felt a breeze caress his neck. It came from the unknown in that tight virgin passage behind him. It was subtle enticement. However, Red crawled on out into the larger passage.

Roger and Denny had felt breezes in some of the other leads, too, but only one of the passages was large enough for the three to survey, and it was just a long cut-around. It had been a good trip, but it had not put them any closer to the connection between the Flint Ridge Cave System and Mammoth Cave.

Then a few weeks later, on the 1972 Memorial Day Q-87 trip, John Wilcox showed his hand. Roger was convinced that the old-timers' chances were fading. He pondered the reports of John's trip, and then telephoned Red to say that he thought the connection was imminent. Roger had some plans of his own for connection, and wanted Red to help.

But Red was once again busy with other things. He was going to New Mexico, where his wife, Patty Jo, was directing an archaeological dig that summer. Then they would leave in early September to spend a year on an island in the Aegean Sea. Red is always going away when we need him, Roger thought.

Where were all the *young* romantics? The engineers were at work. Roger cast about to get his own program started, and realized that the best place to begin was at home.

"Tom," Roger said to his son one day, "I'd like you to take charge of connecting Mammoth Cave."

"Okay," Tom replied. Tom had already appointed himself to the task many months before, anyway. Why else had he painstakingly enlarged the Kaemper map of Mammoth Cave? He had redrafted it square by square on Mylar film, and had enlarged the topographic map to the same scale. The resulting composite map showed all of Mammoth Cave at the same scale as the Flint Ridge Folio map that Roger and Denny had completed many years before.

Tom had been leading trips into Mammoth Cave for two years, since he was sixteen. He knew where every passage went and how each intersection looked. He could project Mammoth Cave three-dimensionally in his mind's eye, just as Roger could do with the Flint Ridge Cave System.

"Where do you think we might connect?" Roger asked.

"Upstream Houchins River," Tom replied.

They agreed that one of the first trips on Roger's 1972 midsummer expedition would be upstream Houchins River in the Flint Ridge Cave System. If that did not go, then they would survey out Marion Avenue in Mammoth Cave to look for leads toward Flint Ridge. The Kaemper map showed Marion Avenue as a wide trunk passage extending north toward Flint Ridge from the Snowball Dining Room in Mammoth Cave. It appeared to end in vague, small leads under the edge of Mammoth Cave Ridge along the edge of Houchins Valley, about 1000 feet from Art's farthest point of penetration in Houchins River from the Flint Ridge side.

Tom and Roger took several trips to John Wilcox's map factory in his home in Columbus. They took along a new Flint Ridge caver, Richard Zopf, who was proving to be very strong. Richard ought to get to know John better, they thought, and the map scene was one more way to infect him with Flint Ridge fever.

John lived in a town house some sixteen feet wide and thirty-two feet long. The living-room wall, eight feet high and thirty-two feet long, was covered from end to end, floor to ceiling, with maps of the Flint Ridge Cave System and Mammoth Cave.

For Roger, the map wall was the most exciting piece of cave outside Flint Ridge. He could spend hours looking at the maps for places where new discoveries might be made. He imagined the thrust of his body through tight rock passages, here smooth and oval and regular, there angular and sharp. Conversation lagged as the visitors got caught up in the map. John was amused at Roger's contortions, and provided a chair for Roger to stand on so he could view the higher passages. The map wall was as hard to ignore as a loud television set in a room.

Later, John took them to the basement to see his collection of gleaming brass and oiled cast-iron gas-engine parts. Gas engines had been John's love long before he started caving.

The engine is a puzzle, Roger thought, like the cave. If the pieces are put together properly, it goes! Roger thought about the mystery of the universe in which gas engines work, and where great caves are carved by the slow action of water in the earth.

Both the engine parts and the cave surveys were numbered and recorded in folders in John's filing cabinets. John was as orderly and neat as Roger was disordered and casual. When Roger had been Chief Cartographer, the map factory was in his bedroom, spread from the shower across the bed, storage chest, and several tables. It appeared chaotic, but Roger could always find what he wanted. So far as is known, he never lost anything of value, perhaps because his wife, Joan, kept watch.

John was reserved, introspective, thorough, and intensely pragmatic. His energy and determination matched Roger's. But where Roger as a child had loved magic and had developed shows to hoodwink anyone who would watch, John had opened up gas engines (and before that clocks), exposing their parts to show exactly how they worked. Roger loved mystery and concealment. John loved objective clarity and openness. Each wanted to connect the Flint Ridge Cave System with Mammoth Cave.

They talked into the night. Roger considered many possible routes at once. John listened patiently, but would venture only a few guarded opinions. Roger's years of persistent, flamboyant, enthusiastic caving were so different from John's organized, efficient, and dogged accomplishment that neither man thought the other would make the connection with Mammoth Cave. Yet John could not help but defer to Roger. It was Roger's example as an expedition leader that had inspired John to plan his own expeditions.

John said that if the Q-survey leads petered out, he favored moving far back up Houchins Valley. The distance between Mammoth Cave Ridge and the Flint Ridge Cave System was much greater there, but prospects looked good.

Roger still liked the Houchins River area. He had surveyed a lot of it. Upstream, Art Palmer had crawled alone on his back with his nose to the ceiling through mud and water for 1500 feet. At the end, the water came within a few inches of the ceiling, so Art had retreated with the feeling that it probably siphoned. This caused Roger some pause. It was difficult to think of pushing beyond where Art had stopped.

However, the upstream Houchins River passage *did* go toward Mammoth Cave, and Art had *not* followed it until it siphoned completely. Roger thought it was headed toward Marion Avenue in Mammoth Cave. He needed a team determined to force beyond the point of Art's farthest penetration. On many other occasions explorers had stopped, only to have a succeeding party find that the low place extended only for twenty or 200 feet before the passage opened up again. Of course, such passages just as often closed off totally, twenty or 200 feet farther along.

Tom would try it. John produced data about Marion Avenue and the U.S. Geological Survey bench marks in Mammoth Cave. For completeness, Roger took a copy of John's eight-page running compilation of unexplored leads in both cave systems. Richard Zopf had said practically nothing the whole evening, but he had listened to it all. He wanted to go along.

Roger had been leading the midsummer expeditions for many years, and many old friends returned in 1972 for the fun. The weather was balmy, and July 1 seemed the perfect day for connecting the Flint Ridge Cave System with Mammoth Cave. Roger began his introduction speech with his usual joke.

"I'd like to announce the connection of the Flint Ridge Cave System with Mammoth Cave," he began, and then went on, "and you know, *this* time, on this very expedition, we're going to do it!"

Roger worked out the party assignments at the end of a picnic table outside the Austin House.

"Who do you want to take, Tom?" he asked.

"Carol Hill and Butch Welch," Tom said at once.

"Good," Roger said, scribbling the names on the assignment sheet.

Butch had been out to the passage before on a biological trip. Carol and Alan Hill had been discovered by Tom Barr years earlier camping near the Great Onyx Entrance on Flint Ridge. After talking to them awhile, Tom Barr had sent them over to the Spelee Hut to talk to Phil Smith and Red Watson. Carol and Alan loved caves, but were not cavers. However, they looked all right, so Phil and Red threw them into Flint Ridge on the spot. Few Flint Ridge cavers have contributed so enthusiastically as Alan and Carol. Over the years their two boys, Larry and Roy, grew up—as had Joe and Gino Austin, the four Brucker children, Anna Watson, and three dozen other cavers' kids—with their caving cousins, playing in the mud and the woods around the Spelee Hut on Flint Ridge.

Cavers hurried through last-minute preparations. Roger had filled them with great enthusiasm about the possible connection, with a minimum of evidence to support his claim that this was *the* expedition on which it would be made. Roger's jokes and techniques were so old by now that they were traditional, but even though they were transparent to everyone, they still worked.

Roger gloried in the role of expedition leader. Even more than for other exuberant expedition leaders like Red, for Roger all these cavers were manifestations of himself. He wanted to go along on every trip with every one of them. It gave him great satisfaction that his son, Tom, was leading a difficult trip on which the connection just might be made. Nearly nineteen years old, lean and taller than Roger, Tom was a born caver, one of the strongest of the Flint Ridge cavers.

Roger sighed. Some of it had not been easy, the last twenty years. In the early days his wife, Joan, had brought the children along, first only Tom, and then Ellen, Jane, and Emily. Joan had dominated as camp manager, bullying the bachelors into doing the dishes. She took only a few trips into Crystal Cave herself, getting almost stuck once in the Crack in the Floor when she was pregnant with Jane. Later, four growing children curtailed these activities, and Joan gave up camp management. As the children got older, she began to feel that Roger was spending too much time with the cave. Joan began to stay home while Roger went caving.

Burnell and Doris Ehman had taken over as camp managers, but now they also had stepped down after many years.

I'm getting old, Roger thought.

Tom Brucker had known caving all his life. Now, in the bustle of cavers converging on the Crystal Cave Ticket Office for surveying equipment, first-aid kits, ropes, carbide, and food, Tom pulled out of the trip-log file a dog-eared manila folder marked *Houchins River*. Flipping through it, Tom found the report he wanted, and read it for the fourth time:

August 27, 1966.
Party: J. Bridge, A. Palmer, M. Palmer.
Objectives: End of Houchins River to survey.
Report:
We went out to continue the surveys at the end of Houchins River, led part way by Fred Dickey's party. The W-survey was continued for three stations where it ended in a gravel fill. The leads at W-24 terminated in a narrow canyon too tight to get through. The lead consisting of three stations of X-survey went on for 1500 feet as a belly-crawl in water terminating in a wide siphon. Several "leads" were checked out on the return trip in Houchins River disclosing no promising passageway. The group felt that the only hope for progress in Houchins River is downstream in the R-survey.
JFB.
In: 10 A.M., 27th.
Out: 4:30 A.M., 28th.
Man-hours underground: 55.5.

It did not sound promising, but Art had said once again that he had not actually followed the passage to a siphon. He had just assumed that it would siphon soon. Well, Tom thought as he clapped the folder shut, he would soon find out.

"Do they all have wool underwear?" Roger asked worriedly.

"Yep," Tom replied. Dad was not to worry.

Tom walked with Carol and Butch down the side of Flint Ridge to the Austin Entrance. Butch had led a party to Houchins River the previous July, but it was new cave for Tom and Carol. They hurried to the Duck-under in Swinnerton Avenue, on through the Candlelight River area, and out to Houchins River. There the U-survey was a crawl in water in an eight-foot-wide passage with intervening gravel bars between long pools. The echo here makes it a very special place. Next came the V-survey, a comfortable walking passage about seven feet high and two or three feet wide. Its waters come from the northeast—possibly from Colossal Cave in Flint Ridge—and its sinuous course lies beneath the axis of Houchins Valley. They passed several low leads that looked more like muskrat bur-rows than passages. Then the survey letter changed to W. At W-34, the walls changed from dark gray to dark brown, studded with knobby cave grape. Mottled pink-and-white pebbles shone on the wet floor.

At the final W-station, they came to the opening in the right wall that contained three X-stations. A cold blast of air came out of the passage, perhaps all the way from Mammoth Cave. Tom leaned over and looked in. It was miserable and wet. And 1972 was one of the driest summers on record. Well, Art had done it. So could they.

They uncoiled the steel tape, got out the compass, and wiggled to X-3. It was belly-flop in water, all the way. The ceiling was too low to permit easy reading of the compass, and the passage was so wet that keeping the survey book dry was impossible. They saw at once why only three X-sta-tions had been surveyed before. Only their backs remained dry. They surveyed eleven more stations, to X-14.

The passage was too cold and too low. None of them, after all, had worn wool underwear. The chilly water sucked away heat, comfort, secu-rity, and finally all sensation.

They stopped surveying at a point where the low passage was joined by a canyon at ceiling level in which they could get up on hands and knees. They crawled 200 feet down the passage to a branch, trying to warm up. To the right the lead squeezed down to another belly-crawl in water, tighter and more grim than the belly-crawl they had just been surveying. They followed the left lead, a drier belly-crawl, around a corner to a large pool.

Tom stopped before submerging himself in the pool. It was probably still 1000 feet to where Art had stopped previously, yet to go on here would require complete immersion already.

The pool extended around the corner, and there was a breeze blowing over it. Tom could hear Carol's teeth chattering behind him. The fast crawling had not succeeded in warming them up after the long period of slow movement while lying in the water surveying. They were completely wet. They turned around for the long trek back to the Austin Entrance.

"We never really got our body heat back until we hit the crawl in

Pohl Avenue, almost to the entrance, hours later," Tom told Roger. Not even the long walk along Turner Avenue had warmed them up.

Later Tom figured out that Art had surely gone through that pool and on for another 1000 feet of cold liquid torture. It was the feat of a superman. Tom admired Art's tenacity, but he was eager to get started on his own alternative strategy for connecting the Flint Ridge Cave System with Mammoth Cave from the Mammoth Cave side. Down some obscure shaft drain along Marion Avenue in Mammoth Cave, one might get just as wet along, say, a 1500-foot crawl that led to survey station X-14 in the Flint Ridge Cave System. One thing was certain, however: It was easier to get within 1500 feet of X-14 from the Mammoth Cave side than it was to get within a mile of it from the Flint Ridge side.

The first part of the Brucker connection plan had fizzled in the low headwaters of Houchins River in the Flint Ridge Cave System. The back-up plan had been outlined in the memo Roger had handed in to the Park Service at the the beginning of the expedition. They would survey Marion Avenue in Mammoth Cave, where Tom would descend some vertical shafts at the far end. One of the pit drains might go north to connect with the upstream reaches of Houchins River in the Flint Ridge Cave System.

During the next few days six parties explored in the Marion Avenue area of Mammoth Cave. They surveyed 8325 feet, but found no leads toward Flint Ridge. It was a solid accomplishment in a part of Mammoth Cave about which little had been known previously. It was not, however, a triumph for the Brucker strategy for connecting the Flint Ridge Cave System with Mammoth Cave.

Now it was John Wilcox's turn again to run the connection show.

Again Q-87

John cons another party into trying to force a way through
the breakdown pile at Q-87. On the way back Pat goes through the
Tight Spot to discover a stream that drains west from the
Flint Ridge Cave System toward Mammoth Cave. However, the Tight Spot
may have separated the woman from the men.

25 Roger's 1972 midsummer expedition decamped, but a contingent of cavers stayed behind. Fred Bögli, the long-time director of exploration in Hölloch in Switzerland, had arrived. For many years Hölloch was the longest known cave in the world, but the Flint Ridge Cave System was growing rapidly. At the IVth International Speleological Congress in Yugoslavia in 1965, Red Watson and Burnell Ehman had been under fierce attack about the length of the Flint Ridge Cave System. This was in the session on the longest and deepest caves in the world, and the intensity of concern about this subject among Europeans is something most American cavers have to see to believe. Red and Burnell were trying to answer questions in their poor French and German when Dr. Bögli stood up, chastised the audience, and then went on to describe the Flint Ridge work. He said—in German—that the Flint Ridge Cave System was then definitely the second- or third-longest cave in the world, and with an emotional catch in his voice he concluded that after so many years of being first, Hölloch was now in danger of being surpassed by the Flint Ridge Cave System. Then he sat down. There were no more questions.

Hölloch is much like the Flint Ridge Cave System turned up at an angle. There is only one entrance, and this is subject to flooding. In the spring, water sometimes shoots out the seven-by-four-foot opening as from a hose under pressure, a maximum of 424 cubic feet per second. Fred and his companions have been trapped in Hölloch by high water nine times for a total of nearly five weeks. In an attempt to avoid high water, they confine their explorations to one session a year in the dead of winter when there is little—but still some—danger that a thaw will send water pouring into the cave. Each year from twelve to forty explorers camp deep in Hölloch for a week or two, supported by as many as 400 volunteers who carry in supplies. It is highly organized and efficient work. It is very European.

Inevitably the question arises as to whether Hölloch is more difficult than the Flint Ridge Cave System. Objectively, no doubt it is. The temperature in Hölloch is 38° F to 41.5° F, while in Flint Ridge it is around 54° F. Obviously, the water is considerably colder in Hölloch than in Flint Ridge, and one does get wet in Hölloch. There is also considerably more climbing up and down in Hölloch than in Flint Ridge, for there are scarcely any horizontal passages in Hölloch. Tubes like those that one could walk in comfortably in Flint Ridge are slanted at angles of ten to fifty degrees in Hölloch, so that cavers must hang on while moving along. The highest point in Hölloch is 2717 feet above the lowest, whereas the Flint Ridge Cave System is confined in a block of limestone only about 300 feet thick.

On the other hand, the elite corps of explorers in Hölloch are relatively comfortable in their camps in the cave, and the work—they say—is pleasant. The only Flint Ridge caver who has been deep into Hölloch is Walter Lipton, who went on a twenty-two-hour supply trip. He says that Hölloch is tough, but remarks that some of the work in the Flint Ridge Cave System is just as difficult.

There are these two big, tough caves. . . .

In 1966, when the Flint Ridge Cave System seemed destined to become the longest in the world—which it did in 1969, pushing Hölloch to second place—Fred Bögli could not rest until he had worked in Flint

Dr. Alfred Bögli and Burnell Ehman. Fred Bögli is the long-term director of exploration in Switzerland's Hölloch, the world's second longest cave, 1972.

Ridge himself. Now Fred was going to spend a week, in July 1972, study-
ing joint patterns in the Flint Ridge Cave System. He was sixty years old,
in excellent physical condition, his hair gray and thinning, and he had a
handsome silver beard. His gray eyes flashed as he shot glances into all
parts of the Flint Ridge camp. Effusive, at home anywhere in characteristic Swiss fashion, he got along well with the predominantly Midwestern
cavers in camp. He clasped old friends about the shoulders, pumped
hands, and smiled, ever apologizing for his rapid, idiosyncratic English.

Tom Cottrell was in charge of the group that was staying to help Fred,
and soon they rushed off to the Crystal Cave Ticket Office to look at the
maps. John Wilcox showed Fred the sinuous line leading to Q-87. Instantly
involved, Fred checked the map scale, asked about the depth of the passage
below the surface, and demanded details about the breakdown plug at
Q-87. John warmed to Fred, losing some of his reserve as the older man
talked of surveying methods in Hölloch. Fred produced a Büchi-compass
used in mapping the Swiss cave. John grasped it with both hands within
a few seconds of its appearance. Fred launched into a detailed explanation
of how it was used.

Years previously, before Fred could speak any English, Roger Brucker
—who knew no German—visited him in his lovely home overlooking a
deep valley and distant mountains in Hitzkirch. With a scattering of maps
and much arm waving and pointing, in two hours they became close
friends, with only one common language: caves.

Fred now opened his attaché case to extract a bulging folder of maps.
He laid them out one by one and told the story of the exploration of the
71.75 miles of Hölloch.

"Ach!" he said. "With your more than eighty-five miles now, poor Hölloch will never catch up. This last winter," Fred went on, "I did not go
into the newly discovered upper regions of Hölloch. My roadrunners did
that."

"Roadrunners?" someone asked.

"Yes, roadrunners," Fred said. "Those birds that were crossing the
road just in front of my car like lightning in New Mexico."

"Beep! Beep!" Roger said.

Everyone laughed. Yes, we had roadrunners. John had his eye on
a pair—newcomers Eric Hatleburg and Mark Jancin, who were staying
to help Fred. Before them, there had been Eric Morgan and Bob Keller,
who ran out great distances in the Flint Ridge Cave System. And, of
course, there had been Bill Austin and Jack Lehrberger.

"Some of our roadrunners almost connected the caves together in
your honor," Roger said. "We'll do it any day."

"When you do, the accomplishment is mine, too!" Fred fired back.
"I am a Member of CRF."

John could not rest for thought of making the connection. Before he
left for home, he asked Tom Cottrell to open the question with Mark and
Eric about taking a big trip the following weekend.

Then at midweek, John discovered that Pat Crowther would be spending that weekend at Flint Ridge, on her way home from visiting relatives.
By now John assumed that Pat was ready for anything. The big trip
could be to Q-87 again.

When John arrived on Flint Ridge on Friday, he told Eric and Mark
immediately that they would be going to Q-87. They could not contain
their excitement, and before John had a chance to ease the news to Pat,
she was startled by a wild Eric who ran into the bunkhouse and shouted,
"Hear you're going to Q-87 tomorrow!"

Pat was stunned, and then irritated. She dropped the sleeping bag
she was unrolling, and marched off to find John.

"This has got to be a joke. I'm never going back to that place," she said
coolly to John, with a clear edge of determination in her voice.

John was already two phrases ahead of her. "It's the ideal time because the water is the lowest it has ever been. This is the best time to
connect. Gordon Smith was going, but now he's sick. And this way we
don't cripple a whole expedition for two days. It's really a good time."

John's enthusiasm began to catch hold. Pat remembered the leads
she had not felt up to pushing last time. She was horrified to find that she
was beginning to feel like pushing them now.

"Okay," she said abruptly, "guess I'd better come along." She went
back to the bunkhouse to try to get some sleep. John was taking her for
granted, so she supposed she had to go.

The next morning, July 15, 1972, the party signed out at 8:30. Mark
and Eric were loaded down. Besides their usual heavy packs of carbide
and food, they carried a longbar, a two-pound hammer, and the heaviest
chisel John could find. They were equipped to break up any of the boulders
in the rock pile at Q-87 that they could not budge.

They moved smoothly to Q-87, with only one stop to set up an experiment. At the junction of Bretz River and Candlelight River, they attached
vertical strips of masking tape to the wall. If the passage were flooded
later, subsequent parties could check for mud on the tape to see how high
the water had risen.

And what if the water should rise while there were cavers in the
rivers? John and the others were themselves a part of the larger experiment. In the depths of the cave, however, you try not to think of every
grisly possibility, just of some of the more amusing ones.

The trip from the Austin Entrance to Q-87 took seven hours and
fifteen minutes. Pat had on an extra wool shirt this time, but still felt
cold. There is no excess fat on her body. To John the passage seemed
damper than he had remembered it. The little trickle of water down
the breakdown pile was now a heavy trickle. Had it started raining outside? When they had entered the cave the day had been bright and
clear, a white haze of heat already rising over Flint Ridge.

These thoughts flew away at the first touch of the longbar to the
pile. The stone wall they had left on the previous trip had turned into

mush. At once some of the largest and formerly tightest rocks crashed down. Four big ones were moved out into the crawlway in rapid succession, and a fifth was loosened. It was so big, however, that the best way to move it out seemed to be to enlarge the edges of the opening.

Chipping away the sides of the canyon did not work. The limestone of the walls was too dense. They tried then to pound the big boulder into smaller pieces. This also was such slow work that they abandoned it.

John started digging at the top of the pile again. With a chilling rumble, all the rocks plugging the end of the canyon collapsed in a shower of mud and stone, and more rocks bulged in to take their place. Everyone shivered involuntarily.

"Well, anyway we got the thing moving again," John said.

They attacked the large boulder on the bottom again. One at at time, they lay on their sides and backs to dig under the boulder in gravelly mud having the consistency of fresh-poured concrete. It was cold. The stone was like a tooth that would wiggle but would not come out. After four and a half hours, they had to stop. John wrote in his notebook that the block might be moved on the next trip if they mined more gravel from beneath it. Perhaps this was the delusion of a cold, tired fanatic.

Crawling warmed the four somewhat as they plowed nonstop back to Q-33. This was the lead in which Gary Eller had abandoned his gloves.

Pat sighed and pushed into the crawlway. It was not all that tight where Gary had turned back, she thought. But it had teeth. She ripped her shirt as she squirmed around a tight S-bend to where she could see blackness about twelve feet ahead. The sound of dripping water was very loud. There really was something there, but the passage ahead was narrower than it had been thus far.

Pat turned on her left side, with her left arm extended ahead. She jammed her right arm up into a ceiling slot four inches wide, exerting pressure with her wrist and elbow to hold herself up just enough to slide along a few inches. Again. Inch forward. Press and pull with your right arm. Rest. Again. It took Pat fifteen minutes to go those twelve feet.

Pat emerged into a very pretty, symmetrical dome with a circular pool of water on the bottom. The pit was ten feet in diameter and fifteen feet high. She had come in through the drain. There were no other openings out of the pit.

It was the first place she had been able to stand up for a long time. And she was alone. Ah! A private bathroom. That taken care of, Pat started out. On her way back to the main passage she fought waves of sudden nausea. She forgot to look for the gloves Gary had lost.

Pat was cold, weak, and sick. The cave had closed in on her and compressed her queasy stomach. For this kind of caving, she was as good as, or perhaps better than, any other caver in the world. She could explore passages so narrow and contorted that most people would not believe that any human being could get through them. She got some satisfaction out of that. But now she had shrunk to one small, sorrowful ball of misery in a cold, endless cave.

They were miles from home. Pat may have felt wretched, but immense stores of strength remained. She followed John to Q-17, where Mark and Eric were chipping at a ledge blocking a lead. Sparks and splinters were all they could remove. Then John shoved the longbar into a crack in the ceiling. He strained, and a chunk of the ceiling and the left wall crashed down. The side passage was now thoroughly blocked by a 150-pound rock. Only one person could reach the block at a time, and it seemed too heavy to move. Then John hooked the longbar around it and heaved with all his strength. After years of cave crawling, he had arms like a high-bar specialist in gymnastics. His muscles bulged, and the block moved slowly out of its muddy socket.

The passage was now large enough for them all. John crawled in. On the last trip Pat had reported that the mud bank could be dug through, but evidently it was underlain by rock. John could not go on.

Watching the work had revived Pat. She sighed and slid past John to look at the rock ledge. This was not her mud bank at all. She had easily slid past this point, and the diggable mud bank was on beyond. But Pat let it go for this day.

They had been in the cave for a long time. Everyone was tired and bruised as they moved on down to Q-1. In the N-survey passage, they sat down in the comfortable room at the junction with the A-survey. The pleasant sandy floor was too fine a place to pass up. Every Q-87 party had rested there.

John looked wearily at the entrance to the A-survey passage, which took off from the room. All the leads in the Q-survey had been examined, if not finally crossed off. The A-survey was next on John's list.

"It's only twelve stations to the end of the A-survey," he said. "We really should go look to find out what we'll be up against next time."

"What's it like?" Pat asked.

John hesitated. "Well," he said, "I guess there's a place where you have to belly-crawl through six inches of water."

Mark and Eric had settled on the sand. No one was eager.

John and Pat crawled down the A-survey. At the end the passage split. This was a surprise to John. The master lead list showed no lead at all at the end of the A-survey—John was looking just to be thorough —but here were two leads. John took the higher passage, a three-foot tube. After thirty feet he came to a rock that he recognized as the one John Bridge had described as closing off the passage. It was just a slab, lying diagonally across the passage. John could see that the passage continued in the same dimensions beyond it. Fatigue was numbing him, but he pulled forward the longbar he had been dragging behind him. He jammed it under the edge of the rock. The floor was deep mud.

It might be possible to dig a way under the slab without moving it. Anyhow, there would be no moving that rock this trip, or perhaps not at all with a bar. John made a note to bring along an automobile jack next time, just in case they could not dig a hole under the slab large enough to get through. Then he returned to the split.

Pat had taken the low, eight-foot-wide squeeze. She could see at once that it was a going lead, and her spirits rose. After 100 feet of virgin belly-crawl that was practically dry, Pat reached a small room with loud water noises coming from a small canyon leading out of its far wall.

Pat managed to slide along the top of the canyon on her side past an awkward tight spot—the Tight Spot—and then dropped into a shallow pit. It was a crossroads, with water coming in from the left and running out to the right. Pat knelt down to look where the water was going, and poked her head out into space. She was at the top of a waterfall, eight feet above the floor of a room that looked huge compared to where she had spent the day. The climb down looked difficult, so she did not try it. Taking out her compass, she sighted over the ample crawlway that was the drain of the pit. It was headed west. Wow! she thought. A real live going lead. The trip was worth it, after all.

On the way back Pat discovered that the Tight Spot was more difficult going out than on the way in. She hoisted herself up to the flat ceiling of the wineglass-shaped opening. The Tight Spot was about fifteen feet long. In it she had to hold herself horizontally against the ceiling with the pressure of her arms to keep from sliding down to wedge in the crack below. She could not push very well with her feet because she had to hold them up to keep them from jamming.

Anyone much larger than herself could not get through the Tight Spot, Pat decided. And if you were not strong enough to hold yourself up against the ceiling, you would be in danger of sliding down into the crack. It would be difficult to move once you got stuck in the crack, and the Tight Spot was not large enough for anyone else to get in to be of much help. Although it was not giving Pat much difficulty, this small stretch of passage—the fifteen-foot-long Tight Spot—was, she recognized, objectively the most dangerous place she had gone through in the cave.

John had followed her down the belly-crawl until his chest was compressed between the bedrock floor and ceiling. His elbows and knees hurt, but it felt good to lie on the hard, wet floor waiting for Pat to return. He could hear her boots scraping and her hard hat clonking.

"Oh!" she said when back in earshot. "It's very tight. But we have cave!"

Pat and John glowed.

"But, John, you and the others might not fit through the Tight Spot," Pat said as an afterthought.

A chill ran through John. He had gone only a short distance up the

100-foot crawl. The Tight Spot was ahead, out of sight. Well, he was too tired to try it now. He would come back. And he *would* get through. Pat said there were several leads out of the pit on the other side of the Tight Spot. This had to be it. He *would* get through the Tight Spot.

It was morning again when the party got out of the cave. There had been rain, and everything looked freshly washed. They had been gone twenty-four hours, but they could not believe that night had come and gone.

John thought unhappily that perhaps the rock pile at Q-87 was at the bottom of a pit close to the surface, and was not connected to Mammoth Cave at all. However, people had been on the surface of Mammoth Cave Ridge above Q-87, and found no pit. Mammoth Cave Ridge was smooth and flat over Q-87. The rock pile had to be at the bottom of a vertical shaft, underground, where it might be connected with passages in Mammoth Cave. Nevertheless, that rainwater had soaked through the pile very rapidly.

"They're back!" someone shouted.

Fred Bögli, wearing a white sweat suit, abandoned his breakfast to rush over to them.

"Did you connect with Mammoth Cave?" Fred asked.

John smiled and shook his head no.

Everyone quieted as they told their tale. Pat had found a lead in an area where new passages were as likely to be a part of Mammoth Cave as a part of the Flint Ridge Cave System. And the passage beyond the Tight Spot was going downstream to the south. The explorers had passed the drainage divide. For the first time in six years there was a going lead toward Mammoth Cave.

"It won't be long now, will it?" Fred asked gaily.

John did not know. He would have to get through the Tight Spot to find out.

Separating
the Man
from the Boys

The Tight Spot strains out the father, but the son
goes on to find a river in the caves beyond. This must be the way
to connect the Flint Ridge Cave System with Mammoth Cave.

26 On that morning of July 16, 1972, Roger Brucker examined the returned Q-87 crew with envy. His own connection strategy had not succeeded, but John's determined work had led to a breakthrough. Roger wanted to hug and kiss Pat for doing such a wonderful thing. He decided not to do it. Pat sometimes seemed like a little girl, but she was also a woman of some reserve who was nearly thirty years old. Maybe Roger should shake her hand. As for John, he was reserved, too. Roger thought about putting John under lock and key for the next few months.

John was scheduled to run the 1972 August expedition. Three Bruckers—Roger, Tom, and Ellen—signed up, as did Richard Zopf. They intended to be there when it happened.

On the first day of the expedition, August 26, John discovered that there were only three parties in camp, and a couple of them were already scheduled. Roger, Tom, and Richard remained, ready to go. They wanted to go all the way to Mammoth Cave. John could not go with them because—aggravating though it is—the job of expedition leader does involve a certain amount of busy work. For one thing, John had to go over to see the Park Service officials.

John thought while the trio stood waiting. Roger and Tom he knew, of course. Richard was almost totally inexperienced, but he had shown himself to be a natural caver and he was strong as an ox. John had not seen the Tight Spot, but Pat had told him he probably could not get through it. If he could not get through the Tight Spot, then Roger and, probably, Richard could not. However, there was a possible bypass route, if they could move the block in the tube. And although Roger was considerably thicker than John, Tom was considerably thinner. Tom was a far-out, going, caving fiend in virgin passage. Tom might get through the Tight Spot, and if he did, he would go. John thought they had at least a fifty-fifty chance of bypassing the Tight Spot through the tube, and if they did, they might very well make the connection without him. Roger deserved it, John thought.

"All right," John said. "See where it goes."

Roger was already dressed in his ragged cave clothes, but on hearing the word he went back to the Crystal Cave Ticket Office to put another meal into his pack. He noticed Tom following him.

"This is it," Roger said.

Tom said, "Yeh."

"You're not sick?" Roger asked testily.

"Dad," Tom said with some embarrassment. "Dad, do you think you're up to it?"

Roger grunted, turned his back, and began to fill his carbide bottle. Tom wrapped some candy bars in aluminum foil, and said no more.

John returned with the cave pass that the National Park Service now required all the cavers to carry. "Here's the Austin Entrance key," he said to Roger. Roger took the pass and key, and then looked more closely. John had given him two keys. Roger had been passed the key to Mammoth Cave that John had been carrying himself since his first Q-87 trip.

The trip from the Austin Entrance to the little junction room in Houchins River where the A-survey took off from the N-survey was uneventful. Roger had been at the junction room several times and remembered it perfectly. It pleased him to have Richard along, for he could tell all his old instructional stories to this newcomer. Also, Richard had carried a heavy automobile jack all the way as though it were a wrist watch.

They ate a meal in the junction room, and then crawled out to the end of the A-survey, and on to the Tight Spot. One point along the A-survey crawlway was a chest compressor for Roger, but he suffered no more damage than the loss of another button and a few more rips in his shirt.

Richard hoisted himself up to the entrance of the Tight Spot and shoved himself through. Good—he was considerably bigger than Pat. Richard shouted back instructions to Tom.

Tom took off his pack and eased himself upright in the narrow canyon that led to the Tight Spot. He shuffled along sideways to the bend in the passage where the lower part of the canyon began to constrict. Here he moved the upper half of his body forward while raising his feet until his body was horizontal. He lay on his right side, his right hand searching for ledges on which he could push himself up, his left hand searching for handholds to pull on.

The Tight Spot.

It was impossible to rest. Tom was larger than Pat, but smaller than Richard, so he did not worry about making it. Inch by inch toward the darkness up ahead, stopping for a breath, then exhaling to make himself smaller, Tom moved forward.

Tom was through. He doubted very much that his father would make it.

Roger pushed himself through the canyon to the Tight Spot. He got his body up horizontal, but it was almost impossible to shove, and there was very little for his hands to grip to pull on. With immense effort he jammed himself solidly into the opening. He exhaled. His heart beat rapidly and he was very warm. He grunted involuntarily, but said nothing.

Should he take off his shirt? He could feel the sharp edges of the walls biting into his ribs and his backbone. He could not get through the Tight Spot. Not this trip, anyway.

Tom had the good sense to say nothing.

Roger rested a moment to get his breath. Then, still lying in the beginning of the Tight Spot, he took out the survey notebook. They had already started a B-survey tied to the last A-survey station behind them.

"You might as well survey on until you're out of earshot," Roger said.

The shots were so short and the passage so contorted that Tom and Richard managed to shout back data for several more stations before Roger could no longer hear them.

"Go see where it goes," Roger shouted. "I'll go back to where we left the jack, and try to move that boulder in the tube."

"Okay," Tom shouted back, and then he and Richard were gone.

Roger had let them go without seeing whether he could get unstuck from the Tight Spot. It was no problem—he had not gotten in very far. He retreated ten feet to the small room that preceded the canyon. It was a space about the size of the interior of the Ford Mustang Roger had driven from Ohio to Kentucky the day before. His left knee crawler had slipped down to his ankle, and the other one was loose. He straightened out his pants legs, and felt the cool cave air ventilate his stinging knees. He pulled knee-crawler straps tight, and felt renewed.

Roger was not overly tired, after all. His carbide lamp shot out a bright flame of yellow light. He was in command, and was not bothered about being alone. He could get out of the cave from the Tight Spot in his sleep, and he had no worries at all about Tom and Richard.

The mud floor was cold, but Roger was still warm from his exertions in the Tight Spot. He slung the strap of his shoulder bag over his head, and swiveled into the passage that had led them to the Tight Spot. The passage was a comfortable four feet wide and three feet high, but very gooey. Ooze squeezed up between his gloved fingers and mud sucked at his knee crawlers. He picked his way across some small pools of water as he crawled along.

Roger had barely noticed the Chest Compressor on the way in because he had been preoccupied with the surveying and thoughts of the Tight Spot. Now, from this other side, the Chest Compressor looked too small.

Roger was practiced in gauging the size of cave passages. He lay on his belly and viewed the Chest Compressor. It was about nine inches high—plenty wide, but only nine inches high. He had squeezed through hundreds of such places in the last twenty years. In fact, he had squeezed through this one without thinking about it, only thirty minutes ago. A faint feeling of irrational fear seeped into Roger's mind. Stupid, he thought, and jammed into the opening. His head went through, but not his chest. He pulled back, and immediately aimed for what appeared now to be a slightly higher spot. His hard hat jammed against the stone ceiling. Pulling back, Roger took off his hard hat and shoved it through the hole. His body was now in a world of darkness. Out there through the Chest Compressor at the end of his hands was another world, a cheery world of light. His objective was to enter that appealing place. He pushed his bag through after the hard hat and light. Willing himself to be as lean as he had been ten years before, he stretched his arms forward and pushed.

His chest stuck.

For a few moments, Roger allowed himself to enjoy the horrible fantasy of being stuck there until Tom and Richard returned, his lamp out of reach, slowly dimming and then going out. Blackness. And then what if they could not pull him out or push him through?

Actually, neither the lamp nor the bag was out of reach. Roger suddenly exhaled all his breath, pushed hard with his feet, and ground his way through the tightest part of the Chest Compressor. He put his hard hat and pack back on, and moved forward to where Richard had left the jack. Grasping the jack and its folded handle, Roger crawled without pausing into the three-foot tube that led to the block.

After about eight feet the tube bent to the left, and then after a few more feet it bent to the right. Roger thought it was surely a cut-around that would bypass the Tight Spot. After thirty feet, he reached the block. He was glad he did not have to carry the jack farther. His admiration rose for Richard, who had carried it in the cave for some four miles.

Richard was twenty-two. Roger was twice that old. Roger had started being a caver just about the time Richard was born.

Moving the slab looked to be a straightforward job. Roger needed only to jack it up, prop it safely with nearby loose rocks, crawl under, and go through. He chuckled as he thought of intercepting the other two a few hundred feet farther on, they having suffered through the Tight Spot while he had taken the easy way around through his newly opened turnpike.

The handle unfolded after a few whacks on the low ceiling. Roger slipped the locking collar over the joint on the jack. He placed the jack under the left side of the slab. The rock was two feet wide, four feet long, and three or four inches thick. It had fallen from the ceiling so that its front end rested against the wall about four inches above the floor and its back end rested against the ceiling. Roger could see that beyond it the passage continued as the same three-foot tube, open into inviting blackness. However, the slab was jammed tight.

The jack was in place, but Roger ate a candy bar first. He would need the energy, and, besides, he enjoyed the anticipation.

The handle rotated easily, driving the jack screw up to bite against the rock. Very easy, Roger thought. There was some opposing pressure, but not as much as he had expected. After about ten turns, he wondered why the slab had not seemed to move. Crawling closer, he saw why. All he had done was screw the jack's base down into peanut-butter mud. He reversed the jack, lowered the screw, and pried the jack out.

The floor was firm under the other side of the slab. The jack extended itself until it was solid against the slab. Then it stopped. Roger could turn the handle no farther. He backed off a turn, then swung around hard. His hand slipped off the handle, bare knuckles scraping the rough ceiling. It hurt. That's what you get for leaving your glove off, Roger thought. He sucked some rock fragments from the bleeding knuckle, then slid his muddy glove back on.

Roger tried a diagonal jack position the next time, but the jack slid away. He continued to work for thirty minutes, but he could not budge the slab. It was stuck under ledges on both sides of the passage. Roger tried to break the slab by pounding it with a loose rock. The slab would neither move nor break. He remembered a conversation he had had with Skeets Miller during the 1954 National Speleological Society C-3 expedition in Crystal Cave in Flint Ridge. In 1925, Skeets had managed to get a jack under the rock that trapped Floyd Collins in Sand Cave, but he could not budge the rock. Roger had been impatient hearing the story, but now in much easier circumstances he had failed to jack up a rock.

Roger lay still and dozed. When he awakened, his lamp had burned nearly out. His body had cooled from the damp floor. He backed away from the boulder to the junction room, fished a candle from his bag, and improved the gloomy place with its light. He changed his carbide methodically, absorbed in the nearly ritualistic act. He had changed carbide at least a thousand times over the last twenty years. He screwed the lamp shut, and then spun steel on flint. The lamp lighted with a loud pop. Now Roger had another four hours of light.

He unwrapped another candy bar, and for the first time wondered about Tom and Richard. They had been gone for two hours. Of course they had found something important. It was too quiet and lonely for total optimism, however. Could they be having trouble getting back through the Tight Spot?

Roger strapped up his bag and went along the tube to the fork again. He ignored the opening that led to the slab, and started instead along the crawlway to the Chest Compressor again. He shoved grimly through it into the little room in front of the Tight Spot. There he heard the nearly inaudible, but unmistakable sound of body movement through a distant passage. His worries were immediately relieved by the joyful noise. He did not bother to shout, for he knew that his voice would not carry as far as the scraping. Roger settled down to wait.

He was cold. He reached up to snap his jacket shut, and found that it was already closed.

"Hello, hello!" Roger shouted.

He could not understand the faint answer. It would be a while.

When Tom and Richard left Roger at the entrance side of the Tight Spot, they had first to negotiate the eight-foot pit that Pat Crowther had not climbed down into. This proved to be not difficult, and they soon stood on the floor of a room ten feet in diameter. The drain was about three feet high, but was divided into two sections, one above the other, as in a figure 8. They squeezed into the muddy top portion and crawled about 100 feet, catching glimpses of a clear-flowing stream in the lower portion. The stream passage then departed at an angle. They continued slithering through the upper tube until it got so tight that they backed up to follow the water.

Tom lowered himself into the water passage while Richard waited, using the time to change his carbide. He thought he might also take a nap. He fell asleep for about an hour.

Meanwhile, Tom was lured into the unknown by an unusual passage. Two-level stream canyons of the type they had been exploring usually meander, with the lower water passage making wide swings, frequently diverging from, but usually rejoining, the upper part. However, the passage down which Tom was following the water had diverged from the general trend at an angle of 160 degrees, turning back on itself almost completely. There was no sign of its rejoining the upper passage. Tom thought with excitement that the stream had been captured by another, lower passage close by. It might be a tributary to a large underground river that would flow into Mammoth Cave.

Tom plunged sideways into the ten-inch-wide drain. It broadened to a foot wide, and was soon six feet high. It wound around like a shoelace dropped on the floor. Soon Tom could hear falling water.

Tom wanted to come to a junction with a larger passage so badly that he found himself saying, "Come to a junction, come to a junction." Around the corner the water fell five feet into a small room, but the drain was as small as the one Tom had just come through.

"But I want a *big* passage," Tom said.

He crawled awkwardly out of the one tight drain, over the pool of water, and into the other drain. He was not quite sure he could get back up into that tiny hole. Mammoth Cave was still at least 2000 feet away. Tom was not expecting to connect on this trip. However, he was a Brucker. As the rock ceiling forced him to his belly in the water, he felt his excitement rise to an almost uncontainable pitch. He was sure that the next corner just had to be the last one before the big discovery. It had to be, for Tom was about to turn around. He had already explored several hundred feet of virgin crawlway.

Tom thought of the years of search, and yet the cave had not yielded the secret of the connection between the Flint Ridge Cave System and Mammoth Cave. He began to feel all the energy, the hopes and disappointments, the collected wills of a hundred strong cavers who had been frustrated time and again by this cave, all of it focused on the next corner.

"No kidding," Tom said later, "I really felt this then."

Tom squeezed around the corner on his side.

It was big cave.

It was the big discovery.

Tom gave a yell of joy, but the noise died quickly in the dank passage. He felt deeply that this was not his discovery alone, but that it belonged to all those who had pushed around just one more corner, whether their discoveries were dead ends, big passages, or just another corner on ahead. He was also slightly pleased that he just happened to be the one of them who was finally good enough to do it.

The new passage was a low crawlway with an elliptical cross-section and a stream flowing across gravel bars. Tom dropped into it quickly and started crawling downstream. This was going to be it. The passage soon became a walkway.

When Tom stood up, he was excited at his find. Deeper in his mind was the thought that the passage might go all the way to Mammoth Cave. But he had already pushed alone through several hundred feet of tortuous, twisting drains. He knew the way back. But . . . can you ever really be sure? The black mud banks loomed up ahead and steam from his heavy breathing obscured his view. The passage went on. Tom knew he was farther out than anyone had ever gone in the Flint Ridge caves. And he was alone. Tom did not pause, but after he thought he had gone 1000 feet, he stopped. He was not tired, but he was panting. The passage continued, grand, gloomy, and peculiar. If it did go to Mammoth Cave it might be a mile or more. It was a long way back to the Austin Entrance. Richard had been alone for a long time. Roger was waiting on the other side of the Tight Spot. Each of the three members of the farthest-out Flint Ridge trip was alone.

Tom drew his initials and the date on a mud bank, and turned back. He went without stopping, almost without thought, to where Richard was waiting. Then Tom's joy knew no bounds and he exploded with the telling of it. It was *the* way to Mammoth Cave. Of that he was sure.

Now that Tom and Richard had returned to the Tight Spot, Roger asked questions impatiently.

"Tom found some kind of river," Richard said. "I'll let him tell you. He's right behind me."

Richard's light glinted through the Tight Spot. Roger moved back against the wall to make room. Without pausing or taking off his hard hat or pack, Richard oozed through the Tight Spot that had stopped Roger cold. Damn him, Roger thought. He had a vision of Richard going through the Chest Compressor back up the passage on his hands and knees, while Roger would be crushing his rib cage to force his way through for the fourth time this day.

Richard's black beard was dotted with globs of dried mud. He had attended Antioch College for two years, but then dropped out to become a carpenter. Well, he was happy. Right now he was humming a lively tune. Roger waited impatiently while Tom pushed easily through the Tight Spot.

The room was too small for three, however, so they set off for the A-survey junction room back in the Houchins River N-survey passage that they had left five hours before. There Tom told his story.

Tom removed his purple hard hat, and his long black curly locks sprang out six inches in all directions from his head. Roger had purchased electric clippers twenty-five years previously, and had run them over his own head at least once a week ever since, keeping a half-inch-long burr-head haircut through all the hair-style changes of the outside world. He was also clean-shaven. His son, Tom, had an immature, wispy black mustache. In the presence of all that hair, Roger wondered whether he ought to let his own grow.

Roger listened impatiently, snorting disgustedly when Tom intoned the line, "Somewhere in the Book of Fate there is an entry which says, 'On August 26, 1972, a big discovery will lie around the next corner.'" Roger was not amused to be getting the same sort of baloney that he continually dished out.

Roger perked up when Tom mentioned that he had seen blindfish and white crayfish in the river.

"What direction does the passage go?" Roger asked.

"Heh, heh," Tom said with some embarrassment. "I, er, didn't take the compass. I left it with Richard."

"You mean you don't know?" Roger screeched, suddenly losing all his awe of the great explorer and feeling like a father again. The kid needed a haircut, too.

"No. I hope it goes toward Mammoth Cave. It must, it's such a big river," Tom replied.

Despite his annoyance, Roger's heart pounded. Tom was right. Any major drainage in this part of the cave would have to follow the general pattern of the known drainage, and that was toward Mammoth

Cave. Roger had the absolute conviction that Mammoth Cave lay at the other end of the new river. Through the Tight Spot . . .

"You really think so?" Richard asked with excitement.

"Yes," Roger said, happy and sad thoughts mingling. Roger was completely in charge now. "No doubt about it. The next party in here will walk into Mammoth Cave." Roger knew that because of the Tight Spot he would not be with them.

Richard was staring transfixed at Roger. The old con man was being forced to change the end of his dream. If he could not take the trip himself—that damned Tight Spot—he could at least guarantee that Tom and Richard would take it.

"Yes, you mark my words," Roger said. "The very next trip out here, you guys will walk right into Mammoth Cave."

Then they started out toward the Austin Entrance, eight or nine hours away. Roger was tired, but relaxed. He moved in what he calls "automatic," a trancelike state which other people call sleep. His view of this behavior is in sharp contrast to how others see it. It was the first time that Richard had observed Roger caving blind, and the experience impressed him vividly:

Earlier Roger had said that he had had a restless night and Tom had asked him whether he was sure he was up for the trip. We were assured he was and I assumed that a spelunker of Roger's experience would have complete control of his faculties at all times. Little did I know the ends dedicated cavers would drive themselves to. I knew of Roger's ability to fall asleep at a moment's notice; I respected this ability and have tried to cultivate the talent myself. But as we finally made walking passage after traveling at a snail's pace through all the crawling passages (I had often felt like a bird dog pointing, my movements being so slow) I realized we were no longer three cavers, but two young men reviewing their knowledge of rock music, and a zombie. As Roger stumbled along the passage, he was awake only enough to ask to stop and rest at any available opportunity. I admit that we all rested and napped, but Tom and I considered starting off again as a necessary consequence of stopping, but Roger evidently did not. Of course, as we finally emerged after 21 hours underground, Roger was wide awake. He had no thoughts of rushing off to bed. I think that somehow he had mystically transferred our strength to his own body while in that trance in the cave. I capped the trip with 16 hours in bed, like a normal person.

"Pete H"

Pat leads a trip down the new river. Tom makes a startling discovery
of a name on the wall and an arrow pointing downstream toward
Mammoth Cave. Wasn't Pete Hanson a Mammoth Cave guide? John
counsels secrecy until they know where they are.

27 News of Tom Brucker's discovery electrified the camp. Did
that river go to Mammoth Cave? It was really going to be
some expedition. Roger bustled around all that Sunday morning wishing he could stay. He finally decided that he had to go back
home to work as planned. Full of advice, he told John that he ought to
put two teams into the new river on a leapfrog survey.

John did not need the advice. He needed the teams. He had already
laid out a schedule of trips, some of which involved meshing plans
with Park Service personnel. There was a wet-suit trip to Roaring River
in Mammoth Cave on Monday, so Tuesday would be a rest day. John
was scheduled to lead a transit survey trip into the New Discovery section of Mammoth Cave on Thursday. The river Tom had discovered
would have to be explored on Wednesday.

Everyone rested on Sunday. On Monday, Steve Wells returned to
camp to report finding a major trunk river passage in a cave below
the Sinkhole Plain, outside the Park. On Tuesday he went out again
with a survey crew and mapped a mile of walking cave. This was part
of Steve's work for his M.S. degree in geology, and it was the most
spectacular hydrological find in many years. John watched silently as
Steve drove off to Indiana on Tuesday night to recruit more cavers to
work under the Sinkhole Plain. Steve could have made the trip out
under Houchins Valley, but he was too excited about his own new find
to be diverted.

This left the expedition with only five members, including John. So
it would be one party, John thought, and it would be on the safety
plan he had originally devised. Greer Price would remain on the surface
to start phoning rescuers if needed. Pat Crowther had fortunately ar-

rived on the bus in time, so on Wednesday, August 30, she, John, Tom, and Richard would go see what they could see.

The evening before the trip out under Houchins Valley to explore the new river beyond the Tight Spot, the four sat in the bunkhouse—Floyd Collins' Home—talking quietly as dusk deepened into complete darkness. The calls of whippoorwills and the chorus of insects only intensified the silence of Flint Ridge. Caving was best when there were only a few on the Ridge. Then one could feel the presence of that enormous sprawling emptiness underfoot. There were long pauses in their conversation. Tomorrow they would be alone, one small party, four specks of light in all those empty miles of the longest cave in the world. It was time to go to sleep.

John awakened the others at six on Wednesday morning. Tom ate many pancakes, and then worried for fear he had eaten so many that he could not get through the Tight Spot. During breakfast, John turned to Pat and said, "Pat, this is yours to lead. Just get me back in time for that transit traverse Thursday."

He must be tired of writing trip reports, Pat thought. However, of the four, only Richard was not an experienced party leader, so it was more or less Pat's turn.

They went again into the Austin Entrance and along that familiar route. At the Duck-under off Swinnerton Avenue, Tom said, as usual, "Quack, quack."

This started them off. Pat knew hundreds of folk songs, and when their boots were filled with Candlelight River, Tom and Richard started composing rock songs of their own. "The Bretz River Shuffle" was sung in the narrowest canyon, where progress ahead is made by shuffling sideways with one's head twisted forward. They connected the stiff neck bone to the strained backbone. There were also "Walkin' Down Candlelight River" and "The Tight Tube Blues." They tried composing a Mammoth Cave song to the tune of "America," but the words refused to come.

At the Tight Tube, Richard entered first. He began shouting a monologue, his voice rising in the cadence of a high-powered evangelist. "Comin' through the Tight Tube!" he shouted. "Now it is well known—it is written—it is written in fiery letters. The Tight Tube is so tight that all who pass through have their sins scraped off. See-uhnn will be scraped from your soul. Hallelujah!"

"Amen!" shouted Tom, tumbling from the Tube right behind Richard.

"Now Sister Pat's a-coming through the Tight Tube! Brother John's right behind," Richard continued.

Pat had elected to shove through the Tight Tube on her back. She came out into the small room at the end with her arms crossed in stately repose.

"Welcome, Sister Pat! You've been born again!" Richard shouted.

Pat quickly rolled out of the way of John, who was coming through, huffing, on his stomach.

"Hallelujah!" John shouted as he reached the room.

The others exploded in laughter at this uncharacteristic outburst from reserved, taciturn John.

"Be-ware. Bee-ware," Richard went on. "For if you pass *back* through the Tight Tube, you'll re-gather those sins. You'll be born again . . . in see-uhn!"

"But what can we do to be saved?" Tom shrieked.

Richard lowered his voice to solemn tones. "The only way to salvation is to connect with Mammoth Cave. Leave through Mammoth Cave." Richard's voice rose. "Leave the Tight Tube forever behind. Then, and only then, will you reach salvation."

They reached Houchins River and followed it downstream. Where its waters pour into the impassable drain, Tom dumped in a bag of fluorescein dye. The bright orange crystals turned the water a brilliant green. Tom hoped to find that the water of Houchins River flowed into the new river he had found three days before.

At the A-survey junction room, they rested on the sandy floor at last to eat a meal.

"Oh," John said suddenly, having reverted to his usual quietness. He had finally figured out that the strange rock he was sitting beside was the jack that he had sent in with Roger's party a few days before.

They crawled down the A-survey to the Tight Spot. John passed through Roger's Chest Compressor on the way, and watched as Richard and Tom went through the Tight Spot. Pat then slipped through, and because there was not space for three in the tiny room beyond the Tight Spot, she and Tom immediately went on to continue the B-survey that Roger's party had started. Richard turned around and crawled back up to the Tight Spot to offer assistance to John if he needed it coming through.

No one had mentioned it, but from the beginning of the trip everyone had known that this would be the crucial moment. John was not so big as Roger, but he was bigger than anyone who had gone through the Tight Spot so far.

John stuck his right arm into the opening, reared his body up flat against the ceiling of the small room, and shoved forward into the Tight Spot. He wedged fast. It was no go. After a struggle, legs flailing above the floor in open air, John backed out. He sat for a moment, and then tried again. This time he put his left arm out in front, with his right shoulder wedged against a wall of the canyon, and his right arm held crossways below his body to keep from slipping down into the wedge-shaped crack. A protruding ledge from the other wall threatened to crush his chest. When he reached the ledge, he paused for a moment, exhaled, and moved a quarter of an inch forward. Then another quarter of an inch. He could feel the skin and hair scraping away from his chest. Another quarter-inch.

"I am maximum size for that passage," he muttered as he dropped into the pit beyond the Tight Spot.

"Yea, John!" Pat shouted back down the passage. John took out his survey notebook and started recording data while Richard explored the other leads out of the pit.

They settled down to what they knew would be a very difficult surveying job. There was no chance that they would exit from Mammoth Cave. They would continue the B-survey, and their reward would be in knowing where the new river was.

The passage twisted and split into a rat's maze on different levels. In 200 feet they set twenty-one stations, some with a distance of only a foot or two between them. In places the passage was so small that Richard and Pat wiped the soot-marked stations off the wall as they passed, leaving the trailing John to guess where the stations might have been for his notebook sketching. Once again, Pat's small size and contortionist's agility were crucial, for only she could have read the compass in that passage.

"What were those dimensions again?" John asked.

"Nonexistent!" Tom shouted back.

The passage is too low for you to stand upright, and too narrow for you to stoop. You go in sideways, shoulders first, legs angled behind for pushing. You avert the danger of falling forward by inhaling deeply. Several times John had to retreat so that his hand holding the pencil could be brought into proximity with the one holding the notebook.

Finally, they had to pass under a small waterfall which soaked the parts of their clothing that had remained dry until then. They passed through a chain of pools in a passage resembling the inside of sausages strung together. They crawled on their sides through a pool to emerge five feet off the floor of a small room. It was difficult not to fall head first into the foot-deep pool below. When the survey was plotted later, it was discovered that this little pool—at B-47—was only eight feet— through solid limestone—from the 160-degree bend at B-30. They had crawled ninety-one feet and set seventeen survey stations to get from the one point to the other. At another place, the passage corkscrewed under itself on its way deeper into the earth.

Tom knew that they were approaching the river, but he kept quiet, wanting to surprise them. He was the one who had been there before, and so far had been behaving like a tour guide.

Then they were there. It was a ten-foot-wide stream passage with water everywhere. Everyone peered to see whether the water had turned green. It had been four hours since they had dyed Houchins River, not very far away. The water was clear. Tom shrugged his shoulders.

It was a real stream, all right, with the water flowing about five cubic feet per minute, in the general direction of Mammoth Cave. However, Pat suddenly noticed that she was still crawling.

"Oh, Tom," she said in disappointment. "Is *this* your wonderful river?"

"Just wait," Tom answered earnestly. "Just wait. I promise you at least six hundred feet of good walking passage."

John peered upstream into an evil-looking side passage, a belly-crawl in the cold water. Fog steaming from their wet clothes was fast collecting, and they could see every breath. They hurried on, and around the bend the ceiling rose. Soon they were setting stations in walking passage nine feet high. They stopped at a tablelike mud bank to eat candy bars and to change carbide.

Now the survey shots stretched out to thirty or forty feet each. Everyone was so enthusiastic that the scene looked almost normal, as though they were re-surveying some routine passage in the known part of the Flint Ridge Cave System.

At B-69 there was a walking lead to the right, back toward Flint Ridge. Tom looked pleadingly at Pat.

"Don't be gone long," Pat said.

Tom ran up the passage 500 feet until he would have had to crawl, and then turned back.

"Wow! What a lead," he told them. "This is real Flint Ridge cave."

"We are under Mammoth Cave Ridge," John said without emotion.

As they set station B-76, Tom pointed into a lead departing from the left wall. "That goes to Mammoth Cave," he said seriously.

Richard squeezed on his side into the narrow canyon passage. He squirmed along for a few feet and then shouted back that it did not get any bigger.

"Come on back," Pat said.

They went on; then John announced that, by his calculations, they had just gone farther from the Austin Entrance than the distance from the Austin Entrance to Q-87. Now that rock pile at Q-87 was no longer the bitter end of the Flint Ridge Cave System. Pat led the way deeper under Mammoth Cave Ridge.

Pat, as leader, had been watching the time. They had entered the cave at 10:00 a.m. on the 30th, and now it was 9:00 p.m. It would take them at least six hours at top speed to get back to the Austin Entrance, and they were due out at 4:00 a.m. on the 31st. Pat decided to survey for another half-hour.

The passage was trending northwest now, and John believed that it was paralleling the axis of Houchins Valley, offset toward the Mammoth Cave side, but not getting any closer to Mammoth Cave. The final station was B-87. They had surveyed 1200 feet in about seven hours.

They put away the surveying equipment, changed carbide, and ate more candy bars. Then Tom led on down the river passage with the others close behind. Just around the corner was the mud bank where he had left his initials three days before, and 100 feet beyond that lay virgin cave. Tom looked back hopefully at Pat.

"We can explore for fifteen minutes, but no more, I think," Pat said.

Without a word, Tom turned to hurry on. He knew that this deep in the System it was good to stick to schedules, but he ached to explore. So did they all.

Just past Tom's initials, the passage split into an upper and a lower level. Tom took the upper tube, sloshing along on hands and knees for a few hundred feet. The pools contained sticks and blind white crayfish. Then the crawlway pinched down to a belly-crawl. The explorers put their elbows ahead and stiffened their arms, sledding along by pushing with their feet and knees. After 150 feet, the tube rejoined the lower passage containing the river.

Tom dropped out of the tube down into the water and crouched over, looking backward, waiting for Richard to follow. Then Tom turned to go forward, and saw something drawn on the waist-high mud bank in front of him. He felt a chill up his back.

"Hey!" Tom shouted. "An arrow!" His knees were momentarily weak.

The arrow was pointing the way Tom was going, downstream, *toward Mammoth Cave.*

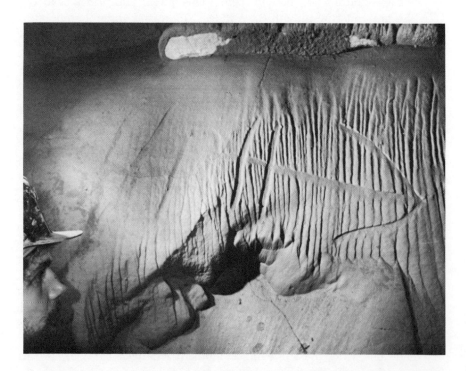

The initials of Pete Hanson and Leo Hunt on mud bank. The arrow points downstream.

Richard tumbled out of the crawlway and turned around to spy something scratched through the mud on the limestone wall above the arched opening of the tube they had just come out of:

PETE H

Pat and John were still in the tube. They could hear shouting up ahead.

"What's going on up there?" John asked.

As John and Pat came out of the tube, Tom and Richard grabbed them and jumped up and down.

"We've done it! We've done it!" they shouted.

"Pete Hanson!" Pat said, as soon as she understood. "He found New Discovery in Mammoth Cave, and explored up Roaring River."

John socked Richard on the shoulder, rather harder than he had intended. Richard was startled, then fiendishly returned the blow with interest.

Suddenly everyone was stunned to silence. It is incredible to come upon an arrow deep in the cave system in a passage you thought was virgin. They were too excited to notice it then, but later examination showed that besides "PETE H" and the initials "PH" by the arrow, there was

Cleve Pinnix and Steve Wells view the autograph of PETE H (Pete Hanson) scratched on the wall. Pete and Leo Hunt probably ended their exploration here in 1938.

another set of initials on the mud bank, "LH," for Pete Hanson's insepa-
rable caving companion, Leo Hunt. They must have explored the passage
from the Mammoth Cave side in 1938, just before World War II. John
and the others named the passage Hanson's Lost River.

"Are we in Mammoth Cave?" Richard asked.

"I don't know," Pat said. "It's ten o'clock."

"Well, at least we know what time it is," John said.

"How does everyone feel?" Pat asked.

Tom and Richard were hyperactive, ready for high-speed exploration.
John had to take Park Service personnel to survey in Mammoth Cave the
next day.

"I'll settle for one hour's sleep," he said.

"Okay," Pat said. "We'll explore another hour."

Tom and Richard immediately set off. Pat and John were close behind.
They went northwest for 1000 feet through knee-deep pools, bending over
in crouchways. There were chert ledges along the passage, just as Tom
and Pat had seen in base-level drainage passages in Mammoth Cave. They
saw blindfish. Then Tom saw a small brown fish that had eyes, and Pat
saw a brown crayfish. Only the white ones were eyeless.

"We must be close to the level of Green River," John said.

They searched for more initials and arrows on the walls, but found
none. The arrow they had seen had obviously been flooded several times.
They were going downstream, and were undoubtedly low enough now
that in high water Hanson's Lost River would be flooded to the ceiling.
Had any more arrows been placed, they would probably have been washed
out by now.

They went another 1000 feet. The passage turned left to the southwest.

"We'll hit Mammoth Cave broadside," John muttered.

Mammoth Cave, here we come!

On and on they went. The water was often waist deep now, sometimes
with only three feet of air space. In places they had to crawl. They estimated that they had gone about a mile from the arrow at "PETE H" when
they crawled out of a three-foot-high passage into a dry canyon. They
fired up their lamps and peered upward some thirty feet. It was very narrow, and would be a tortuous and dangerous climb.

Maybe Pete and Leo had come down from higher Mammoth Cave
passages there. Or maybe they had come in a surface sinkhole that was
now unknown, or even plugged. Maybe the stream flowed to an unknown
low entrance along the Green River. Hanson's Lost River might not be
connected to Mammoth Cave at all.

It was very late, nearly midnight. They had exceeded their hour. John
checked on ahead for 200 feet, and then turned back. The river went on.
They stood cold and shivering, five and a half miles from the Austin
Entrance, farther into the Flint Ridge Cave System than anyone had ever
gone from an entrance, maybe farther than anyone had ever been from

Mud and water in Hanson's Lost River.

any cave's entrance. How much strength did they have for getting back through the Tight Spot?

Pat took out a small copy of the Kaemper map of Mammoth Cave, but they knew that it was impossible to figure out where they were.

"We have to go back," Pat said.

There was silent agreement.

"Why don't we write a note to leave here?" Tom asked.

There was no reply. They all thought of Bill Austin's and Jack Lehrberger's initials that seemed so appropriate in Salts Cave and Unknown Cave. There was "PETE H" back there, and Floyd Collins' name in Crystal Cave, "T. E. LEE" so beautifully drawn in the entrance pit to Lee Cave, "ROY HUNT—TO THE RIVER" in Colossal Cave, and dozens of other signatures going all the way back to Stephen Bishop's in Mammoth Cave. They were conservationists who did not approve of writing names on cave walls. On the other hand, initials were the only history some parts of the cave had.

John increased the water flow in his lamp and the flame shot out three inches. He walked over to the wall, and high up on the light tan limestone he wrote with carbide soot:

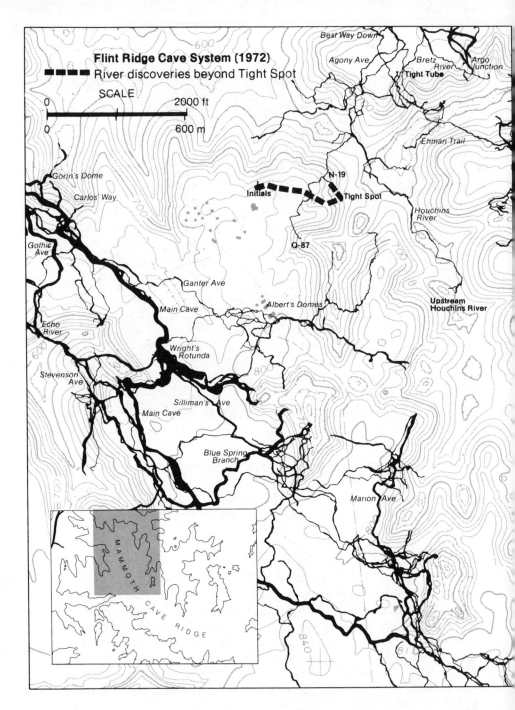

Flint Ridge Cave System (1972)
▬ ▬ ▬ ▬ River discoveries beyond Tight Spot

SCALE

| 0 | | 2000 ft |
| 0 | | 600 m |

Best Way Down

Agony Ave

Bretz River

Argo Junction

Tight Tube

Ehman Trail

N-19

Initials

Tight Spot

Houchins River

Gorin's Dome

Carlos' Way

Q-87

Gothic Ave

Ganter Ave

Albert's Domes

Upstream Houchins River

Main Cave

Echo River

Wright's Rotunda

Stevenson Ave

Silliman's Ave

Main Cave

Blue Spring Branch

Marion Ave

MAMMOTH CAVE RIDGE

Map of the Flint Ridge Cave System showing the location of Pete's and Leo's initials.

PPC   RBZ
TAB   JPW
8-30-72

It seemed much farther back to their end survey point, B-87, than it had on the way out.

"B-87 . . . Q-87," Pat said as they passed by, "the bitter end." They had gone more than a mile beyond them both.

At the Tight Spot, Tom and Richard shoved through without pausing. John found it difficult to climb up the eight-foot wall to get into position. When he got horizontal, there was nothing behind to put his feet against for pushing. Panting and weak, he spent five minutes making no progress.

Behind him, Pat fought down a bubble of panic. Could she slide through beneath John if he got permanently stuck? No. But now Richard had come back to talk John through. He took John's pack.

"I fit through once before, and I can do it again, but it may take some time," John said.

Good, Pat thought.

With his arms free of the pack, John got more leverage, and managed to relax and push at the same time. Almost imperceptibly his chest wedged through the tightest two feet of the Tight Spot, and then he was through.

"I wonder if I'll ever have to go through there again?" he asked. "I would do it if necessary, but not without dread."

Now they were on home ground. They sloshed through pools that they had laboriously avoided on the way in. When they reached the Duck-under to Swinnerton Avenue, Tom said the cave was spinning around him. Pat was limping with a sore knee and hip. John's elbows were rubbed raw. Richard, the devil, was in fine shape, rushing on ahead.

They left the Austin Entrance at 6:15 a.m. as the bats were flying in. They climbed the hill slowly, faces drenched in the golden glow of the rising sun.

"Mum's the word," John said quietly. He would have to think about how and when to announce this discovery.

Later, Pat wrote:

It's an incredible feeling, being part of the first party to enter Mammoth Cave from Flint Ridge. Something like having a baby. You have to keep reminding yourself that it's really real, this new creature you've brought into the world that wasn't here yesterday. Everything else seems new, too. After we wake up on Thursday afternoon I listen to a Gordon Lightfoot record. The music is so beautiful, it makes me cry.

The Tight Spot

Joe aborts the trip of a lifetime.
Everyone is worn out, but they advance not one foot
beyond the point of previous penetration.

28 News of the discovery of Hanson's Lost River and of the imminency of connection between the Flint Ridge Cave System and Mammoth Cave reached Joe Davidson almost by accident. Joe had been president of the Cave Research Foundation for five years, during which time most of his activities—and many of those of his wife, Betty, who was not a caver—were organized around the Foundation. After more than a year of agonizing, he had finally persuaded himself to turn over the presidency to Stan Sides, only two months earlier, in June 1972. Then Joe went mountain climbing for three weeks in Wyoming. After he got back and settled down, on September 1, Joe calculated with surprise that he had not been caving for nine months.

Joe phoned Roger to see about a cave trip. As he listened to the tale of Tom's discovery of the river (Roger did not know about "PETE H" yet), Joe thought that it was just like John not to have let him know. But then he remembered that he was no longer president of the Foundation. There was no particular reason for John to have called him. It was the old Flint Ridge blackout: If you were not on the scene of action, you never knew what was going on.

"They'll probably make the connection this week," Roger said.

Joe arranged to have Roger pick him up at the Louisville airport on Friday night. They arrived on Flint Ridge at about 1:30 a.m. on Saturday, September 2. The entire camp was dark except for a lone light in the kitchen of the Austin House. They were startled by a Floyd Collins-like apparition that materialized between the parked cars. It was John Wilcox.

John walked over to the car noncommittally, but then he could not keep from breaking out into a broad grin.

"You've done it, haven't you?" Roger said at once.

Wordlessly, John led them to the Crystal Cave Ticket Office. There he took out a new plot and superimposed it on a map so they could see.

"The passage has been explored for another mile," he said. "It turns to the southwest, straight toward Mammoth Cave."

Joe began to object that it was not enough. There had been surveys under Mammoth Cave Ridge before, but no connection had resulted. John raised his palm to cut him off. Joe recognized that smile. John went on quietly to tell the story of finding the arrow and "PETE H." He warned that the secret could not be kept for long.

Roger said that he knew that this was the week.

Joe was excited and full of ideas about how to handle the affair. The Superintendent must be informed. Publicity must be handled effectively. It would be an opportunity to speak for wilderness designation for the caves of Mammoth Cave National Park, to point out that the Park was understaffed, to work for conservation, and to publicize the work of the Cave Research Foundation. Who could be brought to the park to make the announcement? The Secretary of the Interior?

But Joe was no longer president of the Cave Research Foundation. He lapsed into a reverie about the irony of the connection being made so soon after he had stepped down. And he had not been on any of the recent hard trips that led up to it, when once he had led all the hard trips.

"Joe, you'll be the fourth member of the party," John said.

Joe was brought up sharply. He had not been listening to what John had been saying.

"Where?" Joe asked.

"To continue the B-survey in Hanson's Lost River," John replied.

"I'm really not in shape for it, John," Joe protested. "I haven't been caving for nine months, and I'm just back from a climbing trip. The first cave trip after the mountains has always been bad for me."

"Well, if you can't go, we won't send a party out there. There just isn't anyone else for the fourth member," John said.

Joe did not hesitate long. He did not want to go, but his commitment had long ago gone deeper than personal desires.

"Okay, I'll try. I know I can get to Houchins River and back okay. It's what's beyond that bothers me," Joe said. John had already told them about the Tight Spot.

Joe immediately gathered up his gear and went into the bunkhouse to get some sleep. He tossed and turned, and was up at 8:00 with the others.

John, Joe, and Richard were ready to go to Hanson's Lost River from the Flint Ridge side, but the fourth member of the party, Gary Eller, had awakened early that morning with a headache. This was unusual for him, but even then he would not have bothered about it had he been going on an ordinary cave trip.

"Give me an extra half-hour," Gary pleaded.

There was little else John could do. There are few far-out cavers who have not felt the strain and ambiguity of both wanting to go on a long, dangerous trip, yet somehow not wanting to go. Gary walked about, relaxed, and the headache mercifully disappeared. Still, he managed to put only two cans of meat into his pack although on trips of such length he always took four, and he put in a can of date-nut roll, which he detested.

It was 10:30 a.m. before the party went through the Austin Entrance into the Flint Ridge Cave System. Richard Zopf raced far ahead, and John Wilcox roamed the middle distance. Joe Davidson and Gary Eller brought up the rear, talking a continuous stream of Cave Research Foundation politics.

Breaking the precedent he had established on the other trips, John called for a rest stop at the beginning of Swinnerton Avenue. They had not even reached the Duck-under where the crawling started. Joe was irritated. He had just begun to feel good, and he hated rest stops. Joe supposed that John thought a rest would make him feel better, for John and the others certainly did not need a rest.

Then they rested only 2000 feet farther on while putting on their knee crawlers at the Duck-under. Joe was still chafing about this when they reached Candlelight River. There John dived into the canyon at a point that was strange to Joe.

"Where you going?" Joe asked.

"This is a better way than we used to take," John replied. "Drier."

"Yeah? Well, the only dry ways I know are awfully hard on the arms," Joe said. However, he followed as the pace slowed.

Joe reflected that newcomers waste a lot of energy because they do not know what to expect next. In the unfamiliar dry crawlways Joe felt like a newcomer, wasting energy.

Despite the rest stops, they reached the sandy-floored A-survey junction room in four and a half hours, which was as fast as it had been done so far. They settled down for a meal. Gary heated a can of beef stew over a small heat-tab stove.

"Beef stew causes gas as bad as beans," Joe said.

"Humpf," Gary replied.

"What is all this trash around here?" Joe went on, waving at the jack, sledgehammer, longbar, and chisel. "It looks like a junkyard."

"You seem to be your old self," Gary said cheerfully.

Joe winced. It was the same cheeriness that he had long been notorious for himself.

"As a matter of fact," he said quietly, "I feel surprisingly weak."

They closed their packs and moved into the A-survey passage. Joe could not remember exactly what it looked like at the end, although he had been on the party that surveyed it years ago.

Gary's mind was working rapidly. He focused on the leads at the end of the A-survey that the old-timers had not pushed, but that Pat Crowther had gone through to find the Tight Spot. Obviously, the A-survey passage boomed on where those guys had stopped. So who had been on that party? How long, and where had they been in the cave? Was the lead man actually too big to get through the Chest Compressor, or had psychological factors discouraged his even trying? Was this generation, Gary's own, more aggressive in exploring leads than the previous ones? Or had the story that the passage terminated just arisen on its own?

The last was closest to the truth. This passage—*the* connection passage—had been lost because over the years, in making up lead lists, the cartographers had failed to notice or had failed to transcribe that the survey booklet showed a going lead at the end of the A-survey. The passage had been explored first by Dave Jones, who said it pinched. John Bridge crawled in to double-check, and decided that it could be pushed. Joe Davidson had gone in later to survey it, and now he remembered the trip in detail. He had been training new recruits, and had had another objective. So when the passage got tight, he had surveyed back out. He had failed to put a note in the survey booklet that the passage went on, although his sketch showed that it did. So if John Wilcox had not been determined to look at every possibility himself, that lead might still be lost.

The A-survey was certainly well traveled now. Richard and John went on through the A-survey to the Tight Spot. John seemed to have made it through the Tight Spot without much difficulty, so Gary ceased worrying about whether Joe could make it. Joe was a little taller than John, perhaps, but certainly no bigger through the chest.

Joe was following John and Richard closely.

"Is this the Tight Spot?" Joe asked. "Oh, yes, I see now."

Joe was thinking about the fact that it had taken John about five minutes to get through that fifteen feet of passageway. Joe watched closely as Richard reached up, leaned forward into a horizontal position, and then all he could see was Richard's feet in his face. He wondered what Richard was doing to keep from wedging down into the deep, tight V of the Tight Spot. But then Richard's feet went out of sight around a corner. His legs would not bend backward, so his heels scraped the wall hard as he went around. That worried Joe, for he was four or five inches taller than Richard.

Richard was through. It was Joe's turn at the Tight Spot.

John had given precise directions: "Stay high and on your *right* side facing the left wall. Keep your left arm ahead and up on the ledge against the ceiling. Pull only with your right arm, underneath."

Joe shoved into the slot on his right side with his pack in his right hand. He had difficulty keeping his feet up horizontal so that they would not wedge down in the slot. He pushed into the Tight Spot up to the corner.

Joe's arms started shaking. The pack was much too heavy, but Richard was there just ahead. Joe tossed the pack up to a point where Richard could crawl in to retrieve it.

Joe's arms were shaking again.

Joe's legs jammed. Richard repeated John's instructions. Joe shook his head. He would have to back out. He shouted back to Gary, who was behind him.

"Gary, pull on my feet and *lift*."

Gary lifted Joe's feet, and pushed hard.

Joe was trying to back up, Gary was trying to push him forward, and Richard was mouthing platitudes.

"Okay, okay!" Joe shouted. "I've got to rest."

Gary stopped pushing and Richard mercifully shut up. The only sounds were John moving around up ahead and Joe's heart pounding. In a few moments the claustrophobic attack subsided. However, Joe's arms were drained and shaking. Today, he decided, he must not go through the Tight Spot. It was comforting to him to know that Roger had not made it through the Tight Spot, either.

Time is important, Joe thought. I must get back out before I get too weak, but I must rest some first.

Richard started repeating those damnable instructions again.

"I just don't have the strength to get through," Joe said.

"If I got through, you can," John yelled back.

"Probably," Joe replied, "but not today."

"We really need you," John said.

"John, I just don't have the strength," Joe said tonelessly. "Even if I could get through, I couldn't get back out after more caving."

"Okay," John said. He watched as Joe raised on his shaking arms again to try to push backward.

"Oh," Gary said, "you wanted me to pull."

Joe knew that this would probably seem funny later, but now he was gradually slipping down into the crack below the Tight Spot. His knees would not bend and he could not get any leverage with his arms. Gary stretched out into the Tight Spot just far enough to grasp the toes of Joe's boots.

They struggled and steamed. Joe moved back an inch. It was excruciating, but he was getting out. After ten minutes—and a total of thirty minutes in the Tight Spot—Joe was safely out, exhausted and sweating profusely.

"It wasn't a matter of size alone," he repeated, "but more a matter of strength. Richard, throw my pack back to me. You guys go on ahead and I'll wait here."

Gary started to protest. He had been a Flint Ridge caver for three years, and the gospel was that you never leave any caver alone for very long, particularly when he is not fit.

John shut Gary off. "Fine," he said. "We'll return in two hours."

Now it was Gary's turn to try the Tight Spot. First he retrieved Joe's pack from Richard. When Gary got a good look at the Tight Spot, he thought, Oops, this is the real thing. Gary got along fine until one of his knee crawlers jammed. He spent several minutes getting unstuck, and then pushed on through the Tight Spot easily.

However, being five foot eight and 140 pounds also has its disadvantages. When Gary arrived at the edge of the eight-foot-deep pit, he could see no way to get down. John said to traverse it by leaning across, but for a person Gary's height that would mean stretching across almost horizontally. Gary decided to inch around one side. Then suddenly he found himself holding loose in his hand a handhold that had been attached to the wall when he had grabbed it. Falling would not kill him, but a sprained ankle this deep in the cave would be crippling. What about a broken leg? Gary decided for the less elegant but certain method. He grabbed the hand that John had stretched out toward him, and was pulled across the pit.

Joe listened to the fading splashes in the distance, and then prepared himself for keeping warm and comfortable. He found a flat rock, carefully placed his gloves on it, and sat down on them. Then he tucked his red kerchief around his neck, buttoned his coat up to the top, put his carbide lamp on the floor between his legs, and held the bottom of the coat forward to catch the heat rising from the lamp.

This was a Palmer furnace, a trick Art had taught the Flint Ridge cavers for warming the body and quieting the mind.

All was silent now, and Joe's thoughts drifted. He thought of how Red had been afraid of the narrow crack in Lee Cave after having been away from caving for a year. Joe had thought it silly at the time, and after going through the crack Red had agreed that it really was not much. Why had Red been frightened? Well, this trip was for Joe a lot like Red's Lee Cave trip.

Here I am, Joe thought, out of shape and in a spot that requires high strength. So I don't make it. Thinking about the Tight Spot, he could remember nothing that seemed impossible, nothing that a pair of good strong arms would not cure.

Joe drifted off to sleep.

He woke with a start, an acrid hot smell in his nostrils.

"Ye gods, what's burning?" he asked. Then he saw his lamp. He had been holding it in his hand when he fell asleep, and now the flame was against his pants leg.

Joe wondered what the others would think about his holding up the party and aborting the trip. He rationalized that it might be good for the Foundation if the old tyrant were to fall flat on his face. To many of the Flint Ridge cavers, Joe's inability to get through the Tight Spot would be more significant than his turning over the presidency to Stan. Ineptness of the old guard is a great stimulant to the aspiring young.

But it's not the end, Joe thought grimly. A lot of cavers had finished their caving careers out here beyond Candlelight River. Joe swore that it would not happen to him. He would be back.

Joe's lamp had burned too low to provide warmth, so he changed his carbide. He figured that John and the others had been gone two and a half hours. This did not bother him. They had probably found something. Joe drifted back to sleep.

He smelled smoke again. This time he had a hole in his coat. Now Joe sat up, wide awake. He thought of Roger. He found himself understanding Roger better all the time. Back in the 1960s it had been partially his belief that Roger was inept that had made Joe push forward. Roger would fall asleep whenever a party stopped moving, and there were spots he would get stuck in where others could get through. Now, Joe thought wryly, he fell asleep himself. And today he could not get through the Tight Spot that had stopped Roger a week ago.

There was the faint sound of water sloshing in the distance. Then silence for a long time. Finally, more sloshing closer in. The others were back.

"Did you find a way around?" Joe hoped they might have found some way to bypass the Tight Spot.

"No," John replied. "We looked out a number of the wetter leads since we weren't to be very long. We're worn out from the cold water."

They had been gone three hours.

As they started back through the Tight Spot, Joe turned, not waiting for them, and crawled toward the A-survey junction room.

Gary was still bright-eyed and bushy-tailed. John had taken them all the way to "PETE H." Every time Gary saw a blindfish he whooped with delight. At Pete Hanson's signature, Gary thought he sensed the vibrations Tom Brucker had told everyone about. He fretted about having to turn around in this booming passage because of lack of time. They had to turn back because Joe was sitting alone on the other side of the Tight Spot.

Nevertheless, John and the others pushed into the side lead that Tom had explored for 500 feet. They took it another 1000 feet back toward Flint Ridge. There was a scattering of sherds of blue pottery in the gravel of the stream bed. Surely this was close to the surface. But they had to turn back before the passage ended.

The party sped out of the cave in five and a half hours. If John was disappointed, he hid it well from the others. Already he was turning over possibilities for parties on the morrow. He had to face the probability that they would not connect the caves on his expedition.

Gary continued to wonder whether Joe could have made it through the Tight Spot. Could he have done it had it been nearer the entrance—say, in Pohl Avenue, or even in Candlelight River? Or on another day? Had adverse psychological conditions been a factor? Gary could not decide. All he knew was that it had been impossible for Joe to make it through the Tight Spot that day.

There was a little banter about Joe's big shoulders, but, as usual, no one in camp even suggested that he should have tried to go on. Joe was content. This was what he had always preached, that one should be open and unembarrassed. Now he was experiencing how important the under-standing acceptance of this openness was to someone who had aborted a trip. No one blamed him, but he felt awful, anyway. Because Joe could not get through the Tight Spot, they had not advanced one foot beyond the farthest point of previous exploration.

"Go up Roaring River, Boys"

Explorers return to Mammoth Cave to see if Pete Hanson had followed
Marty's advice. Perhaps the river containing Pete Hanson's
name that was found from the Flint Ridge side is a tributary of
Roaring River in Mammoth Cave. Explorers fail to find it.

29 For many years Marty Charlet, a retired Mammoth Cave chief
guide, resided in the old Mammoth Cave Hotel. He liked to
talk about exploring in Mammoth Cave, and he always ended
with the admonishment, "Go up Roaring River, boys, go up Roaring River.
That's where you'll find big cave."

Pat Crowther and John Wilcox had been on a wet-suit trip up Roaring River on Monday, and they had noticed several unexplored leads at
water level in the north wall. Maybe one of them was the outlet of Hanson's Lost River. There was no hope of sending another party out from
the Flint Ridge side on Sunday, the last day of the expedition. The Saturday bash on which Joe had been stopped by the Tight Spot had burned
out the strongest cavers. There was still the second half of John's plan,
to send a party up Roaring River in Mammoth Cave to look for the outlet
of Hanson's Lost River. However, he had no people for the trip.

On Sunday morning, September 3, John Wilcox told John Bridge his
troubles. Bridge had been in Seattle at the National Speleological Society
Convention, and had swung down to Flint Ridge for a day on his way
back to urgent work in Ohio. When he detected what was going on, he
stayed. That made three doctors of mechanical engineering on the scene.

Now Bridge puffed on his pipe, thinking of the leads in Roaring River.
He no longer went beyond Candlelight River, but maybe he could be in
on the connection, after all.

"I might be able to go up Roaring River," John Bridge said to John
Wilcox, "but I don't have a wet suit."

"You can use mine," John Wilcox said.

The trouble was that Bridge was six foot four and Wilcox was only
six foot. Bridge was also somewhat wider in the middle than Wilcox.

"I'll go if I can get into your wet suit," Bridge said.

"Do you think you possibly could?" Wilcox asked.

"Well, I don't suppose," Bridge replied, pulling dejectedly on his pipe.

Later, Wilcox looked up in surprise from the reports he was writing. Bridge stood in the doorway, looking like an overstuffed sausage. The black neoprene of the wet suit appeared ready to burst in a dozen places, and a broad strip of Bridge showed around the middle, but he was definitely inside it. Peering from behind him with a big grin and an innocent look was Pat, with her red caving bandana over her hair.

The trip up Roaring River in Mammoth Cave was on.

Two eager people had anticipated the push up the rivers from the Mammoth Cave side, and had brought wet suits for the trip. Ellen Brucker was eighteen, long-legged, with a pert and pretty face under soft blond hair. She was a good rock-climber, and, as befitted Roger's eldest daughter, she was a tough and experienced caver. She and Richard Zopf had been spending time together recently, discussing the caves. He was ready to go, too.

John Bridge, Pat, Ellen, and Richard discussed the possibilities on their way to the Historic Entrance of Mammoth Cave. If Hanson's Lost River headed southwest, it should come into Echo River. If it headed south, it was a tributary of Roaring River. Everyone was under the impression that Jack Hess's survey parties in Echo River had not found any leads, so they were on their way to Roaring River. They were to confine the search to the reach of Roaring River extending from just beyond the divide at Cascade Hall to the first New Discovery boat dock.

The last tourist had left Mammoth Cave three hours before, and no one was around to comment on the quartet carrying wet suits past the Mammoth Cave National Park Visitor Center. However, by now the cavers thought nothing of going into Mammoth Cave. Insects buzzed and flitted around the lamps of the deserted Historic Entrance, and a faint haze distorted the view where the moist cave air collided with the warm night.

They walked down to River Hall. Here, where Stephen Bishop had begun exploring 134 years before, John Bridge again performed his incredible stuffing feat.

"It's somewhat tighter than my skin," he remarked as he rammed his bulk into the neoprene wet suit. "I can sit down in my skin."

Tourists were no longer taken on a boat ride, so a wire fence blocked the passage to Echo River. The cavers in their black wet suits climbed the fence and picked their way along silt-covered walkways. They slid quietly down the bank of Echo River into the dark water, braced against its chill. It was not really very cold, only 54° F, but it looked oily, viscous under the sharp yellow beams of their carbide lamps. It reflected the dripping ceilings and black darkness up ahead. After swimming 200 feet, they climbed up the far shore past stacked aluminum boats. A narrow causeway of rocks spanned a smaller pool of Echo River. A metal-pipe railing extended along the trail. The cavers crossed it and then climbed a sandy slope into Cascade Hall.

*John Wilcox spans with his fingers from PETE H
in Hanson's Lost River toward Roaring River in
Mammoth Cave. Left to right: John Bridge, John
Wilcox, Tom Brucker, Gary Eller, Pat Crowther,
and Richard Zopf (foreground), 1972.*

Walking awkwardly in the wet suits, they turned out of Cascade Hall, over a small divide, to descend to Roaring River. In the water again, they began to check leads to the north—small, low passages that might carry a tributary stream from Flint Ridge.

The explorers paddled along backward on inner tubes. Blindfish a few inches long hung nearly motionless as the explorers passed. Carbide, candy bars, matches, and surveying equipment were sealed in polyethylene boxes inside their packs.

They floated into the first lead for 300 feet. Breakdown blocks lined the walls, and then the water disappeared beneath a pile of breakdown. The passage continued above as a canyon, two feet wide and fifteen feet high.

Richard, awkward in his wet suit, launched himself into the canyon. The floor was of mud that had slumped here and there to create hummocks six to eight feet high. Richard wished he had a trenching shovel to cut steps in the slippery slopes. After forty minutes, he had advanced only fifty feet. The passage went on and he could hear a waterfall, but Richard turned back.

The next lead was lower. Pat checked it. She had to keep one side of her face in the water, because chert beds in the ceiling left only eight inches of air space. After 200 feet she realized that she was trapped between the ceiling and her inner tube. Had she been in the water without the inner tube, she could have gone on to where the passage nearly siphoned, with only two or three inches of breathing space. She kicked and scraped along the ceiling, backing out. She wondered how far Pete Hanson and Leo Hunt would have pushed such a passage.

By then, Pat was shivering and her lips were blue. Her wet suit was sleeveless and only half as thick as those of the other party members. She resolved to buy a real wet suit.

At the next dry spot, Pat's hands would not work, so Richard changed her carbide for her. She managed to tear open a candy bar and ate it from her fist. She was shivering violently.

Nevertheless, Pat followed the others to swim down a long passage with no leads. Then she floated in the main passage while Ellen and Richard crawled through a 250-foot lead, five feet above the water. It ended in a mud fill. Then they found another way into the same lead. The next one was a cut-around. They recognized a lead that John Wilcox had described looking into on Monday, so did not re-check it. Then a short tube led to a siphon.

They swam on, the cold very painful for Pat now. They came to a complicated junction of passages that finally clicked into focus for Pat as the *second* New Discovery boat landing. They were supposed to have turned around at the first one.

"Damn!" Pat said. "The first landing must have been at that dry spot where I began feeling really cold. Turn around."

On the way out, Richard plunged into some very low leads they had saved. All of them siphoned. At last the shivering cavers climbed out on the bank in Cascade Hall. They had not found the downstream end of Hanson's Lost River. They would have to report that prospects looked very dim in Roaring River. As they walked past the Cascade Hall causeway over Echo River, the iron-pipe handrails of the old tourist trail gleamed forlornly in the light of their carbide lamps. They climbed the slope slowly.

The expedition was over.

Leak!

Tom lets the secret out, forcing John's hand.
John must connect the Flint Ridge Cave System with Mammoth Cave
while the rumor spreads that it has already been done.

30 They had found their way into a river deep under the middle of Mammoth Cave Ridge, but the trouble was that they could not say with certainty that they had been in Mammoth Cave. Pete Hanson could not tell them, for he had died in Alaska during World War II. Leo Hunt was dead, too, and no one remembered any stories of their exploring a long river passage northeast in Mammoth Cave. There had been too many disappointed hopes raised over the years for even an arrow and initials on the wall to be enough. Cavers had to go through the connection underground, from a Flint Ridge Cave System entrance to a Mammoth Cave entrance, before it could be said that the connection had been made. The best thing to do was to try to keep quiet until Hanson's Lost River was explored and surveyed all the way.

However, John Wilcox had taken Greer Price aside and told him what they had done. Greer had watched on the surface, and John counted him as much a member of the team as any of those who had gone underground. John cautioned everyone to keep the secret.

Late that morning of August 31, 1972, Tom Brucker opened his eyes slowly and peered at two figures standing in the midday light filtering through the dusty bunkhouse air. He felt vaguely disoriented, but the memory of the arrow in the mud and of "PETE H" on the wall was vivid. It was less than twenty-four hours since Tom had found that arrow.

He had been awakened by two caving friends from Indiana who were passing through.

"Guess what we found," Tom said to them.

They had no idea.

"We've made the connection," Tom said. They obviously did not know what he was talking about, so, excitement spilling out in his words, Tom told them about the discovery of the river, the arrow, and the initials.

" 'PETE H,' " he said. "That's Pete Hanson, an old Mammoth Cave guide who explored a lot. We're in Mammoth Cave!"

From a bunk across the room there came a choked cough. Richard had awakened.

"We've done it!" Tom said, as the Indiana cavers became animated with understanding.

Richard coughed very loudly this time, and Tom gazed at him with a worried expression. Richard shook his head in a flurry of violent disapproval. Tom continued to be puzzled for a moment, and then his face fell. "Oh," he said, "I wasn't supposed to tell, I guess. We don't want it to get out just yet. Can you keep it quiet?"

The Indiana cavers agreed to keep it quiet, and left the room.

"We agreed not to tell," Richard hissed. "Did you forget?"

"I guess so," Tom said. "I guess I shouldn't have told. But they said they'd keep it quiet."

However, the news was out, and cavers in camp began whispering about some sort of big discovery.

On Saturday morning, September 2, there were four of the nine CRF directors in camp. Besides, John, Roger Brucker, and ex-president Joe Davidson had arrived the night before, and Stan Sides, the new CRF president, had driven in that morning with his wife, Kay, and their two young sons. John had told Joe and Roger about the discovery, and Stan had heard rumors about it as soon as he arrived in camp. The four directors met for a few minutes and decided quickly that there was no point in trying to bottle up the news any longer from their own people and the Park Service. John was torn. He had not heard that the story was out until that morning, and he was alarmed and furious. It would spread coast to coast among spelunkers within days, if it had not already. It was the caving news of the century, and now because of Tom's indiscretion, there was no time to plan how to release it. On the other hand, John had worried that morale would drop among the hard-working CRF cavers if the news were kept secret from them. They certainly had a right to know. Well, the tension was too high now. Everyone in camp knew something big had happened anyway.

Roger was dispatched to the nearest pay phone to call the two other ex-presidents of CRF, Phil Smith and Red Watson.

Phil answered the phone absently, but immediately became alert.

"I'm tremendously impressed, Roger. That's wonderful," Phil said. He listened to the story, and then asked detailed questions. Roger belatedly remembered Phil's casual attitude about long distance calls, and wished he had reversed the charges.

"Give them a 'good job' speech for me," Phil said.

Roger hung up, bemused with Phil's love/hate relationship with Flint Ridge. The Cave Research Foundation had been mostly Phil's creation, and as its first president he had lived and breathed it for nearly six years. It had been difficult for him to give it up, but for the permanence of the organization the presidency had to rotate. So Phil had bullied Red into taking the presidency, had pretty much dropped out of active caving himself, and ever since when visited by any of the old cavers would greet them with, "Well, I suppose we have to discuss that damned cave."

Roger did not give Phil's "good job" speech. Most of the people in camp had never met Phil, and many of them would not know who he was. Phil, as an administrator, would have approved of Roger's decision. It still might have hurt Phil a little to face it.

Then Roger telephoned Red.

"I warned you the connection was imminent," Roger said, still smarting because Red had not come to help him in July.

"Yes, yes, I know," Red said, "but you've said that every year for the past fifteen."

Red sounded glad to hear the news, but said he did not know whether he could come to Kentucky right then. Roger thought he detected an edge of sour grapes in Red's voice.

"Well, we'll see," Red said.

When Red told Patty Jo, she said, "Of course we'll go." Within a couple of hours they were on their way to Flint Ridge.

After everyone had eaten breakfast, Stan whistled for attention. The cavers gathered in a circle. Twisting the mustache he had grown the year before, when he was a conscripted medical officer with the U.S. Marines in Vietnam, Stan told the story with forced calmness. Then John told it again. They cautioned the group that silence was essential until the Park Superintendent made a public announcement. That might not be for weeks, so the news would have to be kept for a long time inside CRF, preferably among only those CRF cavers participating in this expedition. Then Stan left immediately to tell Superintendent Joseph Kulesza, smiling at the thought that these Kentucky woods have big ears, and he might already have heard.

Superintendent Kulesza was enthusiastic on hearing Stan's news.

"Yes, when the two caves are connected they will come to something like a hundred and forty-five miles," Stan said.

And with all the passages remaining to be surveyed, it would soon be 150 miles. Yes, Mammoth Cave would soon be known to be honestly 150 miles long, just as the Park Service and previous owners had been claiming for nearly that many years.

"What will we call it?" Red asked as soon as he arrived. The discussion was heated, but it was finally decided that it should be called the Flint Mammoth Cave System. The Flint Ridge Cave System was the longer of the two, and the connection had been made from the Flint Ridge side.

"You'd better call it the Central Kentucky Cave System," Bill Austin said, when he heard. "You know that otherwise everybody will just call it Mammoth Cave."

Red agreed that the Park Service and the general public would doubtless call it Mammoth Cave, but he still argued that it ought to be named according to international cave-naming conventions: The Flint Mammoth Cave System.

Superintendent Kulesza came over to the Flint Ridge camp and spent an hour chatting about the impending connection. Everyone seemed to think that it was a sure thing.

Everyone except John. The expedition had taken an enormous amount out of him. He had spent many hard hours underground, and on the surface he had barely slept at all. The uncertainties had not diminished with the discovery of "PETE H." Instead, the agonies of those uncertainties had increased a thousandfold.

John's frown deepened as the affair slipped from his grasp. Everyone seemed to forget that there was no connection yet. There had been false alarms before. It might be years yet before a connection was made. Nobody knew where Hanson's Lost River was. It might not connect with Mammoth Cave at all. That was John's job. He would get on with it, and leave the politics to others.

John went into the Crystal Cave Ticket Office and rolled out the maps. He used his little finger and thumb to span a measure of approximately 600 feet. Putting his thumb on point B-87 in Hanson's Lost River, John swiveled his hand up to the northwest and then down to the southwest. He ended at a point equidistant—about ten inches, or 1000 feet—from Echo River and Roaring River in Mammoth Cave. Hanson's Lost River was probably a tributary of one of those Mammoth Cave rivers. Where *were* they? John did not know.

There was only one thing to do. John resolved to find out where they were in Hanson's Lost River at once, to connect the caves swiftly, and to get it over with. Tom had forced his hand. John had to connect the Flint Ridge Cave System with Mammoth Cave while the rumor spread that it had already been done.

The Final Connection

John Wilcox, Pat Crowther, Richard Zopf, Gary Eller,
Steve Wells, and Cleve Pinnix make the final connection between
the Flint Ridge Cave System and Mammoth Cave.

31 John Wilcox drove back to Columbus on nervous energy alone after the 1972 August expedition. He had just completed one of the most strenuous and productive weeks in the history of Flint Ridge caving. He had directed the expedition, taken two trips far beyond Candlelight River, and led several lesser trips in the caves, with practically no sleep at all.

"If I'm a walking zombie, what is this?" Roger Brucker asked.

John could not slow down. He had put in motion a course of events that was now sweeping him along. Hanson's Lost River had to be followed to its end, soon, either in a siphon or in known Mammoth Cave passages. John's mind was filled with details, doubts, plans, hopes, fears, and . . . confusion. He needed sleep.

The news of the discovery of "PETE H" would soon reach far beyond Kentucky. John Wilcox dwelt on one thought: They would have to complete the connection quickly if there was to be a formal announcement before the story appeared in the *Louisville Courier-Journal*.

"Sure, go ahead and wrap it up," Stan Sides had said.

"There will be no consideration of releasing any news until and unless the caves are connected and the connection surveyed," Superintendent Kulesza had announced.

"If it were me, I would . . ." Roger Brucker went on unnecessarily.

"We've already done it," Tom Brucker said flippantly when John made a casual check to see if Tom would be available for an attempt to make the connection the following weekend. Tom's classes at Oberlin College were starting, he had a date, and he had been caving a lot that summer.

John was already annoyed at Tom for letting out the news, even though Tom had done it when he was groggy with exhaustion and lack of sleep. Now Tom's attitude galled John. John did not press. Tom was a superb caver. No one but John himself was as hard a pusher as Tom, and probably Roger was the only Flint Ridge caver who knew the total cave system as well as Tom did now. Caving with Tom was fun, and he was a highly sought-after party leader. It was people like Tom who had made Flint Ridge caving—and the Cave Research Foundation—go in the past, and who would make it go in the future.

However, if Tom did not understand that the Everest of speleology was not yet climbed, John was not going to tell him. He would only show him what work remained—an old Bill Austin technique—and let him draw his own conclusions. Tom's discovery of the arrow in Hanson's Lost River *had* shown the way, but someone still had to go all the way. Tom, for all his strength and experience, was just nineteen years old, and he had shot off his mouth when he should have kept quiet. And Tom shrugged off trying for the real connection. So be it, John thought.

No more expeditions were scheduled until late October, so John decided to go back to his original scheme of remote support. On Wednesday, September 6, John had decided that he would try to field two strong parties from the Flint Ridge side on the coming Saturday, September 9, 1972. If he could connect by then, the announcement might still be made in time to beat the rumors to the newspapers.

Furthermore, haste was imposed by the weather. The Green River water level had not been so low in years. In normal times it might be high enough to flood the connection route. If they waited and it rained, it might be a year or more before the water was so low again. Even now the water level might not be low enough, and at the lower end of Hanson's Lost River they might find an impassable siphon. If that happened, they would have to try that tight and dangerous climb where the canyon was perhaps thirty feet high, and they would have to explore all the side leads. That would take months, or even years, if John burned out his best people on this next trip.

Superintendent Kulesza was interested, and said he would monitor the level of the Green River and telephone the Corps of Engineers at the dam upstream to keep them from releasing more water. All right, one man was enough for surface watch. This time John decided that it might as well be the Superintendent of Mammoth Cave National Park.

Joe Davidson was also concerned. He had thought a lot about the connection after returning home. He picked up the telephone and dialed Stan Sides, who was in medical residency in Lexington, Kentucky. Joe felt somewhat embarrassed, remembering how annoyed he had been when ex-president Red Watson used to call him up to give him orders. Now ex-president Joe Davidson was on the phone.

"I think you ought to hold off making the connection until the October expedition," Joe said.

"Why?" Stan asked.

Joe discussed safety, National Park Service politics, and the need to have many experienced people on the scene. Then he remained silent. Although he had his convictions, Stan had to decide.

"Oh, I think it will be all right," Stan said. "We'll make sure we're adequately prepared to bail them out if we have to. I'll be at Flint Ridge early Sunday morning when they're supposed to come out of the cave." He waited. It was Joe's nickel.

Joe finally said, "Well, okay. You're the president."

Other people had thoughts similar to Joe's. About this time Stan received a wistful postcard from Jack Freeman, who was to direct the 1972 Thanksgiving expedition. "Don't make the connection just yet," Jack pleaded. "Wait 'til Thanksgiving."

John had to select the party, provide for emergency backup, clear the trip with the Park Service, study the maps and trip reports, and double-check everything. His plan was to put six explorers in the upper reaches of Hanson's Lost River through the Austin Entrance from the Flint Ridge side. That way, they would be surveying a known going lead, downstream toward Mammoth Cave, rather than searching for an unknown lead into the reaches of Hanson's Lost River that might or might not be found from the Mammoth Cave side. With a party of six, a productive trip was assured, for if one or two could not squeeze through the Tight Spot, the others would still be able to run with a strong river party. If all six made it through, there would be two parties for a leapfrog survey. John made a note to take *four* compasses. Each team would have a spare.

John alerted the backup team by telephone.

"Sounds as though yer knockin' on somebody's back door," Pete Lindsley said from Texas.

John was not telling anyone exactly what was going on, but he could not resist replying, "Yep."

Bill and Sarah Bishop in Albuquerque, Jack Freeman in New York —the scattered cavers would stay close to their telephones that weekend. All could reach Kentucky within eight hours after being called.

The probability of needing a rescue operation seemed remote, but fall was setting in. There could be rain. They had to count on the continuation of dry weather. The water was now abnormally low. Good cavers were ready to go. It might be now or never.

And even if it rained, a river rise might not trap them. Perhaps if they climbed up that untried thirty-foot canyon in Hanson's Lost River, they would find their way into known passages. On the other hand, a one-hour rain of several inches on a local watershed above might snuff them out in a matter of minutes. It had happened before, close by in Indiana, to other cavers.

If an explorer were merely injured beyond the Tight Spot, rescuers would be needed, either for an evacuation attempt or for setting up an underground base for treating the injury. That could be handled.

Don Black had already run several practice rescues in Flint Ridge, and had directed numerous real rescues elsewhere. Don would be near his phone in Tennessee. As for doctors, during the early days of the Cave Research Foundation a deliberate attempt had been made to recruit cavers who were M.D.'s. Now Jake Elberfeld, Jack Grover, and Stan Sides—all M.D.'s—were three of the strongest Flint Ridge cavers. Stan would be on Flint Ridge while the cavers were underground.

These were chilling thoughts, nonetheless. If someone wedged in the Tight Spot, he would probably die from exhaustion long before help could come. And what if five got through the Tight Spot and the sixth got stuck, trapping them? John decided to put those who had not yet been through the Tight Spot between cavers who had. The new people could thus be encouraged and instructed from both sides. If one of them got stuck, trapping someone ahead, there would still be someone behind to go back for help.

John spent the evening on the telephone. Gary Eller would come up from Atlanta. Steve Wells would be there already, doing field work in the park. Richard Zopf was delighted to be asked. Pat Crowther, who was in Columbus working on the cartographic files, was doubtful.

Pat's husband, Will, was out west to avoid the worst of the hay-fever season, and their children were back home in Massachusetts in care of a baby-sitter. School was to start the next week, and there were many things to catch up with after a hectic summer. The children had to go to the doctor and the dentist; they had to be registered in school; their school clothes and supplies had to be bought.

Pat telephoned Will. "Sure," he said in a cheery voice. He understood her desire to be on the connection trip. Never mind, he would get home in time to attend to all the back-to-school hassle for the girls. Pat could go to Kentucky for another trip to Hanson's Lost River. Perhaps this time she would find Mammoth Cave.

John called Mammoth Cave National Park to see if Ranger Cleve Pinnix could join the party. Cleve was as strong and capable as any Flint Ridge caver. The fact that he was a Park Ranger was also important. It seemed appropriate that a member of the Park Service be along. Cleve said he thought he could make it.

Jack and Tish Hess would be there too. They had been living in the Austin House of Flint Ridge camp all summer. Jack was doing hydrological field work for his Ph.D. They told the cavers they had heard footsteps outside at night, and whistling in the dark. When they looked, nobody was there. Jack suggested that it was Floyd Collins' ghost. One night they had heard banging coming from the Crystal Cave Ticket Office.

Floyd's body *was* right there in that coffin in the Grand Canyon of Crystal Cave. There were some who took the ghost talk seriously. One good caver used to forbid talk of Floyd, little men, or disembodied voices in the cave, on threat of aborting the trip if the conversation did not turn to healthier subjects.

Saturday, September 9, was a bright, crisp morning on Flint Ridge. The weather had held. Everyone was optimistic. John filled out the Park Service form and drove to Park Headquarters with it. He was worried that some last-minute bureaucratic nonsense would keep Cleve Pinnix from joining them, but there Cleve was in clothes definitely not standard for a Park Ranger. His shirt had not been washed since its last muddy cave trip. This man was all right.

"Hot damn! We are go," John said.

Cleve drove a four-wheel-drive truck over to Flint Ridge. John Wilcox, Pat Crowther, Richard Zopf, Gary Eller, and Steve Wells piled in. Cleve steered on the abandoned road through the high weeds and brush out to the last promontory of Flint Ridge overlooking the Green River. There the road cut sharply left. The truck plunged down the steep, rutted road, gears grinding as wheels jostled over rocks and fallen tree limbs.

At the Austin Entrance, they put on their knee crawlers. Gary seemed to have only one. Frantically he dumped out his bag. The missing knee crawler was not there.

Damn! I checked, re-checked, and triple-checked before getting on the truck," Gary wailed.

It was not in the truck, either. So Gary and Cleve jumped into the cab to grind back up the hill. There was the knee crawler in Gary's car. They were back at the Austin Entrance in ten minutes.

Gary was chagrined and fretful. Yet, there had never been a Flint Ridge caver who had not at one time or another reached an entrance of the cave only to discover that he had left behind the key, or the compass, or some other vital piece of equipment. You often stupidly felt that it was an unconscious protest against going underground. But on *this* trip? Ah, well.

"Ye gods, are we going in that?" Cleve Pinnix asked.

Cleve had just caught sight of the Austin Entrance, the slanting timbers holding up a sagging boulder, slightly obscured by two other large boulders that had fallen down. Everyone beamed with pleasure at Cleve's reaction. For years attempts had been made to get the Park Service to repair the Austin Entrance. Bill Austin had drawn plans for its improvement more than ten years previously. They hoped Cleve had influence.

Cleve had been a mountain climber before he was a caver, and he was feeling all the anticipation and excitement he had always felt before a major climb. He had discussed all the possibilities, and he sensed the tension generated by the uncertainty of the outcome of the trip.

Also, he had never been in Flint Ridge before. The jumble of collapsed rocks and the dangerous raw face of the cliff over the Austin Entrance did not look enticing. The others were eager to point out that it was dangerous, but on the other hand they had continued using it for years and they did not seem overly cautious now. It was familiar to them. Cleve kept his head very low as he passed under the wooden beams. Yes, well, the Park Service ought to repair the Austin Entrance, but Cleve knew how little money there was for all that should be done in Mammoth Cave National Park.

Underground, all the passages were new to Cleve. He was impressed by the massiveness of Brucker Breakdown, and was delighted with Old Granddad and the other gypsum and mirabilite displays in Turner Avenue. He had studied the maps, and it gave him satisfaction to see those lines on paper come to life.

Gary was impressed with the fact that Cleve's first trip might be *the* trip of all trips in Flint Ridge. He made a rather formal little speech to Cleve about it. Cleve would see more new cave on this trip than was possible on any other trip. And the connection might be made today.

"There are people who have caved with us for years," Gary went on, "who have never seen Turner Avenue. And so many of them would give anything to be on the connection trip."

Gary continued to think about the irony of it. He could quickly list ten old Candlelight River hands who had given their all in repeated attempts to connect the Flint Ridge Cave System with Mammoth Cave. Not a one of them was along today.

Gary's darting mind would not slow down. They were at the Duck-under leaving Swinnerton Avenue already. Several of them left cans of fruit to eat on the way out if, sadly, they had to return this way. Yes, here they were, and no one had taken their pictures, no one had even come to the cave entrance to see them off. Gary had just been reading about the exit of the French speleologist Michel Siffre from Midnight Cave in Texas, where he had spent six months alone as the subject of a physiological experiment. Gary knew, of course, that Michel had to have assistants about, but there were also newsmen and television cameras. It was the French style. But then, after all, Michel was French, and the French cared very much about caving. Caving exploits were covered in detail on the sports pages of French newspapers. Gary was amused at the thought of what American sports-fan reaction would be to the caving news.

After the Duck-under, Cleve began to lose his grasp on the cave. Up to that point he thought he had learned the cave well enough to get out on his own. Then his impressions began to blur, and the cave turned into a long succession of crawlways with little pattern or seeming order.

"Look back now," Gary said. Like many Flint Ridge cavers, Gary was a bit of a pedagogue.

"Yes?" Cleve obeyed.

"See how wide open it looks straight on there?" Gary asked.

"Yes," Cleve replied.

"Well, that's wrong. See? We just came through that little hole. It's important to remember this place," Gary went on. "It's deceptive. You can come roaring along here on the way back and not even notice that little hole."

"Where does the big passage go?" Cleve asked.

Gary shrugged. "Not out."

"We'll let you lead on the way out," Richard sang back to Cleve.

"You might get a scenic trip that way." Cleve chuckled. "But it wouldn't necessarily be in the right direction."

Cleve had heard a lot about Candlelight River, and was pleased to reach that celebrated place, even though they walked right on. Then, at the entrance to the Tight Tube, watching people chimney up the canyon walls to angle into that rathole, he finally got worried about going on. He would have said that there was no way to get into the Tight Tube, but there they went.

When Cleve's turn came, he discovered the main incentive: After chimneying up slippery walls, you either go into the hole or fall. Cleve was impressed with how tight the Tight Tube really was when he got into it, and he moved ahead as rapidly as he could to shorten the ordeal. No one had bothered to tell him about the right-angle turn, and with his head down, he banged his hard hat solidly against the wall. Cleve emerged breathless from the Tight Tube.

"Cave pass, please," Richard said officiously.

"We heard you testing your hard hat," Gary said.

Cleve sat catching his breath. Cave humor was no different from climbing humor. He sat there with a cloud of steam rising from his body. They also had not bothered to tell him to take it easy through the Tight Tube, as all of them obviously had.

"The Loose Tube is easier," John declaimed, "but Roger Brucker says the Loose Tube is four times as long, sixteen times as wet, and sixty-four times as stupid as the Tight Tube." John added, "The trip is half over. Now we start caving."

On they went until they reached the A-survey junction room. That sandy resting place felt like the beginning of the trip to John and Richard. Steve and Gary—young, thin, and strong—were as fresh as when they had entered the cave, four and a half hours before. Pat looked too small and too thin for any of this, but her broad smile and shining eyes betrayed her energy.

"I'm overwhelmed," Cleve said.

The others looked at Cleve with alarm, but were glad to discover that all he meant was that he had, after so many hours through so many small passages from the Austin Entrance, truly begun to understand what it meant to be there. Gary and the others had pointed out seemingly countless points of discovery, side leads that did not go, alternate routes, former Ultima Thules that had then become just names on the route to more distant points.

Later, Cleve wrote of his feelings:

This sense of the countless trips and remarkable individual and group effort that pushed the known limits of the cave to this point cannot be emphasized too strongly. I've heard the discovery and connection trips compared to completing a climb, but the simile just doesn't hold up for me. The effort through such a long period of time in fielding the number of parties it took to push the unpromising-looking leads, survey the hard crawls, and eventually make inevitable the trip when someone would finally be able to penetrate the passage that would lead into Mammoth is incredible to me yet.

Cleve had seen only four and a half miles of the Flint Ridge Cave System's more than eighty-seven miles of passage known at that time.

John was quiet and happy. There were no fat old men to keep them from reaching the goal. And not a member of the team was pessimistic. None was even cynical in the style affected by some of the older Flint Ridge cavers. Morale could not be higher.

"I can't understand why those guys said it was so difficult to get out here," Gary kept exclaiming.

They repacked their gear to crawl in the gooey mud of the A-survey to the Tight Spot, dead ahead. John's tenseness returned, for this would be the crucial test for Steve and Cleve. John shuffled the party so that the newcomers were sandwiched between veteran squeezers.

Richard went first through the Tight Spot, and John followed. Two big men. Cleve, smaller, was next.

Cleve glanced at Pat's expressionless face as he moved up to the Tight Spot. It was a somber moment. "No sensible person would try this lead," Cleve said.

Richard's voice floated back from the far side of the Tight Spot. "That's why we brought *you*, Cleve."

Cleve looked again, then started in. He had been caving enough to know that many of the horrible places were not so bad once you got into them. He perceived at once that the danger of getting stuck in this horrible Tight Spot was real.

Cleve stiff-armed his pack through to Richard's waiting hands. Already Cleve felt the muscle strain. Then he jammed his body into the Tight Spot, flailed his legs out horizontally behind him, and pushed. Which shoulder would snap off first? Anyone who has ever wondered how a snake propels itself, Cleve thought, has never been in a place like this.

"Exhale," John said.

Cleve exhaled and moved forward another few inches. After a few minutes of wiggling, Cleve sank to join Richard and John in the little room on the other side of the Tight Spot.

"You look thinner," Richard said.

"I got through," Cleve said, "using kneecaps, ribs, and obscenities for propulsion."

Pat slipped through, and then Steve, who had not been through before. Steve wondered what the fuss was about. Gary came through last with none of the anxieties he had felt on his first try.

All had made it through the Tight Spot without trouble. John did not dare be elated. He hurried them on to Hanson's Lost River to the last survey point, B-87, and set up the leapfrog survey. Gary, Richard, and Steve began surveying from B-87 at 4:00 p.m. John, Pat, and Cleve went on.

"Catch us if you can," Pat said to Gary's party.

They all liked leapfrog surveying. It is not the lonely caving of a single party way out to hell and gone, but each party is alone for long periods of time. And the reunions are fun, as the rear party finishes its section and then rushes on to take the lead.

The old people—John was thirty-five, Pat twenty-nine, and Cleve twenty-eight—were to move on and count off thirty potential stations before they started surveying. John's count took them past "PETE H."

"Is this the Flint Ridge Cave System or Mammoth Cave?" Cleve asked. John and Pat did not reply. "Or does it matter any more?" Cleve said softly to himself, turning to follow them on down Hanson's Lost River.

John completed his count of passage bends. It had taken only about twenty minutes.

"We start here," John said.

Cleve took the end of the steel tape and walked back down the passage to unroll it. John the Silent. John had disconcerted Cleve at first, but Cleve had caved in Mammoth Cave with John several times and got used to his taciturnity. Cleve had thought John spooky in a cave, never wasting words, stony-faced. What a contrast to Roger Brucker, who joked, played tricks, told outrageous stories, and burst into song or cave-guide monologue. Yet, diametrically opposed as John and Roger were in their styles of leadership, most Flint Ridge cavers were utterly loyal to them both.

Pat was a quiet one, too. Cleve looked at her. She smiled broadly, and her eyes sparkled. It was the same smile that sometimes creased John's face, almost painfully, it seemed. It was in their eyes—all the exuberance and joy that came pouring out of the mouths of Gary, Steve, and Richard was there, beaming out intensely from John's and Pat's eyes.

Their C-survey started at the beginning of a comfortable hands-and-knees crawl in shallow water. Now and then they could stand stooped over. Survey shots averaged forty feet. After twenty-five stations John began listening for the faint sounds that would be the leapfroggers coming. Compass bearings began to shift to the south. The big turn toward Mammoth Cave was near. They had been surveying about two hours. It was 6:30 p.m. when they heard the first distant murmur of voices and the muffled sloshing of feet.

Gary's party hove into view, hooting and steaming like an express train hurtling into a station. They were eager to leap ahead. Water was running off their pants legs, and a gleaming silver survey tape cut an arc into the water behind them.

"Here, here," Gary shouted. "What's this? Picnickers?"

"Count off thirty stations," John said, "and start a D-survey."

The train roared out of the station.

The organizer of a leapfrog survey can do some clever calculating. He sometimes makes the most exciting discoveries, despite the seeming equality of chances for each party. Who makes the discoveries often depends on how the survey is begun. John had estimated rather closely that it was about 5000 feet from the end of the previous survey at B-87 in Hanson's Lost River to where he had left their initials. The first party—Gary's—could be expected to survey around 1200 feet on the first leg. Beyond them, John's party should equal that distance before Gary's party leaped them. Gary's party would go beyond John's last point another 1200 feet or so before starting to survey. John's party would thus do two legs in succession, for a total of about 2400 feet. Meanwhile, Gary's party would be doing its second 1200 feet on past John. The total distance surveyed would then be around 4800 feet, with

Hands and knees crawl in shallow water.

John's party coming ahead for its leap of Gary's party, with only 200 feet to go to virgin cave, and with thirty new stations to count past—in virgin cave—before John's party had to begin to survey again.

John's party would thus spearhead the exploration through perhaps 1000 or more feet of virgin cave, with a high probability of entering Mammoth Cave—or hitting the final siphon—first.

Gary's party did not consist of ninnies. They had this figured out quite as clearly as John did. When they got about 3600 feet down the line, still ahead of John's party, Steve became rebellious.

"Forget the survey!" Steve said. "Let's go on and explore."

"My sentiments exactly," Richard said happily.

Gary continued reading the compass. Before entering the cave, and all along, the three youngsters had been playing the fool. They pretended that they were Pete Hanson and his cronies, and swaggered as they imagined cave explorers of the 1930s would have done. They held their thumbs in imaginary overall straps—Pete and Leo had no more worn overalls into the cave than Gary, Richard, and Steve did—and they talked with exaggerated Kentucky drawls. Steve and Gary had eagerly joined in to support Richard with new verses of "The Bretz River Shuffle" and "The Tight Tube Blues." Gary was an accomplished five-string banjo player, and his voice twanged as though he had his banjo with him in that unlikely limestone cave.

On the way into the Tight Tube, Richard had opened his revival meeting again, stressing greatly that those who had been purified and washed white as snow by passage through the Tight Tube would regain their sins if they returned back through it.

"*If* we return through the Tube . . ." Pat had said quietly.

They had joked a little about a portal-to-portal trip—from the Austin Entrance of the Flint Ridge Cave System to the Historic Entrance of Mammoth Cave—but not much. It was too close, too much desired by them all. They did not want to tempt the real gods of the cave: cold water, darkness, and fatigue.

"I say, forget the survey," Steve said.

"We can't," Gary said.

"You're right, of course," Steve said as they settled down. "But we could run on fast. Let's find Mammoth and then survey back to meet the others. It's all the same, and they'll survey on until they find us."

Gary's heart leaped up at Steve's words, but he kept on surveying.

"If we didn't connect, we wouldn't have anything to show the trend. We have to know where we are." Could that be me talking? Gary thought. Now *he* was the voice of the establishment. He was old at twenty-five. Those two young madmen—Richard, twenty-two, and Steve, twenty-three—really would charge ahead. Or would they? No, they would not.

They could talk, but none of them seriously considered changing John's plan. Impetuous youths had made spectacular discoveries before, but they had also aborted trips and learned nothing, when the plodders and plotters almost always added to the survey and to the growing knowledge of the cave.

Gary, Steve, and Richard surveyed silently. Their spirits had dropped. They figured that this trip was designed basically to survey up to the point of previous penetration, anyway. Then, if they were not too tired, the two parties could explore on ahead in virgin cave for a while together. But there might not be much time for exploration.

Their wild exuberance was also damped by the low crawlway they were surveying. They were on hands and knees in water a foot deep. The ceiling looked solid, but was covered deceptively with a layer of mud three-quarters of an inch thick. They were down close to the source of that mud. Obviously, the Green River flooded back into these passages frequently. The rebellion was quelled in cold mud and cold water. They kept surveying painfully along. They knew that John would follow.

For John's party, the passage had lighted up momentarily in the blaze of the passing leapfroggers' lamps. Then as the noise of Gary's departing crew diminished, all was the same as before. The stations went in quickly and smoothly. Some time later, at a bend in the passage, Cleve shouted that this was the completion of their second section. He had tied into the beginning of Gary's D-survey. It was 8:00 p.m., still early enough to map a lot more cave, but late enough for John to worry about having time to explore. He closed the survey book and tucked it safely away.

As they set off on the leapfrog, John saw at once that Gary's party was carrying the survey through the worst stretch of passage in Hanson's Lost River. Stations were marked in carbide soot on mud banks that completely covered the rock walls. In passages this low and wet, the compass man would get worn out quickly. On many of the bends, there was no way to steady the compass against the wall or a rock. To read the compass, you had to squat in the river, now two feet deep, and grit your teeth to stop shivering long enough for the compass needle to slow its maddening jiggle.

John, Pat, and Cleve found Gary, Richard, and Steve huddled together in a small pool of yellow carbide light, resting on a mud bank with their feet almost out of the water. Gary's young face—despite his twenty-five years, he looked sixteen—showed fatigue. Richard's exuberance had drained. Only Steve responded with enthusiasm to the arrival of the others. He sprang up to stand talking, while Gary and Richard remained stretched out on the bank. They had stopped surveying at station D-23.

"How many feet?" John asked.

They had the figure ready. John added the total. It was about 4000 feet. The initials John had left in August were not much farther ahead. They were surely within 1000 feet of Echo River.

"Don't tell me," Cleve said. "According to your calculations, we've gone right under Mammoth Cave to Joppa Ridge, and we'll have to backtrack."

"I think we'll hit Echo River," John said. "Since nothing went from Roaring River, it must be Echo River."

Cleve thought this improbable. The Echo River area had been well known since the time of Stephen Bishop. How could any passage of consequence be left unrecorded?

John had known what he would have to say even before he had reached Gary's party.

"We'll terminate the survey here. Four thousand feet is a good piece of work for one day." It was 8:15 p.m.

There was a stir of life in Richard and Gary.

"And we'll explore ahead," John said.

The bodies on the bank rose. They had long since put up their surveying gear. They set off close together, all six of them, with John in the lead, moving quickly in an attempt to warm up.

Six hundred feet down the river they came to the initials John had left on the first trip.

"Guts!" Steve cried. "It took guts for you guys to come all the way out here that first time. It is one hell of a distance."

The room with the initials was comfortable, but ahead the ceiling lowered. John walked on. Cleve was worried that the river would soon siphon, and that they would never reach Mammoth Cave. He was wet and chilling fast, despite the activity. The Austin Entrance back the other way was incredibly remote, far beyond a mind-numbing succession of obstacles. And the water was getting deeper.

"Whooee!" Richard said.

They braced against the slippery mud walls, standing bent far over, trying to keep their bellies out of the water that filled the crouchway wall-to-wall. Then they were momentarily crowded together in a ceiling pocket where they could stand upright. Ahead the ceiling was getting lower, dropping to no more than twelve inches from the water. Waves from their movement were making ominous lapping sounds against the ceiling. This was the sign of a possible siphon.

John looked at the deep green water. Brown whorls of silt were billowing ahead from their stirring. The dark, mud-covered walls and ceiling drank in their lights.

Steve was right behind John, and he looked back at the others with such an expression of despair that their hopes vanished. This looked like the abject, bitter end.

Without speaking, John started moving on forward. After all of this, was it going to siphon? There was only one way to find out.

"John, are you really sure you want to get completely wet?" Steve asked.

"Wait here," John replied without looking back. "No point in everyone getting wet if it siphons."

It was the worst place in the world to stand and wait.

John put his hand on the ceiling, moved to the highest side of the passage, and pushed cautiously ahead. The waves were slapping against the ceiling from his movements. If it siphoned, it would no doubt be soon. Then the cold ordeal would be over. Was it possible that Pete Hanson and Leo Hunt had come in another way?

John moved forward. He could still talk to the others, but he was totally alone. If it siphoned, he would be back to look at those other leads, all those other leads, all the way back along Hanson's Lost River. He was prepared for the worst.

John hugged the wall, his body angling down under water. Soon he sensed that there was more open space over his head. He turned slowly to look up and saw that the ceiling had broken upward three feet. He looked back down. An incautious move might dunk him.

"I'm through," he yelled. Then he saw that the ceiling ahead broke upward again. He moved forward. He could see no more ceiling above.

"I've got something!" John shouted.

Upward there was just a black nothing.

"It's big!" John shouted again.

Steve registered what John had said, but what made his heart pound was not the message, but the echo. Steve could hear reverberations from a large room ahead.

"What did he say?" Cleve whispered.

"He's through. Forget the water!" Steve said. He started forward.

John stood stunned. The walls had opened out. He had come into a tremendous void, with a lake ahead. He could hardly sense a wall anywhere, but there was one in the distance, straight ahead. Slowly his eyes became accustomed to the gloom. What was that gleaming, horizontal line against the far wall? Surely it was a pipe! Yes, a handrail! With vertical supports!

John opened his mouth, knowing the effect his words would have. He spoke slowly through his ear-to-ear grin:

"*I see a tourist trail!*"

"We did it! We did it!" Steve screamed, stumbling forward. Cleve tripped and grabbed Steve's shoulders, pushing him down into the water. Richard's shirt caught on a projection, and Pat pounded on his back to get him to move along. Then she rushed out, slipped, and sat down in water up to her neck.

They saw John standing in deep water. He seemed to be in a state of shock. Dimly appearing in the distance was a man-made causeway with a steel railing. When Steve saw the handrail, his elation knew no bounds. He yelled and hooted. The others joined him momentarily, but then they quieted. Some of them fought back tears.

They had done it.

Later, John wrote:

*The lines of a handrail marked the
old tourist trail.*

*My memory of the next few moments is indistinct. Victory is a
feeling of vastness inside the skull. In this case, it is doubly sweet
because it seemed so far away only moments before. It is an end to
the doubts, the questions, the fears of something going wrong at
the last moment. It is the knowledge that we have exercised competence and that this is the instant in which we have won, and that
this triumph will be with us for the rest of our lives. By god, there
is absolutely no question, we have done it!*

Pat knew it was Mammoth Cave, but her deepest feelings could not
accept it. This is Flint Ridge, she thought, we came in the Austin Entrance. Even when she slipped and fell into the water up to her neck, she
was concentrating on making her body accept the transition. Absently,
she reached down to push herself upright again.

John was grinning now. He could not stop. For all of them the feeling
of disorientation was intense, and readjustment was difficult. Only a few
minutes before—it was now 8:45 p.m.—they had been at the end of the
world's longest cave, as isolated from the rest of mankind as though they
had been at the North Pole, or in an ocean abyss, or on the Moon. It was
a nine-hour trip through some of the toughest cave in the world back to
the Austin Entrance. Now, like the rush of a swing and the crack of a
whip, they had been whirled into the world of Mammoth Cave, of tourist
trails, and a short walk home.

Then it all locked into place. The strangeness vanished like a disappearing visual illusion. The sense of remoteness and the sensations of disorientation were gone. In this place they could never be regained.

The entire length of Hanson's Lost River was now open to easy access from Mammoth Cave. That room where last week John had left their initials, where explorers only thirty minutes ago had been as far in a cave from its entrance as any cavers had ever been, a miserably cold long way from home, was only 600 feet from a tourist trail in Mammoth Cave!

Then it was over. They waded cautiously through the deep water and climbed onto the causeway. Their wet clothes weighed them down. John had never been at this place in Mammoth Cave before. He ran to the right along the tourist trail and came to a boat dock. Then everyone galloped to the left, up a sandy slope. Pat said, "We're in Cascade Hall!" Now it seemed impossible that they had come into the cave through the Austin Entrance. They were in Mammoth Cave.

Pat searched briefly, then pointed to a CRF survey station on the back of a boulder. Water streaming from his clothes, John sat down in the sand and said, "Let's eat."

They were quiet and reflective. They dipped into their soggy packs to pull out cans of meat and fruit, can openers, and spoons. Steve was thinking of all the hopes and dreams that had now come true, those of a hundred cave explorers before him. A feeling of humbleness overwhelmed him. It was not this party alone that had made the connection, he thought. It was also all those who had gone before.

Cleve was quieted by the thought of one incredibly huge cave joined with another to make a super-cave so immense that no one would ever see it all.

Gary and Richard talked about just how lucky they had been. There had been years of work extending the survey from the Austin Entrance beyond Candlelight River on under Houchins Valley. It had taken thousands of man-hours underground to put together the route that had led them to Echo River in Mammoth Cave. What a lucky happenstance that, of all the competent Flint Ridge cavers, they had been fortunate enough to be on the connection trip.

They all sat there, thinking of the murky water and the obscure wall indentation from which they had emerged. Pat and Richard had walked on the causeway past that opening twice without noticing it on their wet-suit trip up Roaring River. It was not very obvious. Had the Green River been a foot higher, the connection could not have been made. John would have found a siphon, and now they would have been slogging their dejected way back toward the Austin Entrance.

Gary relaxed his skepticism about old-timers. Pete Hanson and Leo Hunt must have been aggressive cavers to have pushed that miserable lead. Had they known that they had gone completely under Mammoth Cave Ridge to its northern edge? Probably not. So far as anyone knew, they had told no one about the trip on which they had scratched their initials over the continuation of a miserably wet crawl, nearly a mile from where they had entered a tiny tributary of Echo River in Mammoth Cave. They must have simply been looking for more big cave. They could have had no idea of connecting with anything out there, for no cave was known out that way at that time.

Not a single inch of those nearly six miles of the actual connection route from the Austin Entrance to "PETE H" had been seen by any of the old-timers. It had all been explored and surveyed by modern Flint Ridge cavers, beginning with the 1954 discoveries of Bill Austin, Jack Lehrberger, and Jack Reccius, and continued by the Cave Research Foundation.

John, Cleve, and Gary talked over possibilities. John was most reluctant to leave by way of the Historic Entrance to Mammoth Cave. It was only about 9:00 p.m., and they might be seen coming out. Anyone who bothered to check would find that there were no exploration parties in Mammoth Cave that day, and if they went on to figure out that the connection had been made, the news might get out too soon.

"I didn't promise anyone an easy trip," John said. "We're all capable of returning by the Austin Entrance." He paused. "On the other hand, if we could survey another twelve hundred feet, Flint Ridge and Mammoth Cave would be tied together."

"Oh, let's do that!" Pat said.

"If we do, we can't return through Flint Ridge, we'll be drained out," John said.

They all yearned to finish the survey.

"Look," Cleve said, "I can telephone the Chief Ranger from the Snowball Dining Room. I told him we might connect today. He'll radio the night patrolman, who's a friend of mine. He can be trusted. He can drive over to the Elevator Entrance. No one will see us there, and we can ride back to Flint Ridge in the patrol truck."

"Agreed!" John said.

John was ever cautious, but the joy was too much. He could not summon another ounce of doubt. They *would* do it. It was not completely done yet, but just as soon as he finished his can of peaches, they *would* do it. They had discovered the connection, and now they would survey the connection route on one and the same trip.

John's grin threatened to become permanent.

Gary's party was whooping again. They surged up Hanson's Lost River to survey back from the point where not so long before they had sworn that nothing would make them continue the survey another step. Then, after fourteen stations, the compass needle stuck. This was their spare compass. The first one had waterlogged earlier. They could survey no more. Once again, Gary, Richard, and Steve leaned back against a mud bank of Hanson's Lost River to wait.

John's party surveyed quickly from the CRF station in Cascade Hall along the causeway to a point across from the hole in the wall. None of

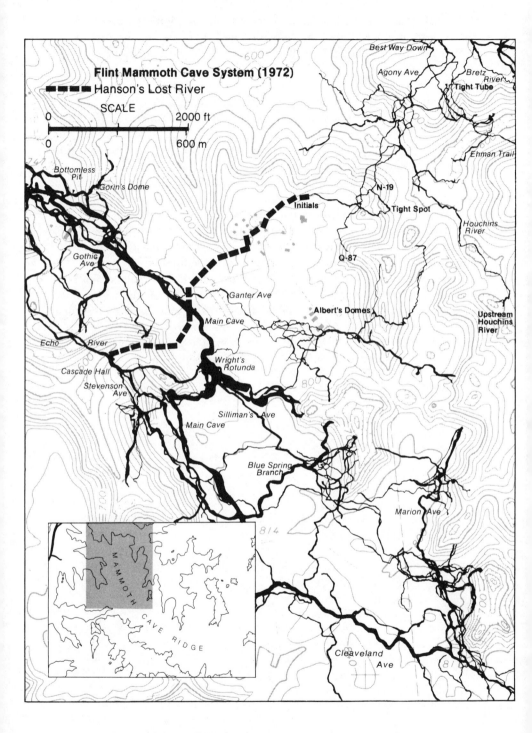

*Map of the Flint Mammoth Cave System
showing Hanson's Lost River.*

them had remembered that Echo River was so cold, but they forced themselves into it. Could they really have made it back out to the Austin Entrance? John still thought so, but he was surely glad they did not have to try.

Pat was cursing softly at her inability to hold the compass steady while shivering. John could hear the needle rattling. He could take notes only by taking a deep breath and holding it to keep himself from shivering. Their survey wound its way from Cascade Hall, across Echo River, into the hole, and up Hanson's Lost River. Finally, Cleve laid the end of the steel tape on Gary's last station.

"The Golden Spike of caving has been driven," Gary said, as Pat surveyed from station E-20 to station D-37, the final tie between the Flint Ridge Cave System and Mammoth Cave.

"Happy, happy," Pat said. She folded up the compass and put it back in her pack. It was the last of the four. The other three compasses had failed. It was just midnight, Saturday night, September 9, 1972.

There was nothing more to do. They turned and filed back down Hanson's Lost River to Echo River in Mammoth—no, in Flint/Mammoth Cave. They plodded slowly up the trail to the elevator that would take them up and out of the Flint Mammoth Cave System into another world. Cleve glanced at his watch. It was now 1:00 a.m. They had been in the cave only fourteen and a half hours. It seemed like a lifetime.

Victory

The longest cave is 144.4 miles long.
The Everest of world speleology has been climbed.

 After they had connected the Flint Ridge Cave System with Mammoth Cave, John Wilcox, Pat Crowther, Richard Zopf, Gary Eller, Steve Wells, and Cleve Pinnix walked out along Cascade Hall, up Silliman's Avenue, through El Ghor, and to the Snowball Dining Room, where the freight elevator would raise them to the surface of Mammoth Cave Ridge. It was Stephen Bishop's old discovery route. They were mostly silent. Only Steve had kept up the high state of excitement.

"Lean over and pinch me. I'm dreaming that I'm walking down El Ghor in Mammoth Cave," he said to Gary.

Gary responded without enthusiasm. They all knew it was not a dream. The dream was over.

"We're the only sinless people in the world!" Richard shouted suddenly. This brought smiles. All were thankful that they did not have to return through the Tight Tube.

"Imagine us sinless!" Steve said. No one tried.

The walk was more than a mile. Their senses were dulled. Water oozed from their heavy shoes, packs, and clothes. Their wet pants legs swished.

Then they entered the eerily lighted underground restaurant. Fluorescent lights hissed, there were flush toilets, drinking fountains, and lingering smells of sandwiches and coffee.

Cleve talked on the telephone quietly. Everything was quiet, quieter than it had ever been in the depths of Hanson's Lost River, it seemed. It was the quiet of an empty restaurant after hours, and of people who did not belong. But there, just past the tables, the blackness opened up again.

They walked over to the elevator and stepped in. As the door shut with a hollow sound, Gary remembered the dull clang of the Austin Entrance door that had closed behind them so many hours before. If they had returned by the Austin Entrance, they would have been in the cave another eight or nine hours. It would have been a twenty-four-hour trip, the hardest trip in the longest cave.

The elevator rose interminably past jagged rock walls, hesitated, jerked, and then stopped. Cleve opened the door and in a minute they were outdoors on Mammoth Cave Ridge.

It was the most brilliant Kentucky night imaginable. The weather was a crisp 50° F., just slightly colder than in the cave. The sky was full of stars. All of them drank in the vegetation-perfumed air and the view. One of the finest moments in caving is that of emerging into the starry night.

The truck arrived. Cleve spoke briefly to the driver, then all six cavers piled into the truck's open bed. The wind was chilly. They huddled together for warmth.

Overhead, the stars pricked through black velvet. Tree limbs loomed in the truck's headlights to pass across the sky and fall away in the distance behind.

Mile after mile the truck traveled northward. The trip seemed to take forever. Had they really come this far underground? It was three miles as the crow flies from the Elevator Entrance on Mammoth Cave Ridge to the Austin Entrance on Flint Ridge, but it was seven miles by road. They had gone more than seven miles from the Austin Entrance to the Elevator Entrance underground.

At the Flint Ridge camp at last, they woke Jack, Tish, and two-year-old Nathan Hess. Dave and Shirley DesMarais also came out to greet them. Everyone was groggy. It was nearly 2:00 a.m. Cleve and Gary walked down to the Austin Entrance to get Cleve's truck. They were gone before anyone remembered that pictures ought to be taken.

"No one change clothes!" Pat ordered.

Steve had worn a nearly new wardrobe into the cave. His white hard hat was still brilliant.

"That won't do," Pat said. She smeared some mud on Steve's hard hat and face.

They stood in the cold until Cleve and Gary returned, and then posed for the flash pictures in front of the house that Floyd Collins' family had built. The sign said: "FLOYD COLLINS' HOME."

Floyd Collins was home. He surely must be now. Whether he ever tried to find connections between the big caves does not matter. He was a caver. He must have thought about it. And the exploration that had led to all the connections in the last twenty years had been made following the footsteps of Floyd Collins.

Then Cleve said he had a bottle of champagne at his house. They drove to the deserted ferry to cross Green River. They had to pull the ferry boat across by hand. The river staff gauge read 1.5 feet. Almost always

Souvenir of John Wilcox's final connection party.
Back row, left to right: John Wilcox, Richard
Zopf, Steve Wells, and Cleve Pinnix; front row:
Gary Eller and Pat Crowther. September 9, 1972.

it was 2.2 feet or above. At 2.5 feet, John would have found a siphon at the end of Hanson's Lost River.

Hamburgers sizzled and the champagne was poured. John lifted his glass:

"A toast to the best party ever fielded in Flint Ridge!"

John, who never drinks, and who certainly had never toasted anyone or anything before, downed his champagne. They had been grinning so much that their faces hurt.

On the way back to Flint Ridge, Steve had a brainstorm. They stopped at a pay telephone at Mammoth Cave National Park Headquarters to call the Grand Old Man of Flint Ridge caving. At 4:00 a.m.

The phone rang in the bedroom in Yellow Springs, Ohio. It was pitch dark outside, and was obviously very late. The ring was shrill. Roger Brucker stirred to consciousness. A relative dead? A cave party in trouble? John Wilcox? By the time Roger had reached the phone and positioned it correctly, he was wide awake.

"This is Roger Brucker."

An echoing voice said, "This is Pete Hanson."

"You've done it!" Roger said for the second time in a week.

At first Roger was stunned, listening to Steve's excited words. As he listened, his own excitement mounted.

"Kiss them all for me," he said, "all of them. Tell them I love them."

After Roger hung up, it was still pitch dark outside the sweeping glass that formed one whole wall of the Bruckers' house. That house had been built partly by an old Flint Ridge caver, Dave Jones. By the time Roger put down the phone, Joan was awake. She said, "Oh, Roger . . ."

He reached for her absently. It was a bittersweet moment. He was pleased that they had made the connection. He had not been along. He was content and happy. But he could sleep no more that night.

When Red Watson heard, he cried, again, sad and happy tears. It had been hard for him to hold back tears a week before, when he had congratulated John on having found "PETE H."

"Well, neat," was all Red could say.

"Yeh," John replied.

At noon the day after making the connection, the explorers were sitting on Superintendent Kulesza's veranda. He looked with some astonishment at John's map showing the connection of Hanson's Lost River to Echo River.

"John, were you up all night?" Joe Kulesza asked.

"Well, yes, Joe," John replied, "but not drawing that map. I drew that last week."

Epilogue

Pat placed the compass on which she had taken the final connection reading back with the others. Unmarked, it would be used again. However, she clutched the sheets of survey data in her hands on the airplane all the way to Boston. Will was waiting. They held each other and talked about the connection. At midnight, Will was asleep, but Pat could not sleep. She got out of bed and went to the computer terminal in their living room to type in the bearings and distances of the survey. By 2:00 a.m. this was done. Then she could sleep at last.

The next day Pat edited the output from the computer and ran the coordinate program. A long paper tape was punched out. Then Pat and Will took the tape to Will's office. There the computer plotter drew the map of the connection route while they watched.

With the map on paper, Pat finally began to unwind. It seemed that she had been wound up since that first trip out to Q-87. Would her children remember this excitement in later years? Now they were young, and caves had always been a normal part of their lives. Oh well. In later years Pat would tell them what it had all been about.

John set to work at once adding up surveys and mileage statistics. There would be many questions, and precise answers would be needed. After several evenings at the adding machine, he had compiled three pages of mileage data from survey books and maps. They boiled down to the following:

September 9, 1972

	miles
Length of Flint Mammoth Cave System	144.448
Length of Flint Ridge Cave System	86.548
Length of Mammoth Cave.	57.900
Length of connection route	7.149
Longest round trip taken by explorers in the Flint Ridge Cave System through the Austin Entrance (August 30, 1972)	10.758
Use of Floyd Collins' Crystal Cave Entrance to the Flint Ridge Cave System would add to the route length	2.797
Use of the Historic Entrance to Mammoth Cave would subtract from the route length	0.717
Length of Hölloch, Switzerland	71.800
Length of Peschtschera Optimistitscheskaja, U.S.S.R.	57.500
Length of Jewel Cave, South Dakota, U.S.A.	43.100
Length of Eisriesenwelt, Austria	26.200

One night John Bridge called Roger excitedly.

"Guess who first discovered Hanson's Lost River. Stephen Bishop! I'm sending you a copy of his 1842 map. The passage is shown as a dry passage taking off just beyond the second boat landing on Echo River."

The map showed the passage extending almost as far as Pete Hanson and Leo Hunt had probed it. Stephen Bishop had been the first person to explore the Mammoth Cave side of the connection route.

Harold Meloy had been busy, too. He had found the passage on the Bishop map, and also on an 1897 map attributed to Ellsworth Call. Then, with John Bridge, Harold found the passage on a 1909 map supposedly drawn by Horace Hovey, but this time it was shown as containing water.

What happened in the interim was that in 1905 a dam had been constructed downstream on the Green River at Brownsville, raising the pool stage—and thus the level of Echo River—six feet. This flooded the lower reaches of Hanson's Lost River.

On Max Kaemper's map of 1908, Hanson's Lost River passage does not appear at all. It was probably under water when Kaemper and Ed Bishop surveyed past it. Hanson's Lost River does not appear on subsequent maps of Mammoth Cave. This is not surprising: The mouth of Hanson's Lost River at Echo River in Mammoth Cave is under water many months of most years.

Thus, the leads at both ends of the connection route had been lost. They were rediscovered, and the connection was made, because of the obsession of a dozen or so people. They wanted to make the connection. They were aided by a few hundred cavers who shared their passion for underground exploration.

On September 9, 1972, when the Flint Ridge Cave System was connected with Mammoth Cave, 144.4 contiguous miles of passages in the new Flint Mammoth Cave System had been surveyed. Since then, exploration and mapping have continued at the unprecedented rate of about nine-tenths of a mile per month. Where will it end?

The caves of Joppa Ridge may be connected at any time with the Flint Mammoth Cave System. Tantalizing segments of the Joppa System reveal patterns that are guiding explorers to new discoveries. The areas of the three main ridges—Flint Ridge, Mammoth Cave Ridge, and Joppa Ridge—in Mammoth Cave National Park under which cave passages have *not yet* been found are fully as large as the areas under which passages are known. The two main valleys—Houchins Valley and Doyel Valley— between the ridges cover an area nearly as large as that of the ridges themselves. As the connection shows, there are passages under the valleys, too. They have been explored barely at all. And Great Onyx Cave on Flint Ridge has not yet been connected with the rest of the system.

Consequently, we predict that the ultimate passage length of the Flint Mammoth (Joppa?) Cave System—*the* Central Kentucky Cave System— will someday be shown to exceed 300 miles.

The "final connection" described in this book is thus not really the final one. Neither is our version of the connection story the last that will be told. The history of Mammoth Cave remains to be written (however, see Appendix I for the most complete story of Mammoth Cave so far), as does the history of the Cave Research Foundation. And some cavers who took part in this exploration disagree with our selection of facts, analyses, emphases, and interpretations. We hope that some of them will publish their memoirs, and that in the future a story can be told that is more comprehensive than the one we tell here.

Non-cavers often wonder why anyone cared. Perhaps most understandable are the lure of the unknown and the thrill of going through darkness to where no one has ever been before. Cavers love the mystery of those hidden places which retain their mystery even when revealed.

We did it because we love the cave and one another. The sharing of this adventure has provided us with some of the most satisfying hours of our lives.

The cave lends itself to a sweet metaphor. For brief moments we have possessed—or have been possessed by—the longest cave. And when in time we become another part of mother earth, know that she has already been a part of us.

R.W.B.
R.A.W.

Unknown Cave.

Appendixes

Annotated Bibliography

Index

Appendix I/Historical Beginnings

1/Prehistoric Cave Explorers

The earliest evidence of man in the Mammoth Cave area dates to Paleo-Indian times, 8000 or more years ago. In Archaic times—3000 to 8000 years ago—people occupied camps or villages along the Green River, a deeply entrenched tributary to the Ohio River. The Green River flows through the Central Kentucky Karst, one of the greatest cave regions in the world. In the Central Kentucky Karst, the Mammoth Cave Plateau contains the longest cave in the world, the Flint Mammoth Cave System.

There is no evidence that the most ancient people visited the caves of the Mammoth Cave Plateau. There is solid evidence, however, that 3000 to 4000 years ago the prehistoric people began serious exploration of big caves in all three of the main ridges of the Mammoth Cave Plateau. Salts Cave in Flint Ridge, Mammoth Cave in Mammoth Cave Ridge, and Lee Cave in Joppa Ridge were explored as far as two miles from their entrances by prehistoric cavers carrying torches made of cane, weed stalks, and sticks. Torches were sometimes tied together with lengths of grapevine, tough grass, or bark, and large bundles of torch materials were cached throughout the caves.

Experimentation in Salts Cave has demonstrated that three pieces of cane a yard long and as thick as your little finger will burn for an hour. Such a torch burns with an orange flame that gives light quite adequate for exploration. It is a softer light than that given by the bright yellow flame of a carbide lamp. A cane torch fills a cave passage with a pleasant glow. It is warming. It also gives off smoke which coats cave walls and ceilings with soot.

You can easily carry enough torch material for a ten-to-twelve-hour trip. Modern cavers discovered to their surprise that it is easy to keep the torches burning, and that with cane torches you can explore comfortably even in the tightest crawlways.

Two experimental trips in Salts Cave were taken with cavers wearing nothing but shorts and carrying nothing but cane. Each trip lasted about six hours, with six cavers going each time more than a mile into Salts Cave and out again. The experience changed entirely the modern perspective on the ancient cavers. Prior to the experiments it had been thought that the caves must have been very spooky for the prehistoric explorers, and their explorations with cane torches were viewed as spectacular feats. Now we know that they were as safe and comfortable with cane torches as are modern cavers with carbide lamps.

Beginning about 3500 years ago, prehistoric people camped inside Salts Cave and Mammoth Cave for extensive periods of time. Even during the winter months when temperatures outside can go below 0° F, cave temperature remains around 54° F to 56° F. Warm during the winter, the caves are cool during the summer, when outside temperatures can rise to over 100° F. The caves were always available to provide shelter, water, and comfort.

The prehistoric cavers also mined the caves extensively. Besides chert, they took out gypsum, mirabilite, and probably epsomite. Using celts, shell scrapers, or convenient rocks from the floors of the caves, they pounded and scraped these minerals off the walls and ceilings. Gypsum forms crusts of sparkling white crystals on walls and ceilings in the dry upper levels of the caves where the passages extend under the sandstone caprock. In Salts Cave and Mammoth Cave the prehistoric cavers mined gypsum under ledges, deep beneath huge breakdown blocks, and from high on walls and ceilings. When heated, gypsum forms a powder that can be mixed with water to produce a white paint or plaster.

Mirabilite is a white-to-clear mineral that grows in various forms and often resembles finely shredded coconut. One form looks very much like gypsum. You can identify it quickly by its salty taste and because it melts in your mouth, which the tasteless gypsum does not do. You can also use a cane torch to test for mirabilite, which sputters and pops in the flame, unlike gypsum. Mirabilite, like epsomite, is a quite effective laxative, and was probably prized as such. It might also have been used—in small quantities—as a salt substitute. Because of the quantity of dried human feces in Salts Cave, modern residents of the area who entered the caves for doses of salts to cure constipation have suggested that the prehistoric people used the salts for the same purpose.

Great quantities of dried human feces—hundreds of deposits in Salts Cave and Mammoth Cave—are the most valuable of the prehistoric remains. Their analyses provide details about the diets and times of year during which the ancient peoples visited the caves. The feces contain weed seeds, sunflower seeds, hickory nuts, numerous kinds of berries, and meat. Of most interest are sunflower and sumpweed, which were cultivated. They make up one of the earliest horticultural complexes in the eastern United States.

The most dramatic evidence of early human activity came to light in the archaeological excavations in Salts Cave directed by Patty Jo Watson (who also directed the torch experiments described above). The entrance to Salts Cave opens from a huge, funnel-shaped sinkhole 100 feet in diameter and 50 feet deep. A waterfall splashes over one wall of the sinkhole to plunge into a black hole underground. Here you descend another sixty feet down a pile of slippery breakdown rocks to a huge entrance room, very dimly lighted through the cave entrance above. On the floor of this room ancient peoples lived and died. From the excavations came hundreds of bone fragments of turkey and deer and smaller animals, and of human beings. The cut marks, splitting, and charring of the human bones strongly indicate that the ancient occupants of Salts Cave were cannibals.

Knowledge of these prehistoric people has been further extended by the discovery of desiccated human bodies in the caves. A middle-aged aboriginal caver, nicknamed Lost John, was found in Mammoth Cave under a ledge of rock that fell on him when he was mining gypsum from beneath it. For many years the body of a nine-year-old child that had been placed on a rock ledge in Salts Cave around A.D. 100 was known as Little Alice, but the results of detailed examinations by Louise Robbins have recently led archaeologists to change the name to Little Al.

There is evidence enough to piece together a speculative account of at least one exploration trip by prehistoric cavers:

The man had planned the trip for some time. Outside of Salts Cave the wintry skies were bleak and bitter cold. Inside, the cave was warm. He and the woman ate a meal of hickory nuts and a few mouthfuls of meat as the glow of a smoky fire danced on their soot-stained faces. The man stuffed a small pile of dried seeds into a woven grass bag and tied it around his neck. The woman picked up a fat pile of cane. They had selected the canes from the main stockpile that they had brought in from outside in the waning days of fall. On this trip they would take only straight, tough canes that were thoroughly dry. The man also scooped up a bundle of cane about a foot in diameter and three feet long. It weighed only a few pounds. Each caver drew out three pieces of cane, and swiftly bound them together with short lengths of grapevine. They thrust the ends of the torches into the fire. The cane burst into cheerful flames that lighted the room. They were off.

They moved along a familiar path over fallen blocks to climb a pile of breakdown blocks that took them up to the ceiling and then down a rocky slope on the far side. Here they entered an immense chamber that seemed larger than it was because the walls were stained brown-black from many fires. The man led the way down a steep slope. They could feel cool dust between their toes. At a junction where several passages met, they took the left-hand lead. It was unfamiliar to the woman, who knew only the route straight ahead where she had gone with others to mine mineral salts.

They came to a seven-foot drop-off. The woman watched as the man picked out the hidden nubbins of rock with his hands and feet. Then it was her turn. She passed her torch bundle down. As she moved forward, her thigh slid along the polished rock at the lip of the drop. The people had smoothed the rough edges with their bodies over the years. She ran her hand over the slick, cool rock in delight at the contrast with the roughness elsewhere. The pair clambered down the rock slope to a hole in the floor of the room. They crawled into the hole and went on.

They came to a pool of water, where they drank deeply. The man said that it would be the last water they would see for some time. They set off along a narrow canyon and walked a long distance. Here and there they could see that they were not the first people to penetrate the lower levels of Salts Cave. There were places where three or four twigs had been burned to comfort earlier miners. Bare feet had flattened the sand trail. They deftly flicked the ends of their torches against the rock wall to trim off the excess charcoal, leaving black marks on the wall as others had before them. They made a game of dancing to the left and right to avoid stepping on hot embers, and then broadened the game to avoid stepping on the long-cold flecks of charcoal from the torches of others who had come this way. This part of the cave was ten feet wide and ten feet high, but projecting ledges and undercuts made it seem much smaller.

After a long trek, the man stopped. The passage, bone dry until now, was changing. The walls glistened with moisture and the sand floor was damp. The woman saw that only a few of the people had been here because there were no scrape marks from mining on the walls, and only a few pieces of charcoal lay on the floor. Around the next corner they came to a pool of water, and beyond it the walls closed in. This was as far as anyone had gone in the cave. They sat down and ate some sunflower seeds.

The man had been to this remote place before. He wondered if it really were the end of the cave. The pair made new torches, and examined the walls of the passage, searching for a way to climb up. The woman thought she saw a place, and led upward, bracing her arms and legs on the walls to keep from slipping. At the top a sandy ledge led onward, trackless. It had never before been seen by anyone. Her heart leaped with excitement as she shouted down to the man. She had found a way! The man climbed quickly after the woman, then close together they walked into the virgin blackness.

Water splashed down from high above. Canyon walls gave way to tumbled blocks of limestone and sandstone, and the floor was mud. They cupped their

hands to shield torch flames from the spray of water, and moved ahead to
where the passage became drier again. Now they felt mud squish up between
their toes. The man's tracks stood out in bold relief in the light of the torches.
The woman's footprints, smaller and shallower, formed a trail beside the man's.
They shivered. There was a breeze. They had been sprayed by the water. And
it was a long way back to where other people had been.

They saw that they had burned a little more than half of their torch supply. They should have started back earlier to be safe, but they had to know
where the passage went. The woman shrugged her shoulders, laughed, and
rushed on. They came to a T-intersection with a much larger passage, twelve
feet high and twenty feet wide, with a dry sand floor. They would go another
half-torch worth, and return with only one torch. They came to a fine, big
room. Ordinarily they would have searched it thoroughly, but now the man led
along the passage departing from the opposite side. Where the ceiling descended
and they had to crawl, they trimmed their torches one more time. Even with
the renewed light, the passage seemed remote and very far from home. The
passage stretched mysteriously ahead, without end.

The two explorers finally turned to retrace their way hurriedly to the
familiar parts of Salts Cave. They left their footprints in the distant mud.
Sometime later, the wet rocks near the muddy place let go with a rumble. Those
blocks sealed the passage until it was found by other explorers, who entered
it from Unknown Cave thirty centuries later. They found two sets of footprints
of bare feet in the mud.

If the man and the woman had continued—and ancient footprints found in
the mud by modern explorers suggest strongly that a man and woman really
did take such a trip—they might have found their way out the entrance of

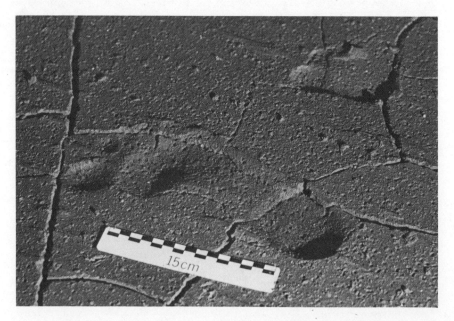

Aboriginal footprints in Unknown Cave.

Unknown Cave, and they would have realized that they had connected two of the immense Flint Ridge caves.

2/The Early History of Mammoth Cave

Today, leaving Mammoth Cave through the Historic Entrance on a muggy night in midsummer can be an unforgettable experience. Fog from the collision of moist, cool cave air and the hot outside air swirls about the top of the long flight of steps, making great white luminous balls around the lights that burn all night. As you walk up the slope, you soon lose sight of the black hole of the entrance in the fog. Much of the history of Mammoth Cave is similarly hidden in the receding past.

Until 1750, Kentucky was a wild, unknown land over the mountains from Virginia. No whites lived there and Indian hunting parties only passed through. Because of the French and Indian War, it was not until 1770 that any serious settlement took place. Then the preoccupation was with survival. Fierce clashes between settlers and Indians went on until Kentucky became a state in 1792. There were other clashes. Much of Kentucky's early history is contained in court records of lawsuits over land claims.

By 1796, the settlers had established the town of Bowling Green in Warren County, about thirty miles from Mammoth Cave. There was good bottomland there, but out toward Mammoth Cave, farmland was marginal. A farmer would file a description of his claim, pay the minimum down-payment, and farm as best he could. At the same time he would scout for a better piece of ground. If he found one, he would move, leaving his previously claimed land in a legal limbo, prey to squatters. Almost from the beginning, however, Mammoth Cave itself was seen as offering opportunities for making money by means other than farming.

The names of the pioneers who found and explored Mammoth Cave are not known. One popular story is that a hunter named Houchins discovered Mammoth Cave in 1797 when he chased a bear into it, but the cave was certainly known well before that time.

The land on which Mammoth Cave is located was surveyed for Valentine Simons in 1798. The 200-acre survey was registered in 1798 at the courthouse in Bowling Green as the "Mammoth Cave tract." Simons soon began digging saltpeter dirt in Mammoth Cave and Dixon's Cave on his land. Saltpeter was important for making gunpowder, not only for personal use, but also to sell. Word of saltpeter deposits in Mammoth Cave and Dixon's Cave was carried back along the trails to the east. Several copies of old sketch maps have been found that show Mammoth Cave as a veritable treasure trove of peter dirt.

Before 1811 saltpeter sold for about seventeen cents a pound. Then the War of 1812 with England broke out, and overseas sources of saltpeter were closed off. Now saltpeter commanded from seventy-five cents to a dollar per pound. In July of 1812, Mammoth Cave and Dixon's Cave were sold to some big-time operators. Charles Wilkins, a saltpeter dealer, was the main partner, and later Hyman Gratz, a wealthy Philadelphian, surfaced as a co-owner. Simons and a string of casual owners who subsequently occupied the land were paid off to clear the title.

Wilkins and Gratz set up an extraordinary mining operation to fuel the guns of war. They brought in gangs of slaves, and at one point had seventy people cutting trees for firewood and boring out logs twenty feet long to make pipelines for carrying water into Mammoth Cave and niter liquid out. Inside the cave, slaves built large collecting vats made of planks sawn by hand.

Ebenezer Meriam wrote of going into the twenty-foot-high entrance to Mammoth Cave in 1813, and along the main passage parallel to the two pipelines. In a circular room, now called the Rotunda, the outfall pipe was joined to a tall pump worked by men on a high platform. The spring-fed fresh-water line turned left, down the passage called Main Cave, where it trickled into the primary leaching vats. Proceeding through the gloom by the light of a single torch, Meriam came upon a blaze of light where the cave earth was being scraped into pans and hauled in ox-drawn wagons to the vats. It looked like a scene from the bowels of hell.

Water in the vats soaked the dirt, turning it to mud. Drains at the bottom of the vats captured the mineral-rich water, which was sent by a second pump back to holding tanks in the Rotunda at the base of the first-stage pump. The water was then hoisted out of the cave by the muscle power of the black men who manned the pumps. The water was boiled down in vats just above the entrance sinkhole.

After the war, demand for saltpeter slacked off, although some of the miners remained. For several years they supplied local gunpowder mills. Profits dwindled, but soon a new interest in Mammoth Cave developed. People would pay money to tour a good cave, particularly if there were something dramatic in it to draw attention.

With all the digging for peter dirt, it was not surprising that the remains of prehistoric people were found where they were buried in the caves. Miners in Short Cave found several mummies—desiccated corpses—that had been in the cave for more than 1000 years. Nearby in Mammoth Cave, human bones were removed from the route of the niter pipes. One of the Short Cave mummies was taken to Mammoth Cave and there placed in the Haunted Chambers (now Gothic Avenue) for exhibition to paying tourists.

Gratz and Wilkins each wrote letters and accounts of Mammoth Cave and its mummy, but the writer who brought the most attention to the cave in the United States was Dr. Nahum Ward, who visited Mammoth Cave in 1815. In 1816, he wrote a letter to a Massachussetts newspaper—copied in newspapers everywhere—about what he had seen. Ward's account of the mummy created a sensation exceeded only by the news that he was bringing another mummy from Short Cave to Massachusetts. By 1817, news of Mammoth Cave and its mummy had spread from the United States to all of Europe.

Hyman Gratz's manager, Archibald Miller, showed a trickle of visitors through Mammoth Cave, and between times continued the mining operation on a small scale. The visitors were mostly local, for in 1819 a financial panic of epic proportions hit Kentucky. Real-estate values plummeted to one-sixth their 1817 boom prices. Many of the settlers left the state in abject poverty. Many sold their slaves.

Wilkins died, and in 1828 the executors of his estate sold his half-interest in Mammoth Cave to his partner, Hyman Gratz, for $200 cash. Gratz struggled along, buying surrounding plots at bust prices, until his holdings totaled 1600 acres. A rustic inn was built, consisting of two log buildings eighteen feet square, separated by a ten-foot airway, with sleeping quarters above. However, in the midst of a deep depression with the frontier moving west and leaving Mammoth Cave in a backwater, the tourist venture was a failure.

In the early 1830s business picked up a little. Edmund F. Lee, a civil engineer, stayed at the Mammoth Cave Inn for four months, and in 1835 he produced the first accurate map of Mammoth Cave. This map shows about eight miles of passages, with a distance of about 2.25 miles to the farthest part of the cave, at the end of Symmes Pit Branch. Despite the survey, Miller told tourists that Mammoth Cave was twenty miles long, starting a tradition of exaggeration that persisted down through the years.

The next map of Mammoth Cave was made by a slave belonging to Franklin Gorin. Gorin lived in Glasgow Junction, eighteen miles from the entrance of Mammoth Cave. It is said that he was the first white child born in the region. He read law and practiced as an attorney. Given the Kentucky penchant for suing, he became as wealthy as any merchant, and looked about for an opportunity to pyramid his money through investment.

In the spring of 1838 Gorin purchased the Mammoth Cave tract from the Gratz brothers for $5000. As tourist manager he retained Archibald Miller, Jr. The Mammoth Cave Inn was enlarged to accommodate forty persons, and fences and stables were built. One of the smartest things he did was to bring in a new cave guide, his own slave from Glasgow, a black man named Stephen Bishop. Stephen Bishop became one of the greatest cave explorers in Mammoth Cave history. His work opened a new era, during which the idea of connection emerged.

3/Stephen Bishop

Of the dozens of individuals who may have been responsible for discovering one or more fragments of the caves that would one day be connected in the Mammoth Cave Plateau, the first about whom we know is Stephen Bishop. He was a slave sent to Mammoth Cave to make money for his master, Franklin Gorin. Before Stephen Bishop left Mammoth Cave, he had acquired an international reputation as a cave explorer.

Slavery in Kentucky reached a high-water mark between 1820 and 1830. Its ebb was the result of large landholdings being split as fathers left farms to sons until farming with slaves on a grand scale was impossible. Slaves were expensive. A promising young black man might fetch $2000 at auction; a black woman, $800. And from 1833, a law forbade the import of slaves into Kentucky. Thus, Franklin Gorin's investment of a slave in the enterprise at Mammoth Cave was somewhat unusual for the times.

Stephen Bishop was a young man of enormous curiosity. He was about five feet four, lean and

Stephen Bishop, the Guide—Mammoth Cave.

Stephen Bishop (1821–1857).

hard, with the build of an athlete. In Mammoth Cave, he moved effortlessly across the tumbled rocks. He was quite intelligent, and was said always to have an assured and tranquil air. He was a slave who had a quiet pride that did not offend his master.

In 1838, Stephen was put to work learning the routes in Mammoth Cave from Archibald Miller, Jr., and Joseph Shakelford, white guides who were sons of former guides. Stephen learned the spiel with no difficulty, and was soon conducting visitors with ease over two or three miles of passages.

The guide uniform of the day was whatever Stephen and the others could get. Stephen wore a chocolate-colored slouch hat, a jacket for warmth, and striped trousers. Over his shoulder on a strap swung a canister of lamp oil. In one hand he carried a basket of provisions for the longer trips—fried chicken, apples, biscuits, and often a bottle of white lightning for refreshment. In the other hand he carried an oil lantern—a tin dish holding oil and a wick, with a small heat shield held above the flame by wires. Above the protector was a ring through which he slipped his index finger.

Cave guiding was fine work. It was fascinating to talk with the people who came great distances to see Mammoth Cave. Stephen never conveyed any boredom with the old trails, but he wondered about the holes leading off from the commercial route. He yearned to explore those beckoning passages.

As the summer of 1838 wore on, Stephen began to probe the obscure by-ways. In a part of Mammoth Cave called the Main Cave, behind an enormous rock, the Giant's Coffin, he squeezed into a small room and down through a crack into a maze of passages beneath. Here he found fragments of burned cane torch and bits of grapevine tie left by ancient aborigines who had explored Mammoth Cave before him. Stephen explored the maze until he came to an awesome well-like hole over 100 feet deep. Here he turned back.

Gorin was pleased with the news. All hands went back to investigate. Gorin named the new pit Gorin's Dome, and sent a letter describing it to the newspapers. The account brought adventurers to Mammoth Cave. Most of them were content to go on the guided tour, but others would pay to be taken deep into the wild cave.

Late in October 1838, the air was crisp and cool. The color of the leaves ignited the forest with blazes of yellow and red. One visitor to Mammoth Cave, H. C. Stevenson, of Georgetown, Kentucky, spent several of those brilliant days touring the cave. Stephen told him the story of the discovery of Gorin's Dome, and of the difficult descent he had made with two others to the bottom. Stevenson asked whether Stephen knew of other passages where no man had ever gone before. Stevenson wanted to see some real cave. Yes, Stephen said, he knew where there was more virgin cave. Did the visitor have guts? Yes, indeed.

Stephen carried a lantern and a basket of provisions. Stevenson brought another lantern. They entered the cave, walked to the Giant's Coffin, and crawled behind it into the low room. Then they squeezed through the crack between the wall and the floor, into a passage where they could walk erect. They continued down the passage, pausing to toss rocks into the depths of Sidesaddle Pit. The oil lamp was miserable for seeing any distance, but they did not need to see the bottom. They could hear how far down it was when the rocks bashed against the walls and exploded into little bullets on the way to a final, distant crash on the bottom. Then they went into the side passage to look at Stephen's discovery of a few months previous, Gorin's Dome.

After returning to the main passage, they soon reached the brink of Bottomless Pit, a gulf that started as a vertical shaft on the left side of the passage. The pit extended across the floor, cutting off further progress. Stevenson tossed a rock into the void, counting to himself. It took 2.5 seconds from the moment of release, by his best estimate, before the rock splattered to rest on the bottom. Stevenson held both oil lamps, lighting the edge of the drop and the six-foot gap that blocked the way to the continuation of the passage on the other side.

The two explorers went into a side passage to return with two cedar-pole ladders. Stephen cleaned off the edge of the hole, scraping the rocks into Bottomless Pit. They thrust the first of the ladders across the pit, jamming it on the other side between two rocks. They rocked the ladder back and forth until the pole ends were seated. This was the result of their talks about what might be found if one could cross Bottomless Pit. The more Stephen had talked, the more Stevenson had wanted to see for himself.

When all was prepared, Stephen sat down, straddling the ladder. He tested it with his weight, rocking forward. It held. Now he was ready to go. He leaned forward on his hands, scooted forward, moved his hands forward, and pushed ahead again. He moved cautiously, a few inches at a time. Then he was over the void. Stephen moved on across quickly. There was nothing below to see except darkness. The pit was in shadow. It might have been more frightening if he could have seen the bottom, but he was interested in the other side. Now he rested his feet against the far edge, quickly scooted forward, and went over on his hands and knees on the other side of Bottomless Pit. Stephen let out his breath, cleared some rocks from the edge, and then told Stevenson to come across. Stevenson bridged the second ladder over the drop. It was longer and seemed safer. For an instant his light showed the depth of the pit, and then he also was across.

While Stevenson rested a moment, Stephen checked around the corner. Here he found another chasm, this one quite narrow. It could be jumped. Stevenson joined him to pause at the brink, then each in turn stepped smoothly across. So far so good. They had certainly moved far beyond the tourist realm. The actual distance back to the tourist trails was not far, but the formidable barriers they had crossed had taken them deep into an unknown underground wilderness.

Stephen was now eager to explore. The last time he had gone off the beaten track he had found exciting new cave, much to the pleasure of his master and himself. There had not been much risk that time. Now he had crossed a deep pit, taking a paying customer with him. The element of risk and the lure of the cave excited them both.

The two explorers set off through an oval passage in which an old stream had left banks of gravel. The passage was tilted downward, taking them deeper into the earth. They went on into a high, dry crouchway. The passage narrowed to only a few feet wide, and the ceiling pushed them down into a stoop. The smoke from the oil lamps was acrid, so they held the lamps at arm's length as the crack narrowed to a very tight squeeze. They forced their way to the top of a narrow canyon, watching to keep their legs from getting stuck. Beyond the tight place the passage opened up again into a fine, sand-floored walking passage. The passage continued to slant downward. Dark mud and the sound and feel of water in the air increased the gloom.

After several hundred feet they entered a mud-floored room that was the junction of several passages. It was an ideal situation for satisfying Stevenson's deepest desires. Stephen stood back as Stevenson walked across first, putting a row of human footprints in the mud of the damp floor. This is what he had come for, to be the first man to enter those Stygian realms.

Then they stood together on the brink of a valley in a passage thirty feet wide and forty feet high. They could see a stream of water below. Stephen stooped over and made a mudball. He hurled it to splash in the dark expanse. They had found the first real river known in Mammoth Cave. They named it River Styx. Stephen looked carefully, examining a possible route downward along one wall to the water. But that would be for later. Stevenson had had his day. It was time to leave the cave. Stephen would come back later.

The two retraced their steps to the junction. Here the right-hand lead led up a sandy slope. Stephen scampered up, pausing to see that the passage at the top led onward to the left and also branched to the right again. The passages were wide open for exploration, and were very tempting. Nevertheless, they had the crossing back over Bottomless Pit to make. Until that traverse became more familiar, explorers would feel very remote in this part of Mammoth Cave. One last look, and then they crossed their rough bridge again.

Back at the inn, they told of their adventures. Stephen stood back while Stevenson talked. Gorin was ecstatic, partly over the contagious excitement of Stephen and Stevenson, and partly because Stephen had—in a few months—discovered more new cave passages in Mammoth Cave than all the other guides together. From the merry, confident look in Stephen's eyes, Gorin knew there was more to come.

Stephen had established himself as a great cave explorer. But if he were promoted as the first great caver in America, he himself knew better. He had found evidence that prehistoric people had explored far back in Mammoth Cave. And he knew that more than eight miles of Mammoth Cave had been explored before he was born, by men whose names he probably knew, but which are now lost to us. Most of the former discoveries, however, were made in easy walking passages. The dimension Stephen added was that of risk in narrow passages and across deep pits. His discoveries and the publicity given to them were the starting point of a kind of tenacious modern exploration. He pushed relentlessly through every passage that was in any way passable, in hopes of finding big discoveries beyond.

That is true. There is, however, something suspicious about this classic version of the first crossing of Bottomless Pit in Mammoth Cave. There are stories that this was not the first crossing. And modern explorers have puzzled about the fact that just down the passage from the traditional cedar-ladder crossing point the gap narrows so that one can easily climb across it. One can wonder. Did Stephen take Stevenson on the first known rinky-dink in Mammoth Cave? It is certainly possible, for Stephen had that kind of sense of humor. And it would be in the best tradition of Kentucky cave showmanship. The paying customer could be given the thrill of a lifetime by allowing him to make a dangerous, but objectively rather safe crossing. He could be permitted to think that he was the first—Stephen would not count—to cross Bottomless Pit. And then he could join Stephen in exploring truly virgin passages. It had all the elements of the best entertainment in caving. These circumstances suggest that the highly touted crossing probably was staged by Stephen for Stevenson, to the thorough satisfaction of them both.

During the slack tourist days that followed the crossing, a bridge of cedar poles was constructed across Bottomless Pit. Now many long trips were made into the new area of Mammoth Cave, to discover, explore, and name Pensacola Avenue, Bunyan's Way, Winding Way, Great Relief Hall, and, deep in the depths, the mud-encased River Styx. Mammoth Cave had exploded. Stephen explored continuously throughout the winter.

Where others feared, Stephen cautiously kicked his heels into the mud slope leading down to River Styx. He could see the bottom through the water after his eyes grew accustomed to the reflections of the ceiling. One remarkable discovery was that the new river contained blind white fish a few inches long. Stephen caught some for exhibit. He waded across the river, climbed the mud bank on the other side, and found an opening leading to a crouchway. It would be another place to explore when the bigger passages had all been looked out.

Gorin was most pleased. In one winter Stephen had doubled the known length of Mammoth Cave. If newspaper editors had gone wild over accounts of the discovery of Gorin's Dome, what would they make of these sensational new discoveries?

Gorin wasted no time. From Thomas Bransford of Nashville, he hired two slaves for $100 a year each. These slaves, Materson Bransford and Nicholas Bransford, were trained by Stephen to guide on the tourist trails of Mammoth Cave. Stephen was a good teacher, and the new guides were soon operating competently.

Tourists began to arrive in the spring of 1839. Gorin's inn was overwhelmed, and the guide force was kept busy with the steady stream of visitors. They came on horseback and by wagon from Bell's Tavern on the main highway. These were good times in Kentucky. Gorin saw the business increasing, and concluded that he had better invest more capital in Mammoth Cave, or perhaps sell it to a favored buyer.

Dr. John Croghan, at fifty, lived well on his family estate in Louisville. He had known about Mammoth Cave for years, but he saw it in a new light on a trip to Europe, where he found many excited reactions to newspaper accounts of the new discoveries. His family had been to Mammoth Cave, and his brother described it vividly. But Dr. Croghan had not focused on it at the time as being one of the great wonders of the world. Now he found himself telling tales about it to his European friends.

Six years after his European trip Dr. Croghan finally visited Mammoth Cave. He met with Gorin several times to hear of Stephen's fabulous discoveries, of the potential of new discoveries, of the growing business, and of the inadequate inn. Abruptly, Dr. Croghan made an offer: $10,000 for the whole works—land, cave, inn, slaves, and all. Gorin accepted, but said he would keep close watch on the place, for he loved it and the people who worked there.

Croghan at once expanded the Mammoth Cave Hotel. Through the acts of a favorable legislature, he got aid to construct one road from Cave City to Mammoth Cave, and another from Rowletts to Mammoth Cave, and from there across the Green River at Grayson Spring. Another branch of the road was constructed to the southwest, toward Dripping Spring. The result was a splendid road system across semi-wilderness country. Croghan's road network also neatly bypassed Bell's Tavern, and thereafter Mammoth Cave took over as the region's main stop on the four-horse stage route. The stage traveled daily, and because the jarring ride from Louisville or Nashville left travelers bruised and tired, they often stopped over at Mammoth Cave. With Stephen and the manager, Archibald Miller, Jr., Croghan worked out a full week's worth of different trips into Mammoth Cave. It was a tempting variety to suit anyone's curiosity and needs, and to delay the weary traveler.

Mammoth Cave rapidly became a celebrated stop on the main route rather than an obscure curiosity off in the wilderness. Croghan offered cave trips from two to nine miles in length. More guides were added. Croghan made more money from rooms and meals than from the cave fees, but it was Mammoth Cave that kept visitors coming and staying and spending. Stephen was also a main attraction. His charm and wit by day, and his determination to unravel Mammoth Cave's secrets on exploring trips at night, made him known and sought after by tourists from all over the world.

In 1840, another visitor confided to Stephen that he sure would like to see some real cave. Did Stephen know where to look? Stephen whisked the man—whose name is not recorded—down to the start of River Hall, where he had made his big breakout the year before. Here the two climbed the right wall up the sand slope. They moved along a seemingly interminable stoopway to a collapse of rocks. Stephen began enlarging the passage ahead so they could get through, passing rocks back to the visitor. It was hard and wet work that no ordinary tourist would stand for, but Stephen had gauged his companion well. After several hours they had dug their way out onto a balcony over the largest void Stephen had ever seen underground. To the left lay a deep pit, ninety or more feet below the balcony, and vaulting over it soared a dome with a ceiling at least another ninety feet upward. Straight ahead, up a teetering rock pile, they scrambled to a second balcony. Stephen chunked rocks into the pit to the left, and into an adjacent pit next to a series of fluted bedrock columns that looked like the portico of a Corinthian temple. Wild with excitement, Stephen charged down the slope behind the columns, then on to where cream-colored flowstone formed a curtain on the wall. He stepped back to take a long look at the top of the dome, shrouded in shadows. Perhaps he and the visitor then drank to the depths, for many tourists reported that Stephen often carried along a little drop to cheer them on their way.

Stephen had shown his worth again. Croghan asked him to spend a few days each week exploring, while Croghan considered ways of publicizing this new discovery, Mammoth Dome.

In later weeks and months, Stephen went back across Lake Lethe, River Styx, and on to Echo River. There was a large walking passage there that branched several ways, and he explored many of the branches. By 1841 everyone who came to Mammoth Cave had heard of Stephen Bishop, and asked that he be assigned as guide. In July, two young men—John Craig and Bruce Patton—showed up and asked if Stephen could take them across the newly discovered Echo River. Stephen sized up the new cavers, outfitted them with lamps and a small sack of food, and took them into Mammoth Cave. At Echo River they pushed off from shore in a skiff that the slaves Matt, Nick, and Alfred had brought in. Stephen pushed the boat into the river passage under a low limestone arch and into a parallel passage that was wall-to-wall river.

What echoes! The three began to shout, then to sing. Any fear they felt was dissolved in revelry as they glided along the deep greenish-black waters.

On the far shore they pulled up the boat, and then Stephen set a brisk pace along a large passage. After half a mile it changed into a sinuous, damp canyon that was ten feet wide and thirty feet high. It extended for another mile. Then the floor rose to near the ceiling and Stephen stopped. This was where they were to explore. No one had been beyond this point. Stephen motioned for the somewhat nervous Craig and Patton to go on ahead. They climbed upward into a canyon much like the one they had been walking in, except that no man had ever been there before.

After about 500 feet they came to a junction where a large passage entered on the left. In cross-section the passage was a perfect ellipse, sixty feet wide by six feet high. The ceiling was like none they had ever seen. It was covered with dazzling white gypsum flowers and snowball-sized encrustations. They named it the Snowball Room, and ate their lunch there. Then they set off, blazing the first trail down what is now called Cleaveland Avenue. They walked for a mile and a half. At one place their footsteps were echoed by the ceiling. They stomped their feet in delight to listen to the reverberations. They stopped their explorations where the passage began to widen to dimensions of nearly 100 feet wide and forty feet high. Stephen had exhausted his companions exploring virgin cave.

Croghan was amazed. He asked Stephen to sketch a map of what he had found. Stephen got the biggest piece of paper he could find, and penciled a fair likeness of the cave. A variety of maps had been made of Mammoth Cave, including several crude impressions drawn by first-time visitors, and one genuine survey. Stephen's map was more detailed than any of these, and Croghan was so impressed that he asked him to make a better one. This was not done until winter, when Croghan took Stephen to his Louisville estate.

It took Stephen more than two weeks to draw the map. He was a perfectionist. He kept walking through the cave in his mind, noting down every interconnection and every lead. He put in several side leads to Echo River that only he had investigated in the heady months after his first discovery. Then the cave map was drawn in ink. It was a thing of beauty. It was also indelibly inscribed in Stephen's mind. He would delight visitors by sketching instant maps on the sand of a cave trail, showing them where they were in relation to the Mammoth Cave Hotel. Copies of the map were made for use at Mammoth Cave, and it was published three years later. Stephen was given full credit for his cartography. His reputation soared.

Stephen's discoveries made everyone think of possibilities even beyond Mammoth Cave Ridge. The Snowball Room, Stephen knew, lay partway down Mammoth Cave Ridge, under which many levels of Mammoth Cave twisted. Mammoth Cave might possibly extend under the flanking valleys. Three years earlier, in 1839, Gorin had ridden over to Salts Cave on Flint Ridge to have a look. Stephen knew the stories that Salts Cave passages were blackened with soot from aboriginal torches, just like many areas in Mammoth Cave, and that the passages were every bit as large as any in Mammoth. He could not help but wonder whether or not Salts Cave and Mammoth Cave were connected. Finally, during a lull in the tourist trade, he went to Salts Cave to look for himself.

He probably concluded immediately that Salts Cave heads north, not toward Mammoth Cave—as some had claimed—but parallel to it. While the two caves might be connected, there was certainly a lot of room and a deep, wide valley between them in the nearly three miles that separated their entrances. And since Mammoth Cave was yielding such discoveries, why stop exploring it in order to seek whatever potential Salts Cave might hold? Stephen went back to Mammoth Cave.

At Mammoth Cave, however, Stephen was in such demand that Croghan made him a showpiece. Stephen was able to do less exploring, and spent most of his time showing the gentry over the tired old trails. Visiting scientists asked for Stephen because he alone knew the geology of the cave, a practical knowledge that he had supplemented by reading. He was doing what he liked, but his reputation confined him, and made him focus more than ever on the fact that he was a slave.

Dr. Croghan became preoccupied with making a part of Mammoth Cave into a sanitarium for tuberculosis patients. Stone huts were erected underground, and patients were moved into them. The guides and slaves brought food to the cave dwellers. Some of the patients died, and all the rest took turns for the worse. Within a year the experiment was abandoned.

In 1849, Croghan himself died of tuberculosis, leaving the cave in trust to a succession of relatives near and far for seventy-five years. The will granted Stephen freedom in seven years. Stephen stayed with Mammoth Cave. He married a maid at the Mammoth Cave Hotel, Charlotte, and had one son. Once he took Charlotte to see the wonders beyond the Snowball Room. They marked the trip by writing their names on the wall.

In 1852, Nathaniel Parker Willis visited Mammoth Cave and asked if Stephen could take him to see the blindfish. On one of the frequent rests on the way, Willis asked Stephen his views on slavery. Stephen told him that, despite the fact that he would be freed in five more years, he was saving money and hoped to buy his freedom and that of his wife and son as soon as possible. They planned to settle in Liberia, which had become a free and independent republic in 1847. While Mammoth Cave meant a great deal to Stephen, freedom meant more. Willis refreshed himself with frequent draughts from Stephen's flask. When they reached River Styx, Stephen dipped up a blindfish with his net. He caught them for display at the hotel.

Stephen did not purchase his freedom, but was set free in 1856. However, he never went to Liberia. In 1857, Stephen Bishop died, a young man of thirty-six. His widow married William Garvin and continued to live at Mammoth Cave. What happened to Stephen's son is not known.

For all practical purposes, Stephen Bishop was the first to connect substantial pieces of the big cave. He was the first to venture beyond the obvious walking passages to explore the crawlways, pits, and rivers of Mammoth Cave.

Stephen's first master and friend, Franklin Gorin, writing many years after Stephen's death, said:

> I placed a guide in the cave—the celebrated and great Stephen, and he aided in making the discoveries. He was the first person who ever crossed the Bottomless Pit, and he, myself, and another person whose name I have forgotten were the only persons ever at the bottom of Gorin's Dome to my knowledge.

> *After Stephen crossed the Bottomless Pit, we discovered all that part of the cave now known beyond that point. Previous to those discoveries, all interest centered in what is known as the "Old Cave" . . . but now many of the points are but little known, although, as Stephen was wont to say, they were "grand, gloomy, and peculiar."*
>
> *Stephen was a self-educated man. He had a fine genius, a great fund of wit and humor, some little knowledge of Latin and Greek, and much knowledge of geology, but his great talent was a knowledge of man.*

4/Lute Lee and Henry Lee

Steamboats in the 1830s, the Louisville & Nashville Railroad, which was completed in 1859, and the connecting stage lines brought many visitors to Mammoth Cave. Then the Civil War stopped tourism as Kentucky families divided loyalties to fight for either North or South. Croghan's will prevented Mammoth Cave from being mined again for saltpeter. The war passed, but Mammoth Cave did not revive as a tourist attraction.

Discoveries in Mammoth Cave became fewer and farther between. It was said that the guide Frank Demunbrun found a remarkable passage beyond El Ghor in the mid-1850s. He described it as a mile and a half long, containing a river ten feet wide. It was named Mystic River. All knowledge of the route faded, and by 1880 no guide knew the way. (It was not found again until 1973.) The cave lay fallow. Not until 1907 did guides explore beyond the Big Break, a giant rock pile, to discover an extension of the cave along what is now called Kentucky Avenue.

Then a further slowdown in exploration perhaps resulted from a new influx of tourists. The Mammoth Cave Railroad had been constructed in 1886 from the Louisville & Nashville railhead at Glasgow Junction to the front door of the cave. The jolting ride in the wooden coaches pulled by the locomotive Hercules gave plenty to write home about.

Mammoth Cave's quiet days were in sharp contrast with the lively exploration across the valley in Flint Ridge. In 1895, Lute Lee and Henry Lee forced their way down through a sinkhole—the Woodson/Adair Entrance—on the south flank of Flint Ridge. Their discovery, Colossal Cave, attracted the attention of officials of the Louisville & Nashville Railroad. In true Kentucky-style legal maneuvering, they negotiated an intricate set of agreements to acquire Colossal Cave. They surveyed it and blasted open a wide entrance only a mile and a half from Mammoth Cave.

Railroad officials were now interested in improving the tourist route in Colossal Cave. There were several long and tortuous passages leading from the Bedquilt Entrance to the north end of Colossal Cave. The railroad thus hired the Lee brothers to find a shorter route from Grand Avenue in Colossal Cave to the Bedquilt Entrance. Lute Lee described their success as follows: "My brother, Henry, told Daniel Breck, who was representing Smith, the L&N president, that if he could have three days, he could connect Bed Quilt Cave with Colossal Cavern. Breck agreed and in two days we made the connection and took him through." This new connection of two segments of Colossal Cave probably captured nobody's imagination except those of the participants. However, Henry and Lute knew exactly what they were doing. They went in the Bedquilt Entrance to the end of a passage where compass readings indicated a heading straight for Grand Avenue. Dusty sand extended almost to the ceiling of the passage. The brothers dug a trench big enough to crawl through, and within a few hundred feet they reached Grand Avenue, the main trunk passage thirty feet wide and twenty feet high in Colossal Cave. This was the work of real cavers.

During multiple legal hassles, L. W. Hazen, Colossal Cave's first manager, was offered $500 by the railroad if he could show a connection to Mammoth Cave. So far as we know, this is the first recorded speculation that any of the caves in Flint Ridge might join Mammoth Cave. Hazen never collected his money. In a passage across the top of Colossal Dome, Hazen's name and the inscription "MAMMOTH CAVE" are scratched on the wall.

By 1910, Colossal Cave was all but dead as a commercial venture. Only six tourists a week visited the cave. Guide Roy Hunt had ample time to explore in the far reaches of Colossal Cave. He found the scratched name of Pike Chapman, who had been killed in an entrance cave-in at the back of Salts Cave in 1897. Of the history of this era, we are left with only a few records. There is a fairly accurate map of 2.5 miles of passages in Colossal Cave, and descriptive accounts of the tourist routes. Colossal Cave was also said to contain a lost river.

While Colossal Cave in the southern part of Flint Ridge was being explored, very little was happening in the northern part of that ten-square-mile upland. A sketch map published in France in 1903 shows Unknown Cave with 1.3 miles of passages north of the Colossal Entrance. Unknown Cave is shown in a squiggle of passage lines as being a back entrance to Salts Cave. Except for the entrance, this map is almost entirely fanciful. A local man, Marty Charlet, had drawn it. From pieces of newspaper and dates scratched on the wall, we know that Unknown Cave was visited by a few people between 1903 and 1913. They knew only a few hundred feet of small crawlways and wet pits, and there is no evidence that the early explorers pushed the tiny, wet-shaft drain passage in attempts to connect Unknown Cave with any other cave.

Thus, the three main caves known in Flint Ridge—Salts Cave, Colossal Cave, and Unknown Cave—seemed not to be adapted for commercial exploitation. They did show potential for exploration, and the idea of connection was in the air.

5/Max Kaemper and Ed Bishop

One of the most remarkable individuals to link his life with Mammoth Cave since Stephen Bishop was a twenty-three-year-old German engineer, Max E. Kaemper. He arrived in New York in 1907, and spent six months there learning English and acquainting himself with American mining and manufacturing methods. Early in 1908, he arrived at Mammoth Cave, expecting to stay a week. The week stretched to a month, then to eight months. The cave bug had bitten him! On a trip to Cathedral Domes in Mammoth Cave he was shown two other routes to the place. He had also led his guide into side passages to make discoveries on his own. However, he was frustrated by the inaccurate guide maps. He resolved to make an accurate map.

The manager thought Kaemper might be learning too much and tried to discourage him, for Mammoth Cave passages might extend beyond Mammoth Cave estate lands. However, H. C. Ganter, a former manager who lived nearby, recognized both talent and zeal in Kaemper. He seized the opportunity, urging the manager to contact the chief trustee of the Croghan estate and pass along

Ganter's endorsement of the project. Back came the favorable reply. Kaemper was given permission to explore and to survey to his heart's content in Mammoth Cave, and he was to be furnished with a guide as an assistant. Young Ed Bishop, a black guide described as Stephen Bishop's grand-nephew, knew the cave well, especially the regions far beyond Echo River where the going was tough. Ed was as enthusiastic about the project as was Kaemper.

Kaemper's plan was well suited to the immense task. He decided to base his map on the very accurate Ganter Avenue survey, and from it traverse the Main Cave and the Long Route, using a surveyor's compass. He paced the distances between survey stations, meticulously wrote the bearings and distances in his notebook, and made sketches of each passage.

Both men loved the cave. Their common undertaking served to help Kaemper learn conversational English from Ed. And Ed learned from Kaemper a kind of patience and persistence that nobody had brought to Mammoth Cave before. Their methodical approach slowly but surely reduced the confusing labyrinth to a pattern of named passages. Stephen Bishop's map had shown some of the passages in Mammoth Cave in their proper relationships. Kaemper's goal was to show them all in scale.

On some days, Kaemper and Ed worked on the surface, making a topographic map. Among the tall trees they sighted level lines, and branched off from the Mammoth Cave Ridge crest into the valleys. Ganter was told that the Main Cave portion of Mammoth Cave ended near his home, but far short of the hillside that he suspected terminated the passage.

Kaemper and Ed discussed the possibility of finding more cave beyond Ultima Thule. The signs were right. In most places where passages terminated, water and mud flowed in from hillsides. At Ultima Thule there was only a dusty, dry pile of slabs and blocks of rock walling off the passage. And now, with survey evidence that they were at least 300 feet from the hillside, they explored.

Kaemper and Ed entered a tiny crawlway, but the jumble of rocks provided smaller and smaller spaces, until Kaemper had to stop. Beyond the final block he could hear the faint falling of water. It was too bad that they could not go on, but the pair returned to mapping for another month.

All that time Kaemper kept thinking of the waterfall beyond the Ultima Thule crawl. At the end of the Long Route another mountain of breakdown sealed a passage, Sandstone Avenue, blocking their progress. This was only about 500 feet from the Ultima Thule termination, so they tried to connect the passages. They had no luck. Frustrated, they returned to Ultima Thule, but with more determination than before. They pried rocks from the floor of the crawlway, scooped out loose pebbles, and pushed the rubble back out of the enlarged crawlway with feet and forearms. They broke through.

They were surprised to find not Sandstone Avenue but an immense oval chamber. It was 160 feet long, 120 feet wide, and sixty feet high. Kaemper named it Kaemper Hall. On the left they found a pit ninety feet deep and named it Bishop's Pit. To the right they found a second room, about seventy-five feet in diameter, and a third room 250 feet long, 125 feet wide, and seventy-five feet high. This was named Violet City, after the chief trustee's wife, Violet Janin. At the end of Violet City was a breakdown pile covered with spectacular stalagmites and drapery flowstone. The manager was so pleased that he immediately enlarged the entry to this part of the cave and built a path to exhibit it.

Kaemper, hero of this triumphal discovery, pleaded to be allowed to use explosives to connect Violet City with Sandstone Avenue. Kaemper's map showed that the two passages ended with one practically on top of the other. Permission was given, so Kaemper and Ed blasted their way forward. However, as Horace Hovey tells it in the May 22, 1909, issue of *Scientific American*, they stopped because the manager feared that they might inadvertently blast an entrance to the outside on land not controlled by the Mammoth Cave estate.

Recent historical detective work by Harold Meloy suggests that Kaemper and Ed indeed *did* blast their way to the surface. Almost in H. C. Ganter's back yard the leaves were seen to lift and settle back, accompanied by smoke. Ganter's knowledge of the location of this "back entrance" to Mammoth Cave would provide another chapter in the connection story six years later.

Kaemper's map was the pride of the management. It was on two sheets. The cave map on one piece of paper was meticulously drafted and lettered by Kaemper. Five levels of passages were delineated in five colors of ink. The map is a masterpiece of cave cartography. Its general accuracy is excellent, considering that the distances were paced. Minor inaccuracies have been found, but the fact remains that Kaemper made a remarkably accurate map of about thirty-five miles of passages. The second sheet showed the topographic contours of the land over Mammoth Cave, and was keyed to the cave map by a set of register marks.

The management wanted to get maximum publicity from the map without revealing it. Their motive was to build the tourist business without getting into problems with adjacent landowners. Kaemper's map had to be kept secret to keep adventurers from seeking new entrances. Thus, they permitted Horace Hovey, author of a celebrated series of guidebooks to Mammoth Cave, to view the map. They knew that Hovey's words would be exciting, and that when sworn to secrecy, he could be relied on to reveal no details. Although Hovey did as told, he nevertheless urged the management to allow the map to be published. That failing, he expressed the hope that the map would be made public after the cave was sold according to the terms of the Croghan will. This, he said, echoing an earlier suggestion, would be possible if Mammoth Cave were made into a national park.

6/Edmund Turner

In 1912, E.-A. Martel, the celebrated French "Father of Modern Speleology," visited Mammoth Cave. Martel declared his absolute conviction that Mammoth Cave and Colossal Cave were connected, reinforcing Hovey's conclusion published earlier. Hazen had opined the same thing, but his statements seemed to be based mostly on financially motivated wishful thinking. Hovey and Martel based their opinions on careful studies of the geological setting and topography.

Martel went out on a long limb in his speculations. He was very specific: "When the connection (for me evident and assured) with Colossal Cavern, Salts Cave, etc. is one day established, the total will be not less than 241 kilometers." That is, Martel just took the management's claim of 150 miles for Mammoth Cave as a prediction that would one day be fulfilled. When the connection was finally made in 1972, the total surveyed mileage of Flint Mammoth Cave System was nearly 145 miles.

Martel joined the chorus, which had included his countryman Max Coupey de La Forest nearly a decade before, in urging that Mammoth Cave be made into a national park.

That same year (1912) a young civil engineer in his early twenties, Edmund Turner, came to Flint Ridge. He was a short man, on the borderline between frail and wiry. His pointed, receding chin was offset by a handsome mustache and fashionably long sideburns. Local people took his fine features to be those of an artist. Turner asked who might show him some real cave, and was soon directed to Floyd Collins. At twenty-five, Floyd was slightly older than Turner, and had been exploring caves for years. He took Turner into Salts Cave, where they went to the far end of the upper passage and left their names on rocks. They made several more trips into Salts, on each of which Turner discovered leads to unexplored passages hidden obscurely behind the huge pieces of breakdown that filled the main cave. Although Turner was not in good health, his frequent visits to Salts and other caves, together with his astute observations about them, gave him a reputation as an expert cave explorer.

One local story goes that in the summer of 1915, Turner approached L. P. Edwards, who owned a farm on Flint Ridge, with a proposition. Turner, it was said, knew of the existence of a cave under Edwards' land as a result of his cave explorations. He now asked whether Edwards would give him an interest in the cave if he could show him where to dig into it. Edwards was interested. Turner thought they had a bargain, so he began digging at a place down the hill from Edwards' house. Pieces of flowstone in the diggings were signs of what was below.

The predicted cave was soon found, and was named Great Onyx Cave because it proved to have the finest cave-onyx columns and stalactites in the area. Beyond the profusion of beautiful formations at the entrance lay a major trunk passage, ten to twenty feet wide and from ten to thirty feet high. One fork led to a fine vertical shaft with water falling into it. Turner continued to explore through the year while Edwards hastened to commercialize his new cave.

Then Turner took a chill and died, probably of pneumonia. He was buried on Flint Ridge. Edwards immediately claimed that he, not Edmund Turner, had discovered Great Onyx Cave. And some say that Turner, who had never benefited from his discovery, had in revenge walled up passages in Great Oynx Cave that connected to nearby caves. It seems probable that Turner did find Great Onyx Cave. As for the legend about hidden passages, an enormous amount of exploration has failed to uncover any way out of or into Great Onyx Cave underground. It remains the only major Flint Ridge cave that has not been integrated into the larger Flint Mammoth Cave System.

7/George Morrison

George D. Morrison, a mining engineer and oil prospector, had arrived in the Central Kentucky Cave Country in 1915. He became fascinated with Mammoth Cave, particularly with speculations that Mammoth Cave extended many miles beyond the boundary of the Mammoth Cave estate. After listening to the stories of the aging H. C. Ganter, Kaemper's friend and the former manager of Mammoth Cave, Morrison took mining options on a large number of acres of land immediately southeast of the Mammoth Cave estate. One of Ganter's stories was about the "secret entrance" to Mammoth Cave that Max Kaemper and Ed Bishop had blasted at Violet City. Ganter pointed out the location to Morrison. The opening was concealed by a pile of branches.

Morrison then ran illegal surveys by night into the southern reaches of Mammoth Cave through the Violet City entrance. After he had determined where passages of Mammoth Cave went under land he had leased, he is said to have sent his employees into Mammoth Cave to listen for the sound of his drilling machine on the surface. He was also accused of sending men into the cave to set off dynamite in passages near the surface in hopes that observers in the woods above might see the ground lift and the smoke vent. He found one entry point by this method, but it was on Mammoth Cave property, not on land he controlled. Furthermore, he was discovered in the act with his cohorts. All were arrested and fined for trespass.

A year later Morrison blasted an entrance to Mammoth Cave on land owned by Perry Cox. He again resumed clandestine underground surveying, and he explored in the Markolf's Dome area. Cox, however, had already sold his underground rights to the Colossal Cavern Company, whose officials promptly went to court and successfully enjoined Morrison from using the Cox Entrance to open another Mammoth Cave commercial operation in competition with Colossal Cavern.

Morrison plotted his maps carefully again, and then went away to raise more money for his venture. He was a shrewd man, confident that he could eventually beat the Mammoth Cave estate. He knew that parts of Mammoth Cave were not under Mammoth Cave estate land, and was determined to gain control of them by finding his own entrance.

In 1921, Morrison returned with a suitcase full of money, or at least with solid promises from backers. He unrolled his maps and resumed his secret surveys in preparation for his next surprise. Soon he had excavated the New Entrance to Mammoth Cave on land he controlled. He built a modest hotel near the entrance and set to work improving underground trails. He constructed a stairway down some of the most spectacular vertical shafts in Mammoth Cave.

What most infuriated the Mammoth Cave management was Morrison's attention to advertising. He erected an extensive network of roadside signs touting the "NEW ENTRANCE TO MAMMOTH CAVE." He represented that he was selling tickets to Mammoth Cave, as indeed he was—to his portion of it. And because his New Entrance was on the road between Mammoth Cave and the main source of tourists, he siphoned off a great deal of business.

To the law! The management of the Mammoth Cave estate went to Federal Court to obtain an injunction forbidding Morrison to advertise using the words "MAMMOTH CAVE." That name belonged to the estate, it was claimed.

With a flair for the dramatic, Morrison produced the maps he had made. They proved to the satisfaction of the jury that what Morrison was showing was indeed Mammoth Cave. The cave was a gigantic labyrinth of passages, extending from the Historic Entrance all the way to the south end of Morrison's holdings. Under cross-examination Morrison was able to show that the Colossal Cavern Company had cave passages extending under lands owned by the Mammoth estate. If Colossal Cave could extend under Mammoth Cave land, then Mammoth Cave could extend under Morrison's land. The court decided that Morrison could use the name of Mammoth Cave.

The Appeals Court sustained the lower court, but did find Morrison's competition to be somewhat unfair. With native Kentuckian attention to the detail of property rights, the court decreed that Morrison must add the following disclaimer to his promotional literature: "We do not show any part of the cave which prior to 1907 was generally known as Mammoth Cave; that portion of the cave can be seen only through the old entrance."

In 1924, Morrison remedied the last defect of his New Entrance venture. He blasted another entrance at the Frozen Niagara area, where the most spectacular cave-onyx drapery and waterfall formations abounded. Here was Crystal Lake—an artificially dammed pool complete with a 100-foot boat ride—and a way to avoid having to take tourists back out of the cave by retracing the route and climbing the exhausting flight of stairs. Morrison then prospered until his land was bought in 1931 for the Proposed Mammoth Cave National Park. Many people spoke well of Morrison. And even the Mammoth Cave estate officials could not help but admire the nerve and success of his maneuvers. George Morrison was an authentic hero of the cave wars.

8/Floyd Collins

During the winter of 1916–17, Floyd Collins was more determined than ever to find a cave. Edmund Turner had found a cave, and the newly opened Great Onyx Cave just down the ridge from the Collins family's farm was beginning to attract cash customers. Floyd found an 800-foot section of passage—Floyd's Cave—on the home farm, but it ended in impenetrable breakdown at both ends. A couple of other holes on the property yielded segments of passage, some with pretty decorations, but no giant cave such as Stephen Bishop, the Lee brothers, and Turner had found. There was some solace in the fact that there was a cash market for cave-onyx. It is said that Floyd and his brothers cracked off dozens of stalagmites and stalactites, both in their own small caves and from other caves as far away as fifteen miles. Two cave-onyx "mines," one in Unknown Cave and one in Salts Cave, have scratched on the walls the names of most of the male members of the Collins family. Formations could be sold to cave-onyx sales stands that lined the road to Mammoth Cave. Sales were brisk because tourists felt that they had not really visited the cave unless specimens were brought home for the mantelpiece.

In 1917, so one version of the story goes, Floyd was making the rounds of the family trap line. When he looked for the trap set in a small sinkhole just down the hill from the family house, it was missing. A raccoon, probably, had pulled the trap into an opening between the rocks. Floyd checked, and felt cool air coming out. Now he forgot all about the trap. He moved a great quantity of rocks from the bottom of the sinkhole to reveal a ledge of sandstone that rested on limestone. The breeze had increased, and Floyd was certain that moving more rock would reveal a cave.

Floyd went back to the house and casually asked his father whether he could have half the profits if he found a cave on the property with commercial possibilities. Lee Collins pondered the improbability of Floyd's finding anything now after all the poor prospects he had turned up so far. He also knew that Floyd just might find a cave through stubbornness and persistence. After thinking about it overnight, he made the agreement with Floyd, and told him to go find the cave.

Floyd dug furiously and on December 17, 1917, he opened a hole into a cave. Whether he found the trap is not known. What he did find after shoving his kerosene lantern ahead of him for about forty feet in a crawlway two feet in diameter was a passage fifteen feet wide and five feet high. He skirted a deep hole by hugging the left-hand wall, then crawled over limestone blocks for a few hundred feet. Here the passage floor plunged into blackness. Floyd climbed down the slope to the bottom of a vast room ninety feet high. He called it the Grand Canyon. A trunk passage twenty feet high and thirty feet wide crossed through the Grand Canyon at right angles at the bottom.

Probing the breakdown around this junction, and exploring the right-hand passage, Floyd found walls encrusted with cream and white gypsum and hundreds of beautiful gypsum flowers up to eighteen inches long. He called his cave Great Crystal Cave, and it was later known as Floyd Collins' Crystal Cave, or just Crystal Cave.

Within a few days, Floyd's father and brothers had looked at Crystal Cave. They made plans to smooth the rough floor into tourist trails as soon as the spring work in the fields was completed. However, the excitement was too great to bear. The family started at once to move rocks and to construct wheelbarrows for moving sand from the side of the passages to fill in holes in the proposed trail. In bad weather they worked inside Crystal Cave. In good weather they worked outside to make a stairway down into the entrance passage. They contracted with a manager and erected signs along the road to Mammoth Cave so that they could capture their share of tourists.

There were very few tourists. World War I had stopped most travel, and competition was stiff for tourists who did find their way to the caves of central Kentucky. Crystal Cave needed a ticket office out on the main road. Smoother trails would be an advantage. Better lighting would make the cave more attractive.

When tourists did come, a handful a week, Floyd might personally fire up the lanterns and take them into Crystal Cave. In the large room, the Grand Canyon, he would hook one lantern on a twenty-foot length of pipe and hold it high in the air to show the vast dimensions of the place. In other passages the lights reflected off the gypsum-encrusted walls.

Floyd sometimes would reward visitors with a souvenir of their visit by breaking off a gypsum flower. The cave must have seemed to be an inexhaustible garden of crystal flowers. Floyd provided specimens that went to university museums, where they became permanent advertisements for Crystal Cave.

For all of the Collins family's backbreaking labor, Crystal Cave was not a big moneymaker, even after the tourists resumed travel in 1919. It was too far from Mammoth Cave. When guides and workers at other caves were asked about Crystal Cave, they would express scorn or ignorance and switch the subject.

With time on his hands, Floyd turned more and more to exploring Crystal Cave. He often went alone. Sometimes he took his little sister, Alice. Floyd's main explorations were through a tight side passage just beyond Scotchman's Trap, a big rock leaning over the passage that marked the end of the commercial tour. Floyd would push for hours through body-tight crawlways. He discovered thousands of feet of lower-level passages.

Once, far back along the main route of the cave, Floyd's lantern went out and he could not get it relit. The Collins family was used to Floyd's exploring

*Floyd Collins in the Flower Garden of Floyd
Collins' Crystal Cave, about 1922.*

for as long as twenty-four hours, so would not miss him until the next day. Floyd found his way to the entrance alone in the dark. It took him eighteen hours. The experience did not deter Floyd from further exploring.

With the prosperity of the mid-1920s, more tourists arrived. There seemed to be enough money now for George Morrison to skim the cream, while the historic Mammoth Cave tour retained a substantial flow of revenue. Then on down the road, across the valley from Mammoth Cave on Flint Ridge, there was Great Onyx Cave with its small hotel. Finally—only for the most adventurous—Crystal Cave could be reached over a rough dirt road. The road was a spring-buster, and there were no overnight accommodations for tourists at Crystal Cave. It was fortunate that the Collins family prospered as truck farmers, because Crystal Cave was not a commercial success.

Floyd had found a fine cave, and had developed one of the prettiest commercial tours in the region, but it became increasingly clear that the situation was almost hopeless. Tourists traveled on the road to Mammoth Cave, but very few of them could be lured over to Flint Ridge. Obviously, George Morrison had done the smart thing. He had moved up the road toward the main highway, between Mammoth Cave and the main towns, to open his New Entrance so he could capture tourists before they got to the historic Mammoth Cave entrance. Tourists had to pass by Morrison's New Entrance, *and* by historic Mammoth Cave, *and* by Great Onyx Cave, before they got to the entrance to Crystal Cave.

Floyd had watched Morrison's moves like a hawk from his perch out on a remote corner of Flint Ridge, envying Morrison's style, backing, and guts. Floyd had guts, but neither the backing nor the style nor the right cave. Then, in 1922, Floyd found a crawlway bypass that took him around a formidable drop that had blocked his previous exploration progress in Crystal Cave. He continued onward into a muddy passage, which he followed. Finally its ceiling gave way to a black, upward void. Floyd climbed up into an immense passage, twenty to thirty feet wide, seven to fifteen feet high. He paced off a solid mile of cave. Perhaps this would draw tourists. However, he could not find a practical way into it, or a place to dig an entrance. And even had he been able to open this passage, Crystal Cave was still too far down the road.

Floyd thought it over, then made his move. He walked down the road past Sell's General Store, following the crest of Flint Ridge past the entrance to Salts Cave. He followed the crest as it cut southwest around the end of Strawberry Valley along the narrow bridge of upland that joined Flint Ridge with Mammoth Cave Ridge. He went on south toward Mammoth Cave Ridge. At the Mammoth Cave Road, he turned left, away from Mammoth Cave, toward town. He was going to do to George Morrison what Morrison had done to the Mammoth Cave estate. Floyd would find a cave up the road from Morrison's New Entrance, develop it, get first chance at the tourists, and laugh at his rivals farther down the road. From being the last in line, Floyd intended now to be the first.

Floyd made a contract with three farmers who occupied the narrow bridge of land that joined the lower ends of Flint Ridge and Mammoth Cave Ridge. There just had to be big cave under that piece of upland. Geologic conditions there were exactly like those on Flint Ridge where Floyd had found cave. They were the same as those of Mammoth Cave Ridge where Morrison had been so successful. Under terms of the contract, the farmers would give Floyd room and board on a rotating basis. Floyd would in turn seek out a cave. If he found one, and if the cave were successfully developed, Floyd would take half of the profits, and the three farmers would share the other half.

Floyd started to work in Sand Cave, a shelter cave opening on the Doyle farm. During January of 1925, he spent every day at the cave, moving rocks, pounding off ledges, and shoveling sand and earth out of tight spots. One story is that Floyd worked his way through a very tight crawlway to the brink of a pit. He tied off a rope and went down. At the bottom he crept about 300 feet through a drain, and then turned to the left. Here—goes the unconfirmed story —he found big cave. If so, it was the culmination of everything he had learned the hard way by himself, by caving with Turner, who taught him the use of pace and compass, and by watching Morrison. Floyd was now the best cave explorer in the area. He planned also to be rich.

It is said that Floyd was putting the final touches on enlarging the crawlway so his partners could view the cave when he inadvertently kicked a fifty-five-pound rock that fell across his ankle and trapped him. The crawlway he was lying in was so narrow that he could not rise or turn around to reach the rock with his hands. It did not take him long to realize that he would have to lie there until someone came to get him.

Floyd had been lost and nearly trapped in caves before, and he was a very stoical person. As uncomfortable as his situation was, he probably was not overly worried. The Doyle family knew where he was—that is, they knew he was in Sand Cave. They would quickly find him once they started looking. For the present, it was mostly a matter of waiting. Floyd waited. He was wet and cold, but he probably managed to get some sleep. During the night his lantern burned out.

It poured rain that night. When Bee Doyle found that Floyd was not in his bed the next morning, he went to his neighbor's house to see if Floyd was there. Edward Estes and his seventeen-year-old son, Jewell, joined Doyel to hike to Sand Cave. Jewell crept into the cave to within twenty-five feet of Floyd, who instructed him to hurry back to the Collins house to bring his brothers. With a longbar, Floyd thought, they ought to be able to get him out.

It was not that easy. Homer Collins sped down 100 miles from Louisville in his Model T. He was almost as knowledgeable about caves as Floyd, but he could not move the rock. They did get food and light to Floyd, and they rigged a raincoat to divert the cold water that was dripping on his face. The problem was that Floyd was stuck in a belly-crawl. He was facing out and his body filled the passage almost completely, leaving no room for anyone to get past to reach the rock that pinned his ankle. Hundreds of pounds of loose rocks were moved from the passage leading up to Floyd, but the massive blocks and bedrock walls around him stopped effort to get over, under, or around his body to reach the rock holding his leg.

By telephone the news went out that Floyd Collins was stuck in Sand Cave. People began to gather at the cave entrance. For several weeks there had not been much exciting national news. Now, when the story of Floyd's entrapment hit the wire services, Sand Cave became a journalistic magnet. Reporters from everywhere rushed to the site. Thousands of sensation-seekers converged with them to enjoy the spectacle.

There were many heroic attempts to free Floyd from his stone prison. William Burke (Skeets) Miller, one of the youngest and smallest reporters on the scene, carried in a small jack to try to free Floyd. He managed to squeeze in far enough to emplace the jack under the rock, but there was not enough room to operate it properly. He tried time and again, but always the jack slipped out from beneath the rock. He could not budge the rock a fraction of an inch.

As the days succeeded one another, the rescue dissolved into anarchy. There were claims that Floyd was dead. Others said that he was alive, but that the entrapment was just a publicity stunt. Work was impeded by celebrants who did not care about Floyd, but who enjoyed the excitement. Skeets Miller reported everything, and his eyewitness stories earned him the Pulitzer Prize for journalism for 1925.

On February 8, 1925, thousands of people were milling around the entrance to Sand Cave. It was a carnival of people seeking morbid thrills. Moonshine was sold openly. Balloons with "SAND CAVE" printed on them were hawked.

Things went from bad to worse as the National Guard commander took over. Direct rescue attempts were stopped while an engineer began planning to drill a shaft down to Floyd. Then the ceiling collapsed in the passage, cutting Floyd off. The last day on which anyone could say for sure that they had heard a sound out of Floyd was Friday, the 13th of February, on his fifteenth day in the hole. After those fifteen days of page-one headline coverage and anxious vigils, the digging of the shaft was finally begun. After three feverish days the shaft broke through into the passage above Floyd. He was stiff and cold. According to Dr. Hazlett, Floyd had been dead between two and four days, of "exposure and exhaustion."

Lee Collins expressed the family's wish that Floyd be left where he was, and that the shaft be resealed so that Floyd's grave would be in that passage in Sand Cave. Photographs were taken of the body in place. Then concrete was poured in the heading, and logs and dirt were thrown in to close the shaft. The entombment, however, was to be short. A couple of months later, Lee Collins had the shaft reopened. Floyd's body was removed, embalmed, and taken to a grave on Flint Ridge near Crystal Cave. The publicity associated with Floyd's ordeal and death had brought hundreds of tourists to Crystal Cave, and they wanted to see Floyd's tomb. Lee Collins had provided a tomb, but now decided to sell Crystal Cave if he got a good offer.

Dr. Harry B. Thomas, a dentist in the town of Horse Cave, Kentucky, owned and operated two commercial caves: Hidden River Cave and Mammoth Onyx Cave. In 1927, he bought Crystal Cave to extend his commercial cave empire. He had already pioneered the showing of caves by electric light, and had set up a hydroelectric generator in the town of Horse Cave. If Crystal Cave were lighted, and if the name were changed to Floyd Collins' Crystal Cave, and if Floyd's mortal remains were on the premises, the cave ought to attract increasing numbers of tourists.

Dr. Thomas purchased a glass-covered metal coffin, put Floyd's body in it, and placed the coffin and the headstone inside the cave, on the floor in the center of the Grand Canyon, the magnificent entrance room Floyd had found.

The new tourist attraction—Floyd's corpse—was a sensation. The guides would lecture solemnly about the exploits of "the world's greatest cave explorer," and then the tourists would file by the body on display. These funereal proceedings went on for many years. However, not everyone was attracted by such morbid goings-on. Roger Brucker, when he was first taken to see Mammoth Cave at the age of eight, would have been wild to see the body, but his mother carefully concealed from him the existence of Floyd Collins' Crystal Cave, and of Floyd under glass.

Floyd's service in the cave wars extended even further beyond the line of duty than having his dead body put on exhibit. One March night in 1929, the resident manager and guides were startled to find that Crystal Cave had been broken into, the coffin opened, and Floyd's body stolen. Sounding the alarm, they fanned out around the entrance with bloodhounds to search for clues. Before long the body was found at the base of the Green River bluff. The thieves had tried to throw the body into the river, where it would have been carried away. No one doubted the motivation. Some rival cave operator felt that showing Floyd gave too much of a competitive edge, and had dispatched body-snatchers to equalize things. The attempt had failed, however, and the body was taken back to the coffin. The incident brought great publicity for Floyd Collins' Crystal Cave, undoubtedly to the great annoyance of the rival who had financed the heist. All that was lost was one of Floyd's legs, which was never recovered.

That was show business. It was at the height of commercial cave-showing in the region, and anything went. At one time there were twenty-two caves on exhibit.

The times and notions of propriety soon changed, and a metal cover was placed over Floyd's coffin. Thereafter Floyd's corpse was not shown, except unofficially by guides who might lift the cover for an ample tip. Until Crystal Cave was purchased and closed to the public by the National Park Service in 1961 the tourist trip included only a dignified pause at Floyd's coffin, with a short lecture on his caving exploits. Some of the latter-day tourists probably did not realize that Floyd's body was actually in the coffin, and many probably doubted that it was. It still is.

9/Follow the Water to Mammoth Cave

Out of the flurry of renewed commercial interest in the caves of the Central Kentucky Karst in the 1920s came a piece of hard evidence about cave connections. The steep-walled gorge of the Green River near Mammoth Cave seemed to be a good site for a hydroelectric dam. In 1925, the Louisville Gas and Electric Company obtained the services of R. B. Anderson, an engineering geologist, to make a survey of the possibilities. The first question asked was whether or not a dam on the Green River would hold water. Given the proximity of Mammoth Cave, it might leak through cave passages to bypass the dam.

Anderson posted observers at three big springs along the Green River—Pike Spring, which drains Flint Ridge, and River Styx and Echo River, which drain Mammoth Cave Ridge—to watch for traces of dye. One by one, he dyed the springs on Flint Ridge and the rivers inside the caves. With one exception, all the water he dyed emerged at Pike Spring. The exception was startling indeed. Dye placed in Three Springs, located on the southwest corner of Flint Ridge, was pushed by heavy rains beneath Houchins Valley and under Mammoth Cave Ridge to emerge in River Styx of Mammoth Cave. This evidence of a connection of cave passages from Flint Ridge to Mammoth Cave convinced Anderson that it was not feasible to dam the Green River near Mammoth Cave. No lake could be backed up behind a dam here because the rock is like a sponge. Water would escape, no matter what anyone tried.

This dye tracing was the first direct evidence that the caves of Flint Ridge were connected to those of Mammoth Cave Ridge. Previous convictions about the eventual discovery of a connection, for all their surety, were based on geological inference and hope. However, if water had traveled from Flint Ridge to Mammoth Cave Ridge, cave explorers might eventually take the same trip.

A dam would threaten Mammoth Cave by flooding it. The cutthroat competition among commercial cave owners, and the stripping of cave-onyx from caves for sale along the road to tourists also endangered the natural beauty of the cave country. Over the years, a group of Kentuckians had organized to protect the area by establishing a Mammoth Cave National Park. One thing that may have made them more active was the increasing arrogance of "Judge" Albert Janin, manager of the Croghan estate at Mammoth Cave. He had been cutting a virgin forest on the land.

The story of what happened is not fully known, but it goes something like this: In 1924, the surviving heirs to the Mammoth Cave estate asked the Edmonson County Circuit Court to appoint three trustees in addition to the two already appointed. They named as their candidates three members of the family. The Court agreed to name three trustees, but instead of the three family members, the Court appointed three leading, public-spirited citizens of Edmonson County. The heirs sued, but got nowhere. Then in 1927, the last original heir died. The succeeding heirs pooled their fractional interests in the estate—comprising Mammoth Cave and 1610 acres of land—and turned it over to a trust company in Louisville that issued certificates in return. It is commonly thought that the Mammoth Cave National Park Association bought these participation certificates one by one from the various secondary heirs until control of Mammoth Cave was acquired. The Association was active in the political arena as well. In May 1926, Congress passed an act stating that when the lands within the Mammoth Cave area were delivered in fee simple to the United States, a national park comprising about 70,618 acres would be established.

The legislature of Kentucky moved to set up the Kentucky National Park Commission expressly to obtain lands—by eminent domain where necessary—for the new park. In 1928, the Commission went to court to condemn property within the proposed park boundaries. The heirs fought, but they were unsuccessful in keeping Mammoth Cave. Altogether, the Association and the Commission put together about 50,000 acres for the new park. The Proposed Mammoth Cave National Park was operated during the 1930s by a joint operating committee consisting of some members of the Association and of the Commission.

Two islands of private property remained within the park as inholdings: Floyd Collins' Crystal Cave and Great Onyx Cave. They were annoyances because they siphoned off dollars from the fledgling Mammoth Cave National Park. And things were getting worse. Hercules, the steam locomotive, pulled the last train to Mammoth Cave in 1929. The Depression and the automobile killed the Mammoth Cave Railroad. George Morrison was forced to sell his interest in the New Entrance to Kentucky in 1931. The Depression set in, and not many people came to see Mammoth Cave.

10/Pete Hanson and Leo Hunt

In 1938, Superintendent R. Taylor Hoskins of the Proposed Mammoth Cave National Park decided that exploration in Mammoth Cave might improve morale, and if anything significant were found, it might stimulate the tourist trade. He did not have to persuade his guides, who were eager and ready. For several years they had taken parties of men and boys enrolled in the Depression-born Civilian Conservation Corps into Mammoth Cave to build trails and to survey. A considerable amount of exploration and mapping was done in the cave by CCC workers. However, the elite corps of explorers consisted of the Mammoth Cave guides Carl Hanson and his son Pete, and Leo Hunt, Young Hunt, and Claude Hunt.

In 1938, the Hansons and the Hunts began to look for two lost rivers. The first was the legendary Stevenson's Lost River, an extensive waterway supposedly found in 1863 at the bottom of Gorin's Dome. The other was Mystic River, the long major river passage whose location had been lost after 1870. They fanned out to examine all the watery places systematically. They probed from Bottomless Pit to Echo River, along Echo River, and beyond the second boat dock. We know enough to conjecture about one such trip:

In the fall of 1938, the weather was dry and the water in the Green River and thus in Mammoth Cave was at a low level. Pete Hanson and Leo Hunt crunched along the path through dry leaves to the Historic Entrance of Mammoth Cave. Both men had two-cell flashlights thrust into the pockets of their short cotton jackets. They wore heavy work shoes and soft, mud-smudged hats. Leo carried a gasoline lantern and a couple of spare mantles. One could count on breaking at least one lantern mantle in tight passages. Pete carried a small can of white gas. They had candy bars and sandwiches. For several days they had been

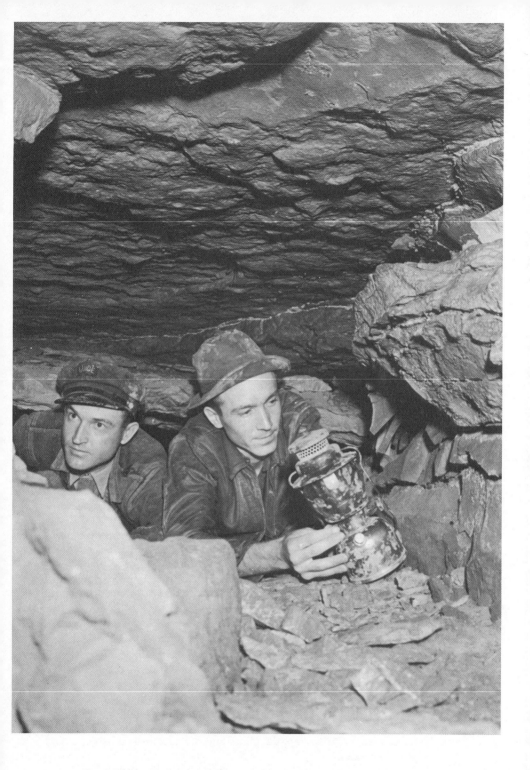

Leo Hunt and Pete Hanson, 1938.

probing crawlways, and today they were tackling the wet lead beyond the second boat dock of Echo River.

Boating in Echo River is always a joy, no matter how often you have done it. The gliding movement in silence under the rock vault ceiling produces a comfortable, yet slightly eerie feeling of detachment. The two explorers came back to reality as their boat bit into the sand bank at the second dock. They gathered their belongings and scrambled out. About eighty feet over smooth sand banks they came to where the trail had been extended over a pool of water on a causeway made of piled-up rocks. Steel-pipe railings protected both sides of the trail. There was a low horizontal slot in the far wall, across the pool.

Pete and Leo carefully picked their way down the bank. Pete sank into the mud over his ankle, but recovered to stand upright. They laughed, but the cold water sobered them. They groaned as cold water inched up on warm bodies, and waded along the wall, up to their waists in the water.

Pete led the way through the low slot. The ceiling was only twelve inches above the water. The bright light of the lantern illuminated jagged, carved stone pendants on the ceiling, producing a mirror image on the water beneath.

Pete and Leo had been this wet before, but usually there had been a more gradual introduction to the unpleasantness. They held their lights and lunch above the water.

The going was easier after the initial thirty-foot stretch of water. They walked and waded part of the time in a comfortable passage five to seven feet high and six feet wide. There was mud on everything, including the ceiling. Now and then the passage split, one branch occupied by the stream, the other being a cut-around. The cut-arounds were larger and less wet than the active stream passage, so they followed them when they could.

The passage had started heading due east, and then had turned north. They guessed that they must have passed under Ganter Avenue, and that they were headed for the northeast flank of Mammoth Cave Ridge. After about 4000 feet the passage got smaller. Now they had to crawl most of the time. In a relatively dry stretch between two reaches of the river, they ate lunch. The passage seemed endless, and they compared it to low, long leads they had been in off Stevenson's Avenue. They decided to continue for a while, although the temptation to turn back was reinforced by knees made tender on the limestone floor. Leo pumped furiously on the lantern's pressure pump and sloshed the lantern to see how much fuel remained. The light was brilliant in the small passage, challenging the gloom of dripping brown mud everywhere.

After another 2000 feet, Pete and Leo agreed that they ought to return. The passage seemed interminable. They were in a muddy room six feet high and eight feet wide. The mud on the walls and ceiling was like cake frosting. It merged with the slick mud banks of the stream. Upstream, water trickled from a triangular black hole.

In case they came into this place from the other direction someday, they decided to mark it. Pete and Leo had left their names in many places in the cave, and in turn had come by some roundabout routes into places marked by earlier pushers such as Ed Bishop, Kaemper, and others. It was a useful record of work in a complex cave.

Pete peeled mud off the arch over the continuing passage with the sharp end of a chert rock and scratched "PETE H." The letters showed the bright color of scratched limestone against the dark brown mud. However, anyone coming from the other direction would probably miss the name, so Leo drew "LH" with his finger on the mud bank. Pete then drew his PH behind Leo's initials, extending the bar of the H to the right to form an arrow pointing the way back to Echo River. They fueled the lantern and turned around for the long trip downstream. They made little of the trip because the passage had not led them to anything big and exciting, nor had they been able to go to its end. They did not return to try.

Instead, through the winter they explored up Roaring River. They had taken a twelve-foot flat-ended skiff over from Echo River. On several trips they found Roaring River to be promising, with wide passages and plenty of headroom. The heavy flow of water was most encouraging. There were riffles and flowstone dams to drag the boat over, but at least the Roaring River passage was not small like the one off Echo River.

They probed each side lead from Roaring River. On one of these trips, Carl Hanson and Leo went out a side passage to the right. It went on and on, with some crawling and stooping. The next day they returned with Pete, who was in front when the passage began to get bigger. He yelled and raced ahead, emerging in a large trunk passage with no footprints on its floor. They had made the new discovery they were seeking.

Superintendent Hoskins listened to their story with pleased excitement. His decision to encourage exploration had paid off. They would have to survey the new passage, and if it proved to be of more than routine interest, they could

Explorers in New Discovery, Mammoth Cave,
1938. Left to right: Carl Hanson, Leo Hunt,
Pete Hanson, and Claude Hunt.

publicize it. But now they would proceed without making much fuss. More than once a couple of hundred feet of new passage had been touted as a great "New Discovery." Hoskins was going to be sure that the new section was worth notice as shown by maps and descriptions before he said anything about it officially.

There was also a problem with the river. Roaring River could rise suddenly during a storm, and a three-foot rise in one hour had been observed. Hoskins did not want his parties to be trapped in some unknown place, so he forbade exploration when it looked like rain.

In 1939, New Discovery was a bright spot in otherwise dull, slow years. There were several miles of large trunk cave passage there, much of it containing rare and delicate gypsum crystals. Gypsum cotton, like batting, hung from the ceiling in one location. It quivered when the explorers walked by. The survey also indicated a place where a new entrance might be drilled into the section right along the road to the Carmichael Entrance that had just been dug to provide ready access to the Snowball Dining Room. There was plenty of CCC manpower available, so the second new entrance was opened. Some efforts to develop walkways and trails were started, but then World War II drew away the workers.

Pete Hanson and others went into the Army. Pete died in the Aleutian Islands. Others came back, not as cave explorers, but as older, less venturesome men with families and responsibilities. Leo Hunt died in 1960. There was practically no more cave exploring in Mammoth Cave for nearly thirty years after the great discoveries of the Hansons and the Hunts in the late 1930s.

11/Jim Dyer

Exploration sagged in Flint Ridge, also, after the death of Floyd Collins in 1925. Then, in 1941, in Floyd Collins' Crystal Cave, the guides Harry Dennison and Ewing Hood found the route to Floyd's Big Room, which had become known as Floyd's Lost Passage. It had never really been lost, of course. It was just that the few people who knew the way there—like Kanah Cline and Ellis Jones—did not have the occasion or the permission to go.

In the Kentucky Cave Country, the way you got to be a cave guide was by knowing someone who was a cave guide who would vouch for you. A long succession of guides at Crystal Cave extended through this old-boy network back to Floyd himself. One long-term guide was Elkanah "Kanah" Cline, from Northtown, eight miles away. Kanah looked fifteen years older than his forty-odd years in 1948, a slim, short man with a Popeye face. He had a shock of unruly sand-colored hair, and he walked with a limp. Kanah was a non-stop talker, with an inexhaustible store of tales about the caves. The farther away his listener came from, the faster Kanah would talk and the heavier his accent got.

Kanah took Jim Dyer in hand in 1946. Jim was fascinated by Kanah's tales of Floyd's Lost Passage, and by the stories of Dennison and Hood, who had been hired by Dr. Thomas to guide, and to explore the cave during the winter. Once, Kanah said, Dennison and Hood had found an enormous passage that was even larger than Floyd's Lost Passage, a cavernous vault that contained a piece of breakdown block as big as a house.

"Where?" Jim asked.

"Don't know," Kanah said, shrugging. "They lost it."

Jim had seen some pictures Harry Wilson had taken in Floyd's Lost Passage. He asked Kanah if he would take him down there. Kanah said sure.

In later years, Jim recalled the trip as a nightmare. They crawled along, seemingly for hours. Whenever they came to a junction, Kanah would allow as how it was mighty confusing, and he would take his chin in his hand and muse about whether he had ever seen this place before. He seemed completely unworried that they had paused at junction after junction where he had expressed uncertainty as to the right way to go. Now and then he would recognize some landmark, cackle to himself, and crawl on slowly and methodically. Not only did Kanah limp when he walked, Jim said, but he limped when he crawled, too.

Jim had worked around carnivals, but he had never seen a con man with Kanah's consummate skill. Years afterward he said that he never did find out whether Kanah actually knew the route to Floyd's Lost Passage well or not, but he did get them down there and back. At one point, Kanah announced soundly that they were lost. He sat down, smoked his pipe awhile, and then got up to go on without a word in the same direction they had been traveling before. Jim could do nothing but follow.

Kanah showed him Floyd's Jump Off, where on an early exploration trip Floyd had jumped down nine feet and then could get back up only by searching far and wide for rocks to pile up. The pile is still there. At one place, they left a wide walking passage to squirm on their bellies through a narrow passage to avoid a big pit. Finally they climbed up out of a muddy passage into Floyd's Lost Passage. After the narrow crawlways, Jim was suitably impressed, to Kanah's satisfaction.

Jim had come to the cave country in 1926. From 1927 to 1933, he worked as a solicitor and guide at Crystal Cave, and then went back north. In 1946, he returned to be resident manager of Crystal Cave. His brother-in-law, Luther Miller, joined the guide force. Jim and Luther spent many hours in the lower levels of Crystal Cave, exploring at night after the tourist day was over. They took no notes and made no surveys. It was real old-time caving that they did, the substance of many yarns about remote and interminable passages in the dark spookiness underground. Their sport was to be changed, however, by a determined young man who was a member of the family that owned Crystal Cave.

Bill Austin was the only grandchild of Harry Thomas. Bill worked in his grandfather's caves from the time he was in grade school, and during the summers when he was in college. Dr. Thomas frowned on his heir's exploring caves. He thought Bill was too young, and that exploring was dangerous. After Dr. Thomas died in 1948, Bill joined Jim Dyer's midnight exploring club. He was twenty years old.

Bill was soon impatient to know where he had been underground. Jim and Luther had considerable confidence in their sense of where the passages were with relation to the upper-level passages and the surface, but Bill wanted to know precisely. Jim encouraged the interest.

Bill was going to the University of Kentucky, where he was taking a degree in civil engineering. Dr. E. Robert Pohl, a geologist who had married Bill's aunt Ruth Thomas, was now managing the Thomas estate. He agreed readily to Bill's plan for mapping the lower levels of Crystal Cave. Bill and Luther then made many surveying trips, chaining their way through the crawlways and

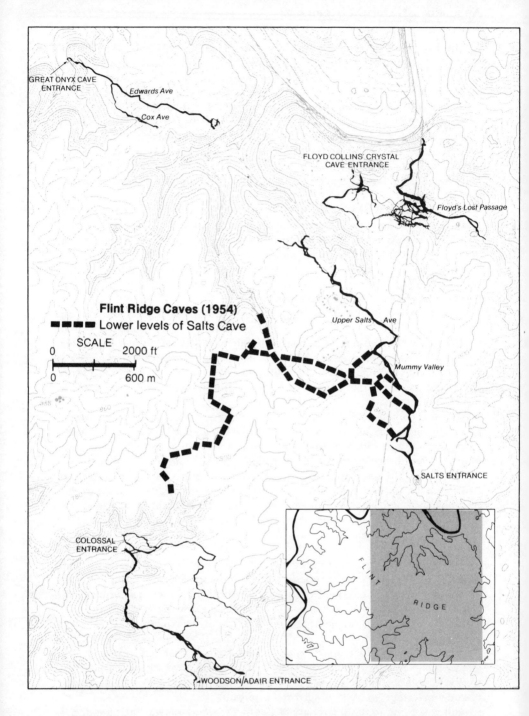

Flint Ridge Caves (1954)
▬▬▬▬ Lower levels of Salts Cave

SCALE
0 ————————— 2000 ft
0 ————————— 600 m

GREAT ONYX CAVE
ENTRANCE

Edwards Ave

Cox Ave

FLOYD COLLINS CRYSTAL
CAVE ENTRANCE

Floyd's Lost Passage

Upper Salts Ave

Mummy Valley

SALTS ENTRANCE

COLOSSAL
ENTRANCE

WOODSON/ADAIR ENTRANCE

FLINT RIDGE

*Map of Flint Ridge caves showing Jack
Lehrberger's discoveries in the lower levels of
Salts Cave in March 1954.*

canyons to Floyd's Lost Passage, working their way to the underside of the bluff facing the Green River to the north, and stopping at the breakdown that closed Floyd's Lost Passage to the south. Bill also began to develop his skills as a photographer. The results were line-plot maps of the underground that revealed very little to anyone who was not personally familiar with the cave passages. However, he also produced photographs of cave features and passages that revealed even to non-cavers the beauty and the lure hidden in the darkness underground.

One thing the survey of the lower levels of Crystal Cave showed was the proximity of some passages to Salts Cave. Bill, Jim Dyer, and Luther had no map of Salts Cave, so they visited it and made one. One thing they could find out was whether there was truth in the rumor that the Collins family had mined onyx at the back end of Salts Cave openly because it was under Collins family property. Bill's map showed that Salts Cave did not extend that far.

Bill also improved the lighting circuits in Crystal Cave with high-voltage transformers. And he showed the lower levels to cavers from other parts of the country. One caver who visited the lower levels with Bill was Joe Lawrence, Jr., who with college friends spent three days camping and exploring in Floyd's Lost Passage. Joe was so impressed by this experience, during which he found the cave too large to understand in such a short stay, that he wanted to run a big expedition into Crystal Cave. He talked with officials of the National Speleological Society, worried that they might not want to support the expedition. However, Charles Mohr, who headed the society, was enthusiastic. He knew Dr. Pohl from trips to photograph cave animals in Mammoth Onyx Cave. He at once asked Dr. Pohl if the Society might run an expedition in Crystal Cave.

Dr. Pohl discussed the idea at some length with Bill. There was some risk in turning loose a bunch of untested Eastern cavers in a long cave like Crystal Cave. If some local veterans could be in charge of the tactical parts of the expedition, perhaps it might work. Jim Dyer was working in Ohio, but was willing to participate in the expedition. Bob Handley, who had visited the cave, was eager to return. Joe Lawrence himself and his friends were competent cavers. Luther Miller certainly knew the cave.

There was also a young man from Louisville, Jack Lehrberger, a clean-cut, soft-spoken, serious fellow who had stopped by to learn what Bill knew about Salts Cave. Bill was impressed by Jack's accomplishments, and in turn was interested in knowing what Jack had found in Salts Cave. Jack's questions had been intriguing. His answers were more so. He had taken several trips in Salts Cave. On one of them he found a crawlway leading out of Mummy Valley, with Edmund Turner's name scratched on the wall. Did Bill know who Edmund Turner might be? Bill did, and soon he and Jack were taking trips together. Jack, too, should be brought along for the big expedition.

There were also some cavers Jim Dyer had recruited in Ohio. Roger Brucker, Phil Smith, and Roger McClure had shown themselves to be competent cavers. Yes, the thing could work. Dr. Pohl gave his permission, and the cavers who would start the modern era of exploration and connection in the caves of the Central Kentucky Karst began to gather.

Appendix II/Chronology

Note: Explorations on the direct line to the Flint/Mammoth connection are marked with an asterisk.

*2500–2000 B.C. Earliest known prehistoric Indian explorations in Lee Cave of Joppa Ridge, Mammoth Cave of Mammoth Cave Ridge, and Salts Cave of Flint Ridge.

1797. A hunter named Houchins chases a bear into Mammoth Cave.

1798. Valentine Simons registers "the Mammoth Cave tract" in the courthouse at Bowling Green, Kentucky.

1812. Charles Wilkins and Hyman Gratz, owners of Mammoth Cave and Dixon's Cave, mine them for saltpeter.

1813. Ebenezer Meriam visits and writes about the Mammoth Cave saltpeter works.

1816. Nahum Ward's letter on Mammoth Cave is published in many newspapers.

1835. Edmund F. Lee makes the first accurate map of about eight miles of Mammoth Cave.

1838. Franklin Gorin purchases the Mammoth Cave tract from Hyman Gratz and Simon Gratz. Gorin's black slave guide, Stephen Bishop, crosses Bottomless Pit with H. C. Stevenson.
*Stephen discovers Echo River in Mammoth Cave.

1839. John Croghan purchases Mammoth Cave and Stephen Bishop from Franklin Gorin.

1840. Stephen Bishop discovers Mammoth Dome.

*1841. Stephen Bishop discovers Snowball Room.

1842. Stephen Bishop draws a map of Mammoth Cave.

1856. Stephen Bishop is set free.

1857. Stephen Bishop dies at the age of thirty-six.

1858. Frank Demunbrun discovers Mystic River beyond El Ghor in Mammoth Cave.

1886. Mammoth Cave railroad from Glasgow Junction is completed.

1895. Lute Lee and Henry Lee discover Colossal Cave in Flint Ridge.

1896. Lute Lee and Henry Lee connect the Bedquilt Entrance to Colossal Cave.

1897. L. W. Hazen crosses the top of Colossal Dome in Colossal Cave.
Pike Chapman is killed digging a back entrance to Salts Cave in Flint Ridge.

1903. Marty Charlet draws a map showing Unknown Cave in Flint Ridge.

1907. Mammoth Cave guides discover Kentucky Avenue.

1908–1909. Max E. Kaemper and Ed Bishop, a black guide, map thirty-five miles of Mammoth Cave.

1912. E.-A. Martel visits Mammoth Cave and predicts an eventual connection with the caves in Flint Ridge to make a system 150 miles long.
Edmund Turner goes caving with Floyd Collins in Salts Cave.

1916. Edmund Turner finds Great Onyx Cave.
George Morrison blasts an entrance into Mammoth Cave, but is enjoined from using it.

*1917. Floyd Collins discovers Great Crystal Cave in Flint Ridge, later known as Floyd Collins' Crystal Cave.

1921. George Morrison opens the New Entrance to Mammoth Cave in competition with the Historic Entrance commercial operation.

*1922. Floyd Collins finds a mile-long passage in Crystal Cave, later known as Floyd's Lost Passage.

1924. George Morrison blasts open the Frozen Niagara entrance to Mammoth Cave.

1925. Floyd Collins is trapped and dies in Sand Cave.

1926. Congress passes an act to accept land for a Proposed Mammoth Cave National Park.

1927. Harry B. Thomas buys Crystal Cave.
Jim Dyer starts working for him.

1928. Land within the boundaries of the Proposed Mammoth Cave National Park is condemned.

1929. Floyd Collins' corpse is stolen from its casket in the Grand Canyon of Crystal Cave, but is recovered.
Hercules, the steam locomotive, pulls the last train to Mammoth Cave.
Dr. E. Robert Pohl marries Harry Thomas' daughter, Ruth.

1930. Cathedral Domes Entrance to Mammoth Cave is opened.

1931. George Morrison sells his Mammoth Cave interests to the Proposed Mammoth Cave National Park Commission.
Violet City entrance to Mammoth Cave is opened.
Carmichael Entrance to Mammoth Cave is opened.

1938. The Hansons and the Hunts explore the rivers in Mammoth Cave.

*Pete Hanson and Leo Hunt explore what later is known as Hanson's Lost River. Carl Hanson, Pete Hanson (Carl's son), Leo Hunt, and Claude Hunt (Leo's cousin) find the New Discovery.

1940. New Discovery Entrance to Mammoth Cave is opened.

1941. Harry Dennison and Ewing Hood rediscover Floyd's Lost Passage in Crystal Cave.

1946. Cave guide Elkanah Cline takes Jim Dyer, the new Crystal Cave manager, to Floyd's Lost Passage.

1948. Harry Thomas dies and Dr. Pohl takes over management of the Thomas cave estates. Jim Dyer, Luther Miller, and Bill Austin explore the lower levels of Crystal Cave. Jim crosses Bottomless Pit in Crystal Cave.
*Jim Dyer, Luther Miller, Bud Bogardus, and Bill Austin explore from Floyd's Lost Passage out B-Trail to the Bogardus Waterfall area.

1950. Jack Lehrberger and other Louisville cavers take an interest in Mammoth Cave National Park caves.

1951. Five Virginia cavers—Roy Charlton, Joe Lawrence, Adam Chow, Ed desRochers, and Jim Gosney—are taken into Crystal Cave by Jim Dyer and Bill Austin; this trip inspired the National Speleological Society C-3 Expedition.

1952. Phil Smith and Roger McClure establish the Central Ohio Grotto of the National Speleological Society in Springfield, Ohio. In February they hitchhike from their homes in Springfield to Kentucky to see Crystal Cave.
February. Lower levels of Salts Cave in Flint Ridge are discovered by Jack Lehrberger and Jack Reccius; they name one of the passages Indian Avenue.

1953, April. Roger Brucker, Joan Brucker, and their three-month-old son, Tom, attend the National Speleological Society Convention in Mammoth Cave National Park. Roger meets Jim Dyer, Dr. Pohl, and Bill Austin. Phil Smith also attends this convention.

November. Jim Dyer arranges for Bill Austin to take Phil Smith, Roger McClure, and Roger Brucker on an introductory tour to the lower levels of Crystal Cave, a trip later referred to as the Rinky-Dink.

1954, February. Joe Lawrence leads sixty-four cavers in a week-long National Speleological Society expedition—the C-3 Expedition—in Floyd Collins' Crystal Cave.

*Jack Lehrberger, Russ Gurnee, and Roy Charlton explore to Bogus Bogardus Waterfall, where Jack finds the Shortcut to the area.

A supply party gets lost on a fourteen-hour nightmare.

Roger Brucker, Earl Thierry, and Roy Charlton survey from Floyd's Lost Passage to Bottomless Pit to connect the map of the lower levels to the map of the upper levels.

April. Roger Brucker shows the composite map of Crystal Cave to Bill Austin at the National Speleological Society Convention. They join forces to explore the cave.

*July. Bill Austin, Jack Lehrberger, Roger Brucker, Phil Smith, and Dixon Brackett explore beyond Bogardus Waterfall. Roger discovers Black Onyx Pit.

*Jack Reccius and Jack Lehrberger enter Unknown Cave and explore the top of the shafts. Then Jack Reccius and Jack Lehrberger discover Pohl Avenue. In September, Bill Austin and Jack Lehrberger enter Unknown Cave and discover Upper Crouchway, Brucker Breakdown, Swinnerton Avenue, and other passages on a series of trips. Jack Reccius, Charlie Fort, Bill Walters,

and John Key spend seventy-two hours in Unknown Cave.

*November. Bill Austin, Jack Lehrberger, Phil Smith, Roger Brucker, Roger McClure, Dave Jones, Jack Reccius, Bill Hulstrunk, Dixon Brackett, Bob White, and Red Watson explore beyond Black Onyx Pit in Crystal Cave. Bill and Phil discover the Overlook, Storm Sewer, and Eyeless Fish Trail.

1955, May. Bill Austin and Henry Porter rescue a man trapped in a cave near Sulfur Well, Kentucky.

*September. Bill Austin and Jack Lehrberger go into Unknown Cave while Roger Brucker and Red Watson go into Crystal Cave, to try to make the connection. They fail.

FIRST CONNECTION: Two weeks later Bill and Jack connect Pohl Avenue in Unknown Cave with Eyeless Fish Trail in Crystal Cave to integrate Unknown/Crystal Cave.

Jack Reccius is apprehended at the Unknown Entrance by Park Service Ranger Joseph Kulesza.

November. Bill Austin takes Roger Brucker, Roger McClure, and Dave Jones on a tour in Unknown Cave; they agree to help survey to locate a place for a new entrance—the Austin Entrance—on Crystal Cave property.

December. Brother Nick Sullivan, President of the National Speleological Society, reads Roger Brucker's paper at the American Association for the Advancement of Science annual meetings, claiming that the twenty-three-mile-long Flint Ridge Cave System is the longest in the world. Dr. Alfred Bögli counters with Hölloch in Switzerland, which is thirty-five miles long. National Park Service literature claims that Mammoth Cave is 150 miles long; actually, about forty-four miles of Mammoth Cave are known and mapped.

Phil Smith, Roger Brucker, and Roger McClure organize

the Flint Ridge Reconnaissance as a project of the Central Ohio Grotto of the National Speleological Society.

Superintendent Perry Brown of Mammoth Cave National Park refuses permission to explore in Salts Cave.

Phil Smith and Roger Brucker field several teams Left of the Trap in Crystal Cave, but after more than a mile of surveying in a continuous crawlway, the failure of attempts to camp in the cave, and a thirty-six-hour trip by Jack Lehrberger and Dave Jones, they are defeated by the endurance barrier.

Bill Austin, Roger Brucker, and Dave Jones take Coles Phinizy and Robert Halmi of *Sports Illustrated* to see Eyeless Fish Trail; Roger slips and plunges head-first into the dangerous river.

The Caves Beyond: The Story of the Floyd Collins' Crystal Cave Exploration, by Joe Lawrence, Jr., and Roger W. Brucker, is published by Funk & Wagnalls.

*1956, May. Austin Entrance into Pohl Avenue of Unknown/Crystal Cave is completed.

1957, August 13. The Cave Research Foundation is incorporated with Phil Smith as president and a board of directors consisting of Roger Brucker, Jim Dyer, Burnell Ehman, Dave Huber, Dave Jones, and Jack Lehrberger.

1959. The National Park Service signs an agreement permitting the Cave Research Foundation to undertake cooperative research in Mammoth Cave National Park.

August. Bob Rose gets lost and found in Unknown/Crystal Cave.

*November. In Unknown/Crystal Cave, Roger Brucker discovers the Duck-under, a shortcut to passages off Swinnerton Avenue.

*1960, January. From the Duck-under, Jake Elberfeld discovers Jake's Breathing Trail extending southward from Gravel Avenue to Shower Shaft.

*May. Roger Brucker discovers Candlelight River out of Shower Shaft.

*August 22. SECOND CONNECTION: Jack Lehrberger, Spike Werner, and Dave Deamer connect Colossal Cave with Salts Cave to integrate Colossal/Salts Cave.

*September. Bill Hosken discovers Agony Avenue beyond Candlelight River in Unknown/ Crystal Cave.

1961, January. Joe Davidson takes John Wilcox to Carter Caves, Kentucky, but John is not impressed.

Great Onyx Cave and Floyd Collins' Crystal Cave are sold to the U.S. Government to become a part of Mammoth Cave National Park.

August. Dave Jones falls forty-four feet down a pit, but is able to climb back out.

August 21. THIRD CONNECTION: Dave Deamer, Judy Powell, Spike Werner, and Bob Keller connect the Lower Crouchway in Unknown/Crystal Cave with Indian Avenue in Colossal/Salts Cave to integrate the Flint Ridge Cave System.

1962, August. Spike Werner, Dave Deamer, and Sandy Irwin survey passages on the other side of the top of Colossal Dome.

1963. Cavers attempt without success to connect Great Onyx Cave with Unknown/Crystal Cave; Tom Brucker, ten years old, does some of the exploring.

Bill Hosken explores two long leads in the Candlelight River area.

*1964, July. Denny Burns, Jake Elberfeld, Bill Morrow, and Dick Burns survey to the end of Agony Avenue and discover Houchins River; this is the first penetration under Houchins

Valley. The next day, Roger Brucker, Spike Werner, Alan Hill, and Claude Rust survey in Houchins River.

Joe Davidson, Red Watson, and Bob Hough explore a long lead in the Houchins River Area; they survey back leaving a hanging survey. Fred Dickey, John Bridge, and Denny Burns go out to complete the hanging survey, but return after only three hours because they lose their pencil; Joe Davidson takes Fred and John right back into the cave to complete the survey.

*August. John Bridge, Fred Dickey, and Kim Heller survey in Agony Avenue beyond the Candlelight River area in Unknown/Crystal Cave.

*November. Dave Jones and John Bridge explore the A-survey passage off the N-survey Junction Room.

Congress passes the Wilderness Act. Attempts are made to get the Flint Ridge Cave System protected as underground wilderness.

*1965, January. Joe Davidson, Bill Coe, Chris Metzger, and John Lindsay explore and survey in the N-survey and put in the A-survey in the Houchins River area.

Red Watson, Stan Sides, and Dennis Drum explore and survey in the Houchins River area.

July. John Bridge, Fred Dickey, and Scooter Hildebolt start the Q-survey.

August. John Bridge, Stan Sides, and Tom Conlan take supplies out to Shower Shaft as protection for Q-survey parties.

September. Walter Lipton, Roger McClure, Dave Roebuck, and Dick Chappell take more supplies out to Shower Shaft. John Bridge, Roger McClure, and Art Palmer complete the Q-survey at Q-87, a point set at a sandstone breakdown closure of the passage under Mammoth Cave Ridge; this is the first crossing under Houchins Valley from Flint Ridge.

Joe Davidson, Dave Roebuck, and Pete Barrett attempt to dig through the Q-87 breakdown without success.

Red Watson succeeds Phil Smith as president of the Cave Research Foundation.

The claim of Red Watson and Burnell Ehman about the length of the Flint Ridge Cave System is defended by Alfred Bögli at the IVth International Speleological Congress in Yugoslavia.

1966, May. Joe Davidson, Art Palmer, Red Watson, and Bob Fries, supported by John Bridge, Fred Dickey, Bill Hosken, and Terry Preston, make another futile attempt to break through at Q-87.

August. Art Palmer, on a trip with Peg Palmer and John Bridge, pushes a tight belly-crawl in Houchins River alone 1500 feet.

October. Red Watson, Stan Sides, Roger Brucker, and Alan Mebane explore and survey in Flint Ridge Cave System.

November. Joe Davidson, Scooter Hildebolt, Pete Anderton, and Mike Goodchild discover the Northwest Passage.

Denny Burns succeeds Roger Brucker as Chief Cartographer.

1967, July. Joe Davidson succeeds Red Watson as president of the Cave Research Foundation. Spike Werner quits Flint Ridge caving.

1968, October. Joe Davidson takes John Wilcox to Flint Ridge, where John is immediately captivated by the immense cave system.

November. Gordon Smith and Judy Edmonds discover Lee Cave in Joppa Ridge.

1969, March. The Flint Ridge Cave System becomes the longest in the world, with about sixty-five miles surveyed. Hölloch has 64.5 miles, and Mammoth Cave forty-four miles.

May. After an absence of over a year, Red Watson leads a

sixty-man expedition, and vows that it is the last expedition he will lead.

Joe Davidson persuades Red Watson to return. Red sends ten parties into Mammoth Cave to begin Cave Research Foundation surveying and exploration there.

1970. Joseph Kulesza becomes Superintendent of Mammoth Cave National Park.

John Wilcox succeeds Denny Burns as Chief Cartographer.

1972, January. John Wilcox asks Joe Davidson for permission to try to connect the Flint Ridge Cave System with Mammoth Cave. Joe says to go ahead.

May 27. John Wilcox, Gary Eller, Pat Crowther, and Will Crowther try without success to penetrate the Q-87 breakdown.

July 1. Tom Brucker, Butch Welch, and Carol Hill explore in Houchins River, but turn back because of the cold water.

July. Alfred Bögli spends a week doing field work in the Flint Ridge Cave System, with the assistance of Tom Cottrell, Eric Hatleburg, and Mark Jancin.

*****July 15.** John Wilcox leads Pat Crowther, Eric Hatleburg, and Mark Jancin for another futile attempt at Q-87, but on the way back Pat pushes past the Tight Spot beyond the A-survey off the N-survey Junction Room to discover a going lead.

August 15. Stan Sides succeeds Joe Davidson as president of the Cave Research Foundation.

*****August 26.** Roger Brucker cannot get through the Tight Spot, but Tom Brucker and Richard Zopf go on, and Tom finds a river.

*****August 30.** Tom Brucker finds an arrow with the initials "PH" and "LH" beside it in Hanson's Lost River, and then Richard Zopf notices "PETE H" on the wall; with John Wilcox and Pat Crowther they explore a mile farther, but have to return to the Austin Entrance, on the longest trip ever taken in the Flint Ridge Cave System, 10.758 miles.

August 31. Tom Brucker leaks the news about finding the initials and arrow in Hanson's Lost River.

September 2. Joe Davidson cannot get through the Tight Spot, but John Wilcox, Richard Zopf, and Gary Eller go on to explore side leads in Hanson's Lost River.

Stan Sides tells the Cave Research Foundation expedition crew and Park Superintendent Joseph Kulesza about the discoveries in Hanson's Lost River.

September 3. John Bridge, Pat Crowther, Richard Zopf, and Ellen Brucker explore leads off Roaring River in Mammoth Cave.

*****1972, September 9.** *THE FINAL CONNECTION:* John Wilcox, Pat Crowther, Richard Zopf, Gary Eller, Steve Wells, and Cleve Pinnix connect the lower end of Hanson's Lost River (which so far had been known to the explorers only through the Austin Entrance of the Flint Ridge Cave System) to Echo River in Cascade Hall of Mammoth Cave in Mammoth Cave Ridge to integrate the Flint Mammoth Cave System, 144.4 miles long; they survey the connection and exit through the Elevator Entrance of Mammoth Cave. The Everest of world speleology is climbed.

Appendix III/Glossary of Cave Terminology

ARROW, n. A pointed line indicating the way out of a cave. Explorers sometimes mark arrows on walls or rocks with a rock, a finger, or soot from the flame of a carbide lamp.

BASE LEVEL, n. The elevation of the bed of the largest surface stream in a region. The Green River is the base-level river in the Central Kentucky Karst.

BEARING, n. The angular direction measured from one survey station or point to another with reference to magnetic north. In cave surveying, the bearing is read from a compass and is written in a survey notebook.

BELAY, v. To secure a climber on a rope for protection in case of a fall.

BELLY-CRAWL, n. A cave passage so low that you can travel through it only by squirming along in a prone position. Ceiling height is typically twelve to eighteen inches or less.

BENCH MARK, n. A surveyor's mark cut in a brass disk placed on a permanent base to indicate a point in a line of levels for determining altitude and sometimes longitude and latitude.

BLOWHOLE, n. An opening in bedrock or between boulders on the surface from which air flows perceptibly from underground.

BREAKDOWN, n. A jumble or pile of rocks in a cave passage produced by ceiling and wall collapse. A **BREAKDOWN BLOCK** is an angular piece of rock with dimensions ranging from a few inches to tens of feet on a side. **TERMINAL BREAKDOWN** is a pile of breakdown blocks that closes off a cave passage.

BREATHING CAVE, n. A cave passage in which air moves perceptibly in one direction and then in the other.

CABLE LADDER, n. Two metal cables—usually made of stainless steel—with rungs of lightweight metal tubing such as aluminum, six or eight inches wide spaced about eighteen inches apart. It can be rolled into a compact, lightweight bundle.

CANYON, n. A cave passage in which height exceeds width, usually by two or more times.

CARBIDE, n. The chemical compound calcium carbide, which produces acetylene gas when it reacts with water. **MINERS'-LAMP CARBIDE** is in quarter-inch chunks. **TO CHANGE CARBIDE** is to recharge a carbide lamp by removing the spent carbide and putting in fresh carbide; with the carbide lamps used in the Flint Mammoth Cave System, this must be done every three or four hours.

CARBIDE LAMP, n. A two-part container —usually made of brass—for generating acetylene gas. Water which drips from the top part onto carbide in the bottom part generates gas that escapes through a tiny hole in a tip or nozzle directed horizontally from the front of the container. The gas is ignited by sparks from a small steel wheel spun over flint, mounted on a reflector 2.5 inches in diameter. The lamp is mounted on the front of a hard hat, so that light is directed in whichever direction the wearer turns his head.

CARTOGRAPHY, n. With reference to caves, exploring and surveying with attention to passage detail, recording data, plotting, and drafting to produce a map showing the contours and details of the cave passages, names, north arrow, topographic contours, and other details.

CAVE, n., v. A natural cavity beneath the earth's surface that is long enough and large enough to permit human entry. Caves in the Central Kentucky Karst are produced by solutional action of groundwater draining through natural openings and flowing to the base-level Green River. Also called **CAVERNS. TO CAVE** or **TO GO CAVING** is to explore a cave. **BIG CAVE** usually means walking passages. **TO LOOK FOR CAVE** or **TO LOOK FOR MORE CAVE** is to look for new cave passages.

CAVE CONSERVATION. A policy of managing caves to protect and preserve their natural appearance by minimizing the adverse environmental effects of man's activities.

CAVE MEAL, n. A small can of meat, a small can of fruit, and some candy bars.

CAVE-ONYX, n. Banded flowstone.

CAVE RESEARCH FOUNDATION. A non-profit organization founded in 1957 for the purpose of supporting cave science, interpretation and education, and conservation. CRF is not a membership organization. Further information is available from: Cave Research Foundation, 445 W. S. College Street, Yellow Springs, Ohio 45387.

CAVE RIVER, n. Any stream of water in a cave is called a river. A cave river may be a few inches (Houchins River) or tens of feet (Roaring River) wide and deep.

CAVER, n. A cave explorer.

CEILING POCKET, n. A concave opening in the ceiling of a cave passage.

CENTRAL KENTUCKY KARST, CENTRAL KENTUCKY CAVE COUNTRY, n. An area of karst terrain centered about 100 miles south of Louisville. It is bounded on the north by the Hilly Country and on the west by Barren River, and includes the Mammoth Cave Plateau and the Sinkhole Plain. It contains about 200 square miles.

CHAIN, n. A metal tape measure used in cave surveying, usually fifty or 100 feet long. Also called a **TAPE.**

CHEST COMPRESSOR, n. A low horizontal belly-crawl passage that you can get through only by squeezing and often only by exhaling to reduce the size of your chest.

CHIMNEY, v. To climb the walls of a narrow cave passage or vertical cleft in a cave wall by bridging the opening with opposing arms and legs, or by putting back and hands against one wall and feet against the other.

CLAUSTROPHOBIA, n. A morbid dread of being in closed, narrow, or tight places.

CLOSED SURVEY, n. A survey in a loop from one point back to that same point.

COLUMN, n. A pillar-like deposit extending from ceiling to floor of a cave passage, formed by the joining of a stalactite and a stalagmite.

COMMERCIAL CAVE, n. A cave containing trails and lights that is exhibited to the public for an entrance fee. Mammoth Onyx Cave in the Central Kentucky Cave Country is a classic commercial cave.

COMPASS, n. A pocket-sized surveying instrument or hand-held transit consisting of a magnetic needle and folding sights for reading bearings, and a spirit-level clinometer for reading vertical angles. The Brunton compass is used in surveying the Flint Mammoth Cave System. (This is a local usage for what is more generally called a **HAND TRANSIT.**)

CONNECT, v. To find and survey through a natural cave passage that joins what were previously known as two independent caves.

CONNECTION, n. In caves, the naturally formed passage between two or more caves previously known through separate entrances.

CRACK, n. A narrow opening in the wall, floor, or ceiling of a cave passage.

CRAWLWAY, n. A cave passage so low that you can get through it comfortably only on hands and knees. Ceiling height is typically two to three feet.

CROUCHWAY, n. A cave passage so low that you can get through it only in a stooped or duckwalk position. Ceiling height is typically three to five feet.

CUT-AROUND, n. A section of secondary cave passage that departs from a main passage and returns to it after a short distance.

DOLOMITE, n. Calcium-magnesium carbonate. Dolomitic limestone contains around 20 percent or more dolomite.

DOME, n. A vertical, usually oval opening in the ceiling of a cave passage, closed at the top, produced by collapse or by the solutional activity of descending water. Vertical shafts viewed from below are often called domes.

EPSOMITE, n. A white-to-colorless mineral, hydrated magnesium sulfate, occurring as crystals, and useful as a salt substitute and as a cathartic.

ERROR OF CLOSURE, n. The calculated difference between the survey point at the beginning and the survey point at the end of a big loop that starts and ends at the same place. If the beginning survey point and the end survey point are coincident, there is no error of closure.

ESCARPMENT, n. A steep slope or long cliff resulting from erosion or faulting that separates relatively level areas of different elevation. In the Central Kentucky Karst the Big Clifty Sandstone often forms the resistant upper rim of cliffs along the edges of the major karst ridges, such as Flint Ridge.

EXFOLIATE, v. To come off in sheets, flakes, or layers parallel with the surface; for example, pieces from shaly rock walls and ceiling of a cave passage.

EYELESS FISH, n. Blind cave fish of several species that have such degenerate eyes and pigment that they appear to be white and eyeless. They are found in base-level cave streams such as Echo River. They seldom exceed four inches in length.

FLINT RIDGE, n. The northernmost karst ridge on the Mammoth Cave Plateau in Mammoth Cave National Park, containing Floyd Collins' Crystal Cave, Unknown Cave, Great Onyx Cave, Salts Cave, and Colossal Cave. The four connected caves (all except Great Onyx Cave) were known as the Flint Ridge Cave System from 1961 to 1972. In 1972 the FRCS was connected to Mammoth Cave to form the Flint Mammoth Cave System.

FLINT RIDGE CON, n. A pattern of speech and body language calculated to win the confidence of the listener and to promote expectations in him of wonderful underground discoveries. As perfected by Jim Dyer, it fascinates and disarms the listener and makes him eager to explore the place described, even if he is aware that he is being manipulated by an artist.

FLINT RIDGE SMILE, n. A broad grin constituting the only answer to a serious question about the Flint Ridge caves, calculated to reveal the existence of mysteries without revealing any substantive information. As perfected by Bill Austin, the Flint Ridge smile can produce feelings of intense determination to learn for oneself, feelings of fury, or feelings of inferiority, or all of these at the same time.

FLOWSTONE, n. A secondary deposit in caves of calcium carbonate, usually in the form of the mineral calcite, precipitated from groundwater. It occurs in the form of sheets, drapery, dams, lily pads, etc. (This is a local usage for what is more generally called **TRAVERTINE**.)

FLUORESCEIN DYE, n. A concentrated non-toxic chemical that colors water a vivid green. It is used to trace the course of underground streams. It can be detected in concentrations of a few parts per million.

FLUTES, FLUTING, n. Vertical striations, grooves, or alcoves in cave passage walls or vertical shaft walls, produced by the solutional activity of concentrated streams of descending water.

GO, v. Cave passages that go are passages open for exploration, or **GOING CAVE**. A lead that looks passable is said **TO GO**. One also says of a passage that has been explored that **IT WENT**.

GRAPE, n. A secondary deposit of calcium carbonate on cave walls, with surface projections consisting of globules ranging from pea size to grape size.

GYPSUM, n. A white-to-colorless mineral, calcium sulfate, deposited in caves in a variety of crystalline forms resembling needles, flowers, cotton, feathers, and wood shavings, and as faceted crystals. Gypsum is sometimes colored yellow, orange, or brown by impurities.

HANGING SURVEY, n. Cave passages that are surveyed, or the data of that survey, which are not connected to the main grid of surveyed passages in a cave.

HARD HAT, n. A protective head covering of fiberglass or strong plastic, held on the head by a chin strap, and containing an attachment on the front for the placement of a carbide lamp.

HELICTITE, n. A secondary deposit of calcium carbonate. Helictites look like stalactites that twist, spiral, and branch. They resemble chicken claws, twisted fingers, twigs, etc.

JOPPA RIDGE, n. The southwesternmost karst ridge of the Mammoth Cave Plateau in Mammoth Cave National Park, containing Proctor Cave and Lee Cave.

KARST, n. A characteristic landscape produced by solution and underground drainage in areas of soluble bedrock such as limestone and dolomite. Karst landforms include sinkholes and sinking streams, irregular ridges such as Flint Ridge, and blind or closed valleys such as Houchins Valley.

KARST RIDGE, n. A cavernous upland, bounded by karst valleys, base-level rivers, or sinkhole plains. Flint Ridge is an example.

KARST VALLEY, n. A cavernous lowland, usually formed by solutional activity of water along the course of a former surface stream, the waters of which have been captured underground. Sometimes called a **SINK VALLEY.** Houchins Valley is an example.

KNEE CRAWLERS, n. Molded rubber pads, each with two straps for fastening on your knee. They protect your knees from bruises and abrasions when you crawl on rocks in caves.

LAY-BACK, n.; LIE-BACK, v. A method of rock climbing in which your fingers pull one way and your feet push the opposite way in a crack in a wall, with your body out in space parallel to the wall like an inchworm. The vector of opposing forces effectively neutralizes gravity to permit you to move your hands and feet alternately to climb vertically.

LEAD, n. A cave passage that looks big enough for you to explore. (Some leads turn out not to be big enough to get into.)

LIMESTONE, n. A sedimentary rock composed principally of calcium carbonate. The caves of the Flint Mammoth Cave System are developed primarily in the Girkin Limestone and the Ste. Genevieve Limestone of Carboniferous age.

LINE PLOT, n. A map made by drawing straight lines between survey points.

LONGBAR, n. A steel crowbar three or four feet long.

MAMMOTH CAVE NATIONAL PARK, n. Established in 1941 as a part of the National Park System. For information, write to: Mammoth Cave National Park, Mammoth Cave, Kentucky 42259.

MAMMOTH CAVE PLATEAU, n. An upland in Mammoth Cave National Park in the Central Kentucky Karst containing three major karst ridges separated by two major karst valleys. Flint Ridge (ten square miles) is separated by Houchins Valley (two square miles) from Mammoth Cave Ridge (3.5 square miles), which in turn is separated by Doyel Valley (2.7 square miles) from Joppa Ridge (four square miles).

MAMMOTH CAVE RIDGE, n. The central karst ridge of the Mammoth Cave Plateau in Mammoth Cave National Park, containing Mammoth Cave, which has been known since 1798 and which became a part of the Flint Mammoth Cave System in 1972.

MIRABILITE, n. A water-soluble colorless mineral, hydrated sodium sulfate, occurring in a variety of crystalline forms, often resembling needles, coconut, or powder. Useful as a salt substitute and as a cathartic.

NATIONAL PARK SERVICE, n. Founded in 1916 as a branch of the U.S. Department of the Interior. For information write to: National Park Service, Washington, D.C. 20240.

NATIONAL SPELEOLOGICAL SOCIETY. A membership organization founded in 1941 and affiliated with the American

Association for the Advancement of Science. The purpose of the NSS is to encourage scientific study and conservation of caves, and it maintains extensive conservation and publication programs. Local branches of the NSS are called Grottos, and are located in all areas of the USA. For further information, address queries to: National Speleological Society, Cave Avenue, Huntsville, Alabama 35810.

PASSAGE, PASSAGEWAY, n. A horizontal opening in a cave large enough for you to enter.

PINCH, PINCH-DOWN, n., v. Where a passage gets so low or narrow that you cannot penetrate it. Then you say the passage has **PINCHED OUT.**

PIRATING, v. In caves, the capturing of a stream of water from a higher-level passage by a lower-level passage. Stream piracy in caves often produces dry higher-level passages.

PIT, n. A vertical opening in the floor of a cave passage produced by collapse of rock, slumping of breakdown, or the solutional activity of descending water. Vertical shafts viewed from above are often called pits.

PLOTTING, v. Use of a protractor and scale to draft a cave map. Sometimes a computer plotter is used to draft the map automatically from data supplied in digital form.

POINT, SURVEY POINT, n. A marked, numbered location in a cave passage on which the compass is placed for measuring the bearing, and on which the chain is placed for measuring the distance to the next point. (This is a local usage for what is more generally called **STATION** or **SURVEY STATION.**)

POINT MAN, n. The first person in a cave survey party, who selects and marks each survey point.

POPCORN, n. A secondary deposit in cave passages, characterized by rough, globular surface projections.

PSYCHING UP, v., n. In caving, mental preparation to go through tight passages and long distances in cave exploration; when in this state you are said to be **PYSCHED UP.**

RAPPEL, n., v. The act or method of descending a vertical shaft or cliff by means of a rope passed around your body through any of a variety of devices that produce friction, permitting a controlled rate of descent.

RINKY-DINK, n. A cave tour calculated to confuse, entertain, and fascinate the victims by taking them on a circuitous route over obstacles and at a dizzying pace. Sometimes called a **RUN-AROUND.**

SHORT BAR, n. A steel crowbar about twelve inches long with a curved end, used for digging and prying at breakdown rocks in caves.

SINKHOLE, n. A closed, often oval, basinlike depression on a karst land surface through which water drains underground. A **SINKHOLE ENTRANCE** is an opening into a cave from a sinkhole.

SINKHOLE PLAIN, n. A geographic unit of the Central Kentucky Karst consisting of a gently rolling land surface with drainage carried underground through sinkholes and sinking streams. Water drains underground to base-level Green River and Barren River. The Sinkhole Plain is bounded by the Mammoth Cave Plateau on the north, the Green River on the east, the Warsaw Limestone on the south, and the western divide of the Barren River on the west.

SINKING STREAM, n. A stream that flows in a valley that terminates in a headwall beneath which the stream plunges underground.

SIPHON, n., v. The place at which a cave passage is drowned by the ceiling's extending underwater. A passage that is closed off by water in this way is said **TO SIPHON.** (This is a local usage for what is more generally called a **SUMP.**)

SODA STRAW, n. A hollow stalactite having the shape and dimensions of a soda straw, sometimes exceeding eighteen inches in length.

SPELEOLOGIST, n. A scientist who specializes in speleology.

SPELEOLOGY, n. The science encompassing the origin and nature of physical and biological features and processes of the karst and cave environment.

SPELUNKER, n. A caver.

STALACTITE, n. A pendulant conical or icicle-like deposit. Stalactites are usually of calcite, and hang from the ceilings of cave passages. They are precipitated from mineral-bearing solutions dripping from cave ceilings.

STALAGMITE, n. A cylindrical, conical, or mound-shaped mineral deposit. Stalagmites are usually of calcite, and project from floors of cave passages. They are precipitated from mineral-bearing solutions dripping from the ceiling, and often form under stalactites.

STATION, SURVEY STATION, n. See **POINT.**

SURVEY, n., v. In caves, the systematic process of, or the data from, measuring and recording the bearings and distances of passages, using a compass and chain for the purpose of making an accurate map. Often a cave passage is referred to by the name of the survey in the passage, such as the Q-survey. Thus "the Q-survey" can refer either to the survey data or to the passage from which that data is taken.

SURVEY PARTY, n. Usually consists of three or four cavers, one of whom reads the compass or other surveying instrument, one or two of whom set points and read the chain, and one who takes notes in a survey book.

SURVEY SHOT, SHOT, n. In cave surveying, one of a series of bearing and distance measurements between points.

TAPE, n. See **CHAIN.**

TERMINAL BREAKDOWN, n. A pile of rocks that have fallen from ceiling and walls to form a barrier that closes a cave passage.

TRANSIT, n. An instrument for taking horizontal and vertical sighting in surveying.

TRAVERSE, v. To move laterally across the face of a wall, such as a wall of a pit, using climbing techniques.

TRUNK, TRUNK PASSAGE, TRUNK STREAM, n. A major conduit for existing or former drainage. Dry trunk passages are the largest in the Flint Mammoth Cave System. Hanson's Lost River is a tributary to the trunk stream, Echo River.

TUBE, n. A horizontal cave passage roughly cylindrical in shape, and usually no more than two or three feet in diameter.

UNDERGROUND WILDERNESS, n. Consists of cave passages that generally appear to have been affected primarily by the forces of nature, with the imprint of man's work substantially unnoticeable. At least 150 miles of the known cave passages in the Flint Mammoth Cave System are underground wilderness, and geological indications are that there are at least 150 more miles of virgin, unexplored underground wilderness cave passages in Mammoth Cave National Park.

VERTICAL SHAFT, n. A vertical, well-like, usually oval cave opening underground, produced by the solutional activity of water seeping downward at the intersection of joints or fractures in limestone or dolomite. Shafts range in diameter from a few inches to more than forty feet, and in height from a few feet to more than a hundred feet. In the Central Kentucky Karst, vertical shafts are the underground heads of drainage, and their drains are tributaries to underground trunk streams.

VIRGIN CAVE, VIRGIN PASSAGE, n. A cave or passage that has not been explored.

WALKING CAVE, n. A cave passage high enough for you to walk in.

WILD CAVE, n. A cave that has not been developed with trails and lighting.

Appendix IV/Picture Glossary of Cave Maneuvers

CHIMNEYING OR CHIMNEY-STEMMING

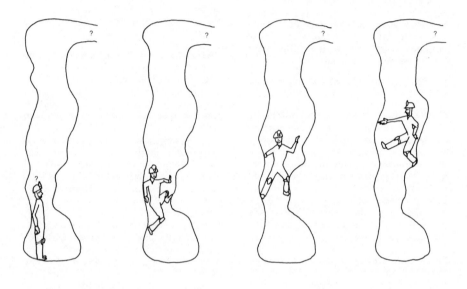

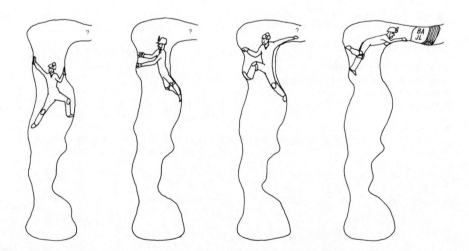

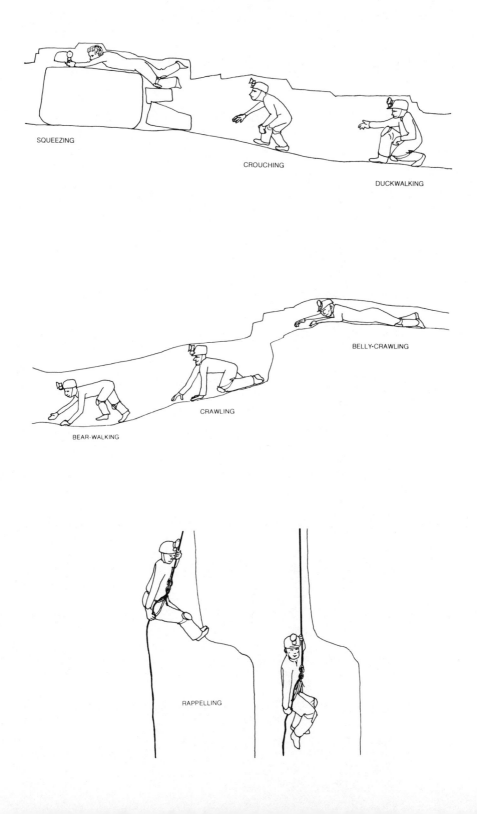

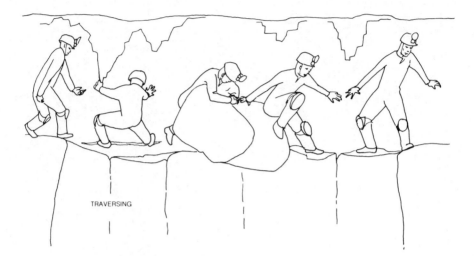

TRAVERSING

Appendix V/List of Participants

1947–1972

Adams, Russell James
Algor, John Robert
Ammon, John D.
Amsbury, Carlene
Amsbury, Wayne Paul
Anderson, Jennifer Anne
Anderson, Richard B.
Anderton, Peter W.
Apschunikat, Ken
Armstrong, Charles H., Jr.
Atkinson, Gordon E.
Atkinson, Heike Brigette
Austin, Jacqueline Frost
Austin, Joseph W.
Austin, Virginia
Austin, William Thomas
Backstrom, Neil
Balfour, William M.
Ball, Alan J. S.
Ball, Caro M.
Ball, Lawrence
Banks, Marie Cathey
Barr, Thomas C., Jr.
Barrett, Andrea Rose
Barrett, Peter John
Barringer, Vernon C., Jr.
Bassett, John L.
Bastien, Thomas Walter
Beatty, Ronald Les
Bell, Richard A.
Benington, Bernice
Benington, Frederick
Benington, Michael
Benington, Phyllis
Berglund, Donna Lou
Bergmann, Frithjof W.
Bilbrey, George Richard
Binney, Frank H.
Bishop, Sarah Gilbert
Bishop, William P.
Black, Donald F.
Black, Herbert L.
Black, Mary C.
Black, Paul R.
Black, Ruth A.
Blackburn, William K., Jr.
Blakesley, Audrey E.
Bloch, M. G.
Blum, H. Thomas
Bockstiegel, William C.
Bogardus, Bud
Bögli, Alfred W.
Bohl, Phillip Bruck
Bon Durant, Robert E.
Borin, Charles R.
Bouton, Carl E.
Brackett, Dixon
Brandel, Catherine Terry
Bridge, John F.

Browman, David L.
Brown, Michael C.
Brown, Ray D.
Bruce, C. N.
Bruce, William A.
Brucker, Ellen
Brucker, Emily Lenore
Brucker, Jane Corson
Brucker, Joan Wagner
Brucker, Roger Warren
Brucker, Thomas Alan
Burns, Ayleen Hilt
Burns, Denver P.
Burns, Richard G.
Cake, Marna
Cake, Stephen
Camvlos, Joshua F. B., Jr.
Campbell, Frank Edward
Cantwell, John C.
Carmazzi, H. Joseph
Carwile, Roy Hudson
Case, Janet
Cate, William C.
Chappell, Richard L.
Charlton, Royce E., Jr.
Chase, Dale John
Chester, James M.
Chow, Adam
Christensen, Kenner Allen
Christensen, Margaret J.
Clark, C. Robert
Cline, Elkanah
Coe, William D.
Colson, John G., Jr.
Colson, Peggy
Comes, Edward P.
Conlan, Thomas L., Jr.
Connally, Martin
Cottrell, Thomas Edward
Coulson, Rhoda M.
Cournoyer, Donald N.
Coward, Julian
Coyle, Jenny L.
Craig, Connie Lou
Cronk, Caspar
Cronk, Jean C.
Crowther, Patricia Page
Crowther, William R.
Culver, David Clair
Culver, Naomi Greitzer
Curl, Rane L.
Curl, Shirley Anne
Curtsinger, William R.
Dale, Kimberly
Daunt, Diana O'Neil
Daunt, Jonathan G. T.
Davidson, Elizabeth West
Davidson, Joseph K.
Davis, John Tootle
Davis, Judith Ann
Davis, Nevin W.

Davis, Roy A.
Davis, Samuel
Deal, Dwight E.
Deal, Sandra G.
Deamer, David W.
Deamer, John W.
Deike, George Herman, III
Deike, Mickey
Deike, Ruth G.
Denny, Carlene
DePaepe, Duane
DePaepe, Veda
DesMarais, David John
DesMarais, Shirley Lee
desRochers, Edward
Dewart, Gilbert
Dial, Billy Gower
Dial, Donald W.
Dickey, Frederick J.
Dickson, Richard
Distler, George E.
Dobson, Herbert J.
Doerschuk, David Cappel
Doggett, George Carson
Douglas, Henry H.
Drake, John
Drum, Dennis Edgar
Drum, Mary Elizabeth
Drummond, Ian
DuChene, Harvey Richard
Duffield, Lathel F.
Dyer, James W.
Dyer, June M.
Dyer, Robert Guy
Dymond, Louis T.
Easterday, Richard L.
Eaton, John W.
Eggers, Robert O.
Ehman, Burnell Frederick
Ehman, Carol G.
Ehman, Doris Daugherty
Ehman, Michael Frederick
Eidemiller, Donald Roy
Elberfeld, Jacob H.
Eller, Phillip Gary
Ellis, James G.
Ellis, Sonny
Emlen, John M.
Etkin, Nina L.
Evans, Keith W.
Ewers, Ralph Owen A.
Fagerlin, Stanley C.
Faller, Adolph, III
Farrar, Glen
Faust, Burton S.
Feder, Ned
Feldman, Lee
Fenton, M. Brook
Finkel, Donald
Fischer, Dianne Kay
Fischer, Fred William

Fish, Johnnie E.
Fisher, James Russell
Fiske, Alan Page
Fitzsimons, John G.
Foote, Wayne H.
Ford, Derek Clifford
Fort, Charles B.
Freeman, John Paul
Friel, Veronica Mary
Fries, Robert Hugh
Friley, Eugene Brooks
Frushour, Samuel Stuart
Gardiner, Ann H.
Gehring, John R.
Geil, Earl H.
Geiser, Albert G., Jr.
Gelfius, Steven Charles
Gell, Elizabeth Morris
George, Donald Irwin, Jr.
Gerwig, John
Gibler, Leonard Stockman
Gieseke, James Arnold
Glock, Elizabeth M.
Goldberg, Paul
Goodchild, Fiona M.
Goodchild, Michael F.
Goodlett, Collier W.
Gosney, James
Gourley, Robert N.
Gram, Larry E.
Green, Edward
Greene, Richard W.
Grover, Joan
Grover, John Watt, Jr.
Grover, John W., III
Grover, Richard W.
Grunwald, Thomas A.
Gurnee, Russell H.
Haggard, Mark Harold
Haines, Barbara Jean
Haldemann, René R.
Hall, Denise Martha
Hall, James W.
Hall, Robert L.
Hall, Thomas H.
Halliday, William R.
Halmi, Robert
Handley, Robert H.
Hanna, David
Harding, John David
Hargrove, Eugene C.
Harker, Donald F., Jr.
Harker, Meta Silvey
Harmon, Karen S.
Harmon, Russell Scott
Harsha, Edward Houston
Harsham, Philip
Hartle, Andreas
Hatleburg, Eric Warner
Hauser, James L.
Hawkinson, Edward F.

Heddon, Chester S.
Hedges, James A.
Heinze, Donald L.
Helfman, Sheldon S.
Heller, Kim Allyn
Hennessy, Thomas R.
Henshaw, Robert Eugene
Hensley, Kenneth L.
Herel, Carolyn E.
Herel, Edward A.
Hess, John Warren, Jr.
Hess, Letitia Jean
Hildebolt, Charles Floyd
Hill, Alan Eugene
Hill, Carol Ann
Hobbins, James L.
Hobbs, Horton H., III
Hobbs, Susan Krantz
Hodonsky, Angelika
Hoffmaster, Richard E.
Homan, Paul Thomas
Hosken, Barbara A.
Hosken, William H.
Hough, Robert Lee
Hough, Susan Lee
Houser, Connie Craig
Houser, Robert W.
Howard, Alan D.
Hoxie, Dwight Thomas
Hoxie, Janelle Ann
Hubble, Dorothy M.
Hubbs, David L.
Huber, David A.
Hufford, Judith Powell
Hulstrunk, Alfred
Hulstrunk, William
Irwin, Ethel
Irwin, Sandra L.
Jackson, Bernard V.
Jagnow, David Henry
Jagnow, Rebecca Ann
Jancin, Mark Douglas
Jones, David B.
Jones, Ellis
Jones, Kenneth Morgan
Kacsur, Charles
Kane, Debra Kay
Kane, Thomas
Kappeler, James Ernest
Kastning, Ernst H., Jr.
Kastning, Marjorie June
Keester, Kenneth Lee
Keith, James Hilton
Keller, Robert Martin
Kelley, David Lee
Kendall, John P.
Kessler, Sandra Rife
Kessler, Thomas Melvin
Key, John
Keys, David
Kibbey, Richard E.

Kiefer, Charles Robert
Kihara, Deane Hajime
Klawitter, Ralph A.
Klein, Marguerite M.
Klekamp, C. Thomas
Klippert, Rolf
Knight, John C.
Knudstad, James Escher
Koeneman, Paul William
Krauss, Raymond E.
Krisko, John
Kulesza, Joseph
Lamb, Peggy J.
Lambert, Nicholas E.
Lasiewski, Robert G.
Lawrence, Joseph D., Jr.
Lawson, William
Leblanc, Stephen A.
Lehrberger, John J., Jr.
Leissa, Arthur W.
Leochler, Edward L.
Lesko, Russ
Levy, Janet E.
Lewis, Franklin E.
Lindberg, Ralph DeWitt
Lindsay, John Francis
Lindsay, Kay Lorraine
Lindsley, Karen Louise
Lindsley, Robert Porter, III
Lingenfelter, Ann C.
Lipton, Barbara L. Nack
Lipton, Walter A.
Lloyd, Mary Jane
Lloyd, Robert B.
Long, Richard D.
Longsworth, Charles R.
Lotz, Theodore V.
Louch, William J.
Lowe, Geoffrey David
Loyd, Betty
Loyd, Samuel A., Jr.
Lucas, Ann Jane
Luken, Ralph A.
Lutz, Robert L.
Lutz, Ulysses E.
MacLeod, Barbara
MacLeod, Ross H.
Maggard, Glay E.
Malone, Christopher
Manhart, Larry DeLain
Mann, Charlotte W.
Mann, William F.
Marks, Franklin J., Jr.
Marquardt, William H.
McAdams, Raymond L.
McClure, Roger Earl
McCrady, Allen D.
McGhee, Charles R.
McIntosh, John Thomas
McLarnan, Charles Walter
McMillan, Roger L.

Mebane, Robert Alan
Medville, Douglas M.
Meloy, Harold
Melton, Carl
Merchant, Dean C.
Metzger, Chris William
Metzger, Radley
Miller, Charles Luther
Miller, Edward McC., Jr.
Miller, William B.
Miotke, Franz-Dieter
Mix, Michel
Mohr, Charles E.
Moody, Michael Ray
Morgan, Eric L.
Morgan, Keith
Morgan, Robyn Mary
Morris, David Marriot
Morris, Jonathan White
Morris, Trevor
Morrison, Fred J., Jr.
Morrison, Mary E.
Morrow, William Henry
Moses, Dwight Douglas
Mueller, Albert C.
Mueller, Margaret
Mullett, Frank C., Jr.
Mullett, Lawrence Paul
Munson, Leonard
Murphy, Cecilia M.
Murphy, Thomas Joseph
Myers, Delbert E., Jr.
Mysz, Fred
Neidner, James H.
Nelson, Wayne C.
Nicoll, Robert S.
Niemeyer, Douglas J.
Norton, Russell N.
Ochsenbein, Gary Dean
Ogden, Albert E.
Ogle, Thomas F.
O'Hara, Frank Joseph, III
O'Leary, Robert Dennis
O'Leary, Sharon Lee
Ondroušek, Oto
Palmer, Arthur Nicholas
Palmer, David Clifton
Palmer, Heather E. Z.
Palmer, Margaret L. Vogel
Palmer, Robert
Papenberg, Hans
Parker, John D.
Parker, Judith
Paul, Goldie H.
Payne, Kathryn Lynn
Peck, Stewart B.
Perry, Kenneth M.
Peters, Donald
Peterson, Georgia D.
Peterson, Gilbert M.
Peterson, Gilman P., Jr.

Petranoff, Theodore V.
Petterson, Karl E.
Pfau, Louis H.
Phillips, John Michael
Phinizy, Coles
Pinnix, Cleveland F.
Pohl, Erwin Robert
Potter, Frank Walter, Jr.
Poulson, Elizabeth M.
Poulson, Thomas Layman
Poulton, Guy John
Povirk, Raymond Anthony
Powell, Betty M.
Powell, Ralph E.
Prescott, Lionel H.
Prescott, Martha C.
Prescott, Virginia B.
Preston, Terry Eugene
Preston, Vivian Diane
Price, L. Greer
Purvis, Earl W.
Purvis, Gregory
Quinlan, James F., Jr.
Quinlan, Patricia H.
Rauch, Henry W.
Redman, Charles L.
Redman, Linda E.
Redpath, J. Lesley
Reid, Frank Spaulding
Replinger, Jean Sanford
Replinger, Randall J.
Reccius, Jack C.
Rhodes, Douglass
Rhodes, Linda Starr
Richter, Robert J.
Rigg, Richard H.
Rightmire, James R., II
Rindt, William
Ritter, Parker
Robbins, Louise M.
Robertson, Alicia Sorenson
Robertson, William, IV
Robinow, Meinhard
Rodcay, Bonnie Lynn
Roebuck, David Bradford
Rogers, Jean Muir
Rogers, Nancy
Rogers, William L.
Roleff, Edmund Frank
Roleff, Sandra Cleo
Rolingson, Martha Ann
Rose, Robert Hugh, II
Ross, Claude Osborne, Jr.
Ross, Paul J.
Ross, Quentin Everett
Rothfuss, Edwin L.
Roudabush, Charles E.
Roy, Charles R.
Roy, Kathleen E.
Ruder, Richard A.
Rust, Claude Charles

Rutford, Robert H.
Rutherford, John M., Jr.
Sanders, Richard S.
Sawtelle, Ida V.
Schaumburg, John R.
Schiller, Jerome P.
Schmitt, Stephen
Schneider, Elizabeth
Schulze, William Lee
Schwartz, Henry P.
Schweri, William F., II
Scott, John C.
Seiser, Felix
Sell, Gerry
Sell, Marvin J.
Sell, Edna
Setty, Paul T.
Seward, Hubbard A.
Shamel, David R.
Shamel, Roger Ervin
Shamel, Susan Richter
Shoptaugh, A. Glenn
Shroyer, Robert Edward
Shufeldt, Robert C.
Shuster, Anne Marie
Shuster, Evan Thomas
Sides, Linda Kay
Sides, Stanley David
Siegel, Frederick Richard
Silver, Jan Maurice
Simpson, Morris James
Sims, Richard
Sloane, Erwin
Smiley, Lawrence K.
Smiley, Rachel Regnier
Smith, A. Richard
Smith, David Glenn
Smith, Gary
Smith, Gordon Lynn, Jr.
Smith, Jan G.
Smith, Judith Ann
Smith, Lawrence D.
Smith, Philip Meek
Smith, Sheryl
Snell, Alden E.
Snider, Robert Austin
Snider, William W.
Snively, John Allen
Sorell, Henry
Sorenson, Alicia
Spaulding, Carl
Spaulding, Gregory D.
Spence, John L.
Stairs, Edith
Steele, C. William, Jr.
Stellmack, John Arnold
Stephens, James Patrick
Stephenson, William J.
Stewart, Robert B.
Stoltz, Michael D.
Stone, Ralph W.

Annotated Bibliography

There is a large literature on Mammoth Cave, the exploration of the caves of Mammoth Cave National Park, and the work of the Cave Research Foundation. Only a few of these items are cited below, and their bibliographies should be checked for further reading. Two good sources are Wilkes's *Bibliography* (1962), cited immediately below, and the bibliography of the *CRF Personnel Manual* (1975), cited below in section 1.

Wilkes, Frank G. *Bibliography of the Mammoth Cave National Park, Kentucky*. Louisville: Potamological Institute of the University of Louisville, 1962. 63 pp.
This work is far from complete even for the years covered, but it is invaluable particularly for the citation of unpublished items.

1/GENERAL WORKS ABOUT CAVES AND CAVING

Anderson, Jennifer Anne. *Cave Exploring*. New York: Association Press, 1974. 128 pp.
This is a good introduction to caving techniques.
Freeman, John P. (ed.). *CRF Personnel Manual* (second edition). Columbus, Ohio: Cave Research Foundation, 1975. 109 pp.
This manual embodies the experience of more than twenty years of work in Flint Ridge caves. It describes the philosophy and techniques of organizing and managing large-scale expeditionary caving in a big cave system.
Freeman, John P., Gordon L. Smith, Thomas L. Poulson, Patty Jo Watson, and William B. White. "Lee Cave, Mammoth Cave National Park, Kentucky." *NSS Bulletin*, Vol. 35, 1973, pp. 109–25.
This is a general description of the archaeological, historical, biological, and geological features of a large cave in Joppa Ridge.
Halliday, William R. *American Caves and Caving*. New York: Harper & Row, 1974. 348 pp.
This is the best American guide to general caving.
Moore, George W., and G. Nicholas Sullivan. *Speleology: The Study of Caves* (second edition). Teaneck, New Jersey: Zephyrus Press, 1976.
This is the best American introduction to the science of speleology.

Poulson, Thomas L., and William B. White. "The Cave Environment." *Science*, Vol. 165, 1969, pp. 171–81.
The life histories and evolution of caves and cave life are described.
Smith, Philip M. *Speleological Research in the Mammoth Cave Region, Kentucky: Elements of an Integrated Program*. Yellow Springs, Ohio: Cave Research Foundation, 1960. 18 pp.
This document outlines a research program that has since resulted in several hundred scientific publications.

2/ARCHAEOLOGY OF THE MAMMOTH CAVE AREA

Meloy, Harold. *Mummies of Mammoth Cave*. Shelbyville, Indiana: Micron, 1971. 40 pp.
Meloy traces the discovery and subsequent history of several desiccated prehistoric Indian bodies found in the caves.
Robbins, Louise M. "A Woodland 'Mummy' from Salts Cave, Kentucky." *American Antiquity*, Vol. 36, 1971, pp. 200–06.
This is a detailed study of the desiccated body of a prehistoric Indian.
Schwartz, Douglas W. *Prehistoric Man in Mammoth Cave*. Eastern National Park and Monument Association Interpretive Series No. 2, 1965. 28 pp.
This is a popular study of what is known of the prehistoric use of the caves.
Watson, Patty Jo. "Prehistoric Miners of Salts Cave, Kentucky." *Archaeology*, Vol. 19, 1966, pp. 237–43.
This describes "action archaeology" experiments in which modern cavers explore barefoot with cane torches to see how it was done in prehistoric times.
———. *The Prehistory of Salts Cave, Kentucky*. Illinois State Museum

Reports of Investigations, No. 16, 1969. 86 pp.
This is a detailed account of prehistoric utilization of the cave.
———— (ed.). *The Archeology of the Mammoth Cave Area.* New York: Academic Press, 1974. 255 pp.
This is a detailed account of prehistoric utilization of the caves.

3/HISTORY OF THE MAMMOTH CAVE AREA, OF EARLY CAVE EXPLORATION, AND OF FLINT RIDGE CAVING

Bridwell, Margaret M. *The Story of Mammoth Cave National Park, Kentucky.* N.p., 1959. 64 pp.
This is a good, popular account.

Brucker, Roger W. "Recent Explorations in Floyd Collins' Crystal Cave." *NSS Bulletin*, Vol. 17, 1955, pp. 42–45.
This is the story of the first breakout.

Bullitt, Alexander. *Rambles in Mammoth Cave* (facsimile of the 1845 edition, with a new introduction by Harold Meloy). New York: Johnson Reprint Corporation, 1973. 136 pp.
This is a very popular and colorful account, much plagiarized by later writers.

Crowther, Patricia P. "Into Mammoth Cave the Hard Way." *National Parks & Conservation Magazine,* Vol. 47, No. 1, 1973, pp. 10–15.
One of the main participants tells what it was like.

Faust, Burton. *Saltpetre Mining in Mammoth Cave, Kentucky.* Louisville: The Filson Club, 1967. 96 pp.
This work tells about mining saltpeter for gunpowder.

Halliday, William R. *Depths of the Earth.* New York: Harper & Row, 1966. 399 pp.
This work contains three chapters of interest: "The Story of Mammoth Cave," "The Story of Floyd Collins," and "The Story of Flint Ridge."

Hovey, Horace C. *Celebrated American Caverns, especially Mammoth, Wyandotte, and Luray* (facsimile of the 1896 edition, with a new introduction by William R. Halliday). New York: Johnson Reprint Corporation, 1970. 228 pp.
This is the classic account.

———— and Richard E. Call. *The Mammoth Cave of Kentucky* (facsimile of the 1912 edition, with a new introduction by Stanley D. Sides). Teaneck, New Jersey: Zephyrus Press, 1976. 131 pp.
This guide book contains descriptions of most features and areas shown today, and continues to be of interest and value.

Lawrence, Joe, Jr., and Roger W. Brucker. *The Caves Beyond: The Story of the Floyd Collins' Crystal Cave Exploration* (reprint of the 1955 edition, with a new introduction by Roger W. Brucker). Teaneck, New Jersey: Zephyrus Press, 1975. 283 pp.
This describes the first major modern attempt to open up the Flint Ridge caves, and as the first American book on a cave expedition, it has exerted an enormous amount of influence on the course of cave exploration in America.

Mohr, Charles E., and Howard N. Sloane (eds.). *Celebrated American Caves.* New Brunswick, New Jersey: Rutgers University Press, 1955. 339 pp.
See particularly three chapters: "The Death of Floyd Collins," by Roger W. Brucker; "Mammoth Cave's Underground Wilderness," by Henry W. Lix; and "Medicine, Miners, and Mummies," by Howard N. Sloane.

Sides, Stanley D. "Early Cave Exploration in Flint Ridge, Kentucky: Colossal Cave and the Colossal Cavern Company." *Journal of Spelean History,* Vol. 4, 1971, pp. 63–69.
This gives all the details now known.

———— and Harold Meloy. "The Pursuit of Health in the Mammoth Cave." *Bulletin of the History of Medicine,* Vol. 45, 1971, pp. 367–79.
This describes an early attempt to cure tuberculosis by confinement in Mammoth Cave.

Smith, Philip M. "Discovery in Flint Ridge, 1954–1957." *NSS Bulletin,* Vol. 19, 1957, pp. 1–10.
This describes continuing exploration.

———— "The Flint Ridge Cave System: 1957–1962." *NSS Bulletin*, Vol. 26, 1964, pp. 17–27.
This describes the integration of the caves.

———— and Richard A. Watson. "The

Development of the Cave Research Foundation." *Studies in Speleology,* Vol. 2, 1970, pp. 81–92.
This is a short history.

Wells, Stephen G., and David J. Des-Marais. "The Flint Mammoth Connection." *NSS News,* Vol. 31, No. 2, February 1973, pp. 18–23.
This is the first detailed account.

Wilson, Gordon. *Folkways of the Mammoth Cave Region.* N.p.: National Parks Concessions, 1967. 64 pp.
This describes the lifestyle of many of the people who were moved off the land that is now Mammoth Cave National Park.

4/THE BIOLOGY OF THE MAMMOTH CAVE AREA

Bailey, Vernon. "Cave Life of Kentucky Mainly in the Mammoth Cave Region." *American Midland Naturalist,* Vol. 14, 1933, pp. 385–635. Reprinted as a book by the University of Notre Dame Press, n.d., 256 pp.
This work lists every animal ever suspected of inhabiting the area.

Barr, Thomas C., Jr. "Ecological Studies in the Mammoth Cave System of Kentucky. I. The Biota." *International Journal of Speleology,* Vol. 3, 1967, pp. 147–204.
See next item.

———— and Robert Kuehne. "Ecological Studies in the Mammoth Cave System of Kentucky. II. Ecological Communities." *Annales de Spéléologie,* Vol. 26, 1971, pp. 47–96.
These two articles give the most complete coverage of the fauna and descriptive ecology, particularly of the aquatic environment.

Mohr, Charles E., and Thomas L. Poulson. *The Life of the Cave.* New York: McGraw-Hill, 1966. 252 pp.
This is the best general introduction to the biology and ecology of cave organisms; it is lavishly illustrated with color photographs and diagrams.

5/THE GEOLOGY OF THE MAMMOTH CAVE AREA

Brucker, Roger W. "Truncated Cave Passages and Terminal Breakdown in the Central Kentucky Karst." *NSS Bulletin,* Vol. 28, 1966, pp. 171–78. See next item.
————, John W. Hess, and William B.

White. "Role of Vertical Shafts in the Movement of Ground Water in Carbonate Aquifers." *Ground Water,* Vol. 10, No. 6, 1972, pp. 5–13.
These two articles tell much about the evolution of caves in the area. They also contain the key to many Flint Ridge exploration hopes and techniques.

Deike, George H., III. *The Development of Caverns of the Mammoth Cave Region.* Pennsylvania State University Ph.D. dissertation, 1967, 235 pp.
This is the most detailed work on the evolution of the cave passages.

Lobeck, Armand K. *The Geology and Physiography of the Mammoth Cave National Park.* Kentucky Geological Survey, Series 6, Pamphlet 21, 1928, 69 pp. Reprinted in *The Pleistocene of Northern Kentucky,* Kentucky Geological Survey, Series 6, Vol. 31, 1929, pp. 327–99.
This classic account contains many striking three-dimensional diagrams.

Miotke, Franz-Dieter, and Arthur N. Palmer. *Genetic Relationships between Caves and Landforms in the Mammoth Cave National Park.* Hannover, West Germany: n.p., 1972. 69 pp.
This work contains summary conclusions on important recent work.

Pohl, E. Robert. *Vertical Shafts in Limestone Caves.* NSS Occasional Paper No. 2, 1955. 24 pp.
A classic work on the origin of this feature that also illustrates exploration possibilities where randomly developed vertical shafts connect different levels of horizontal passages.

———— and William B. White. "Sulfate Minerals: Their Origin in the Central Kentucky Karst." *American Mineralogist,* Vol. 50, 1965, pp. 1461–65.
This describes the origin of a major type of cave mineralization in the area.

Quinlan, James F. "Central Kentucky Karst." *Méditerranée,* No. 7, 1970, pp. 235–53.
This is a general description of the physical geography and distinctive karst features of the area.

Swinnerton, Allyn C. "Origin of Limestone Caverns." *Geological Society of America Bulletin,* Vol. 43, 1932, pp. 663–94.

This classic work describes how the caves of the Mammoth Cave area were formed.

Watson, Richard A. "Central Kentucky Karst Hydrology." *NSS Bulletin*, Vol. 28, 1966, pp. 159–66.
In this work, water flow relations in the area are described.

White, Elizabeth L., and William B. White. "Processes of Cavern Breakdown." *NSS Bulletin*, Vol. 31, 1969, pp. 83–96.
This describes the origin of a major cave feature in the Mammoth Cave area.

White, William B., Richard A. Watson, E. Robert Pohl, and Roger W. Brucker. "The Central Kentucky Karst." *Geographical Review*, Vol. 60, 1970, pp. 88–115.
This is the standard description of the general origin, evolution, and nature of the physiography and karst features of the area.

6/CAVE CONSERVATION

Davidson, Joseph K., and William P. Bishop. *Wilderness Resources in Mammoth Cave Natonal Park: A Regional Approach.* Columbus, Ohio: Cave Research Foundation, 1971. 34 pp.
This work describes a regional approach to protecting a large area of natural karst terrain that is surrounded by developed areas.

Smith, Philip M. "Some Problems and Opportunities at Mammoth Cave National Park." *National Parks Magazine*, Vol. 41, No. 233, 1967, pp. 14–19.
This work examines major problems that have not yet been solved.

Watson, Richard A. "Mammoth Cave: A Model Plan." *National Parks & Conservation Magazine*, Vol. 46, No. 12, 1972, pp. 13–18.
In this article the master plan for Mammoth Cave National Park is examined.

———— and Philip M. Smith. "Underground Wilderness: A Point of View." *International Journal of Environmental Studies*, Vol. 2, 1971, pp. 217–20.
In this article the concept of underground wilderness is developed, and a plea for protection of caves is made.

7/POETRY AND PHILOSOPHY

Finkel, Donald. *Answer Back.* New York: Atheneum, 1968. 38 pp.
In this highly acclaimed book-length poem, by the 1974 recipient of the Theodore Roethke Memorial Award, the poet explores the motif of the cave, taking off from his experiences as a Flint Ridge caver.

Watson, Richard A. "Notes on the Philosophy of Caving," *NSS News*, Vol. 24, 1966, pp. 54–58.
This is an initial attempt to tell what it is all about.

Afterword

Immediately after the connection in 1972, exploration continued on several fronts. Many trips were taken out leads heading southwest from Mammoth Cave toward Doyel Valley and toward Proctor Cave in Joppa Ridge. Despite their promise, all these leads pinched out, ending in mud and wet cobble fills. Even Miller Avenue and Snail Trail, which felt right, fizzled.

John Wilcox, Richard Zopf, and others led trip after trip out Bransford Avenue and Cocklebur Avenue. They made formal surveys of the passages until they crossed under the Mammoth Cave National Park Boundary, and then continued with (secret) compass and pace surveys. These discoveries extended more than half a mile south of the park boundary line. Some passages went near Sand Cave, where Floyd Collins died in 1925, but no connection was found.

Proctor Cave yielded seven miles of passages, and by 1978, almost all leads had been explored.

In Morrison's Cave, Don Coons, Sheri Engler, and John Branstetter found several miles of canyons and crawlways.

Then the simultaneous discovery of two great rivers in May 1979 created great excitement. Don Coons's party found a wide underground river in Morrison's Cave, and Richard Zopf dangled on a rope into a pit in Proctor Cave to find another wide underground river. In July, Don Coons came across his own footprints in the mud along the river in Proctor Cave. They were the ones he had made in May in Morrison's Cave. The two rivers were one: the explorers had connected Proctor Cave and Morrison's Cave.

Upstream it was called Logsdon River and downstream, Hawkins River. It constitutes the master southern drainage of Mammoth Cave. The newly found river opened up more than twenty miles of new passages.

By late summer of 1979, the map showed that the end of the A-survey in East Cocklebur Avenue in Mammoth Cave stopped just short of Logsdon River. A connection attempt was quickly fielded from the Frozen Niagara Entrance of Mammoth Cave. John Wilcox and Tom Gracanin left their clothes in a dry spot and snaked through a near-siphon to make the connection between Mammoth Cave and Proctor-Morrison's Cave in August 1979.

Meanwhile, Jim Bordon and Jim Currens, exasperated with the old guard and regimented caving of the Cave Research Foundation, decided to find their own cave. Besides Flint Ridge, Mammoth Cave Ridge, and Joppa Ridge on the Mammoth Cave Plateau, there is also Toohey Ridge. The other ridges contained great caves: Why not Toohey Ridge? They formed the Central Kentucky Karst Coalition (CKKC), and after a lot of ridge walking and poking into holes, they found a going cave and named it Roppel Cave. By the time of the Mammoth-Proctor connection they had surveyed seventeen miles of Roppel Cave.

But now the Cave Research Foundation and Mammoth Cave had penetrated into the west edge of Toohey Ridge. Would the CRF cavers try to connect Roppel Cave into the Mammoth Cave System? Are you kidding? Those guys are insatiable. CKKC cavers were outraged and let everyone know it. They wanted time to develop Roppel into a cave of great length before what was probably inevitable, its incorporation into the Mammoth Cave octopus. And luck was on their side. A siphon blocked the end of Logsdon River, 1500 feet from known passages in Roppel Cave. But just to make doubly sure, the CKKC cavers initiated meetings with CRF members in which an agreement was reached that CRF would not try to connect Mammoth Cave with Roppel Cave.

How long do you think an agreement like that could last? In no time, explorers from CRF on one side of the siphon and explorers from CKKC on the other side of the siphon were making clandestine trips in the area. The CKKC explorers extended a river passage southwest and reported they had run out of leads. The CRF explorers found upper-level passages that might bypass the terminal siphon, but it was tough going. Then came the real shocker.

In May 1981, Tom Brucker found a cave entrance outside the park, but it was not thoroughly explored. Then in July, some cavers affiliated with neither CRF nor CKKC went in this Ferguson Entrance and found that it provides relatively easy access to Hawkins-Logsdon River. What's more, the outsiders leased the entrance. After a flurry of meetings, CRF gained permission to use the leased entrance and helped to erect a stout gate over it to restrict access.

Now you could almost drop right down in front of the siphon. It was not quite that easy, but close enough that CRF cavers could see no way to exploit the exploration potential of this new entrance without alienating the CKKC cavers. The solution was to ask the president of the CKKC to lead parties through the new entrance into the siphon area. Several CRF cavers had helped in exploring Roppel Cave all along, and now cooperation increased.

In February 1983, Pete Crecelius, CKKC's president, found that the siphon was no longer a siphon. The water had lowered to reveal a ceiling, one inch of air space, and a blast of cold wind. Pete kept quiet about the discovery because he was concerned about the sensitive feelings of many CKKC cavers who strongly opposed any connection.

It was a dry summer in 1983. Pete led a mid-August trip to the siphon. Now, instead of a one-inch air space, he found that the water was down eighteen inches, and a gale of air was blowing through the opening. The explorers set off to the northeast, wading for 1000 feet upstream to a breakdown. They went on another 800 feet before Pete turned the party around, afraid that the next step might produce a connection between Roppel Cave and Mammoth Cave that the CKKC had been trying to avoid.

Some of his party members argued that they should connect it, and to hell with politics. They turned back, but the next day rumors began to fly that the connection had already been made, or that in any case, several area cavers intended to make the connection without the approval of anyone. Officers and board members of the CKKC ran up several hundred dollars worth of long-distance phone bills as they discussed the crisis. They decided—by no means unanimously—to ask CRF to join them in an effort to connect Roppel Cave with Mammoth Cave.

Ten explorers were chosen to represent the groups who had been most active in exploring the world's longest cave and the sixth longest cave: Jim Borden, Bill Walter, Roberta Swicegood, Dave Weller, and Dave Black from CKKC; Don Coons, Sheri Engler, and John Branstetter from the Uplands Research Laboratory, CKKC, and CRF; and Roger Brucker and Lynn Brucker from CRF.

In September 1983, eleven years from the date of the connection between the Flint Ridge Cave System and Mammoth Cave, a cave party entered Roppel Cave's Weller Entrance on Toohey Ridge, and another cave party entered the new Ferguson Entrance to Mammoth Cave on Mammoth Cave Ridge. By 1:30 p.m., members of the two parties could hear each other shouting. They met in a muddy room floored with breakdown, to celebrate, eat, and pose for photos. Then each party continued on the way it had been going, each leaving the cave by the entrance the other party had entered.

With the addition of the Roppel Cave survey, the Mammoth Cave System was then 294.4 miles long.

Exploration continued. No one had noticed the day the Flint Ridge Cave System many years before had become the longest cave. And no one noticed when the Mammoth Cave System passed 300 miles. The longest cave is now well over 300 miles long.

In 1976, the longest cave was under 150 miles long, and we predicted that it would one day be 300 miles long. Some people said that was silly. And it was silly. We should have said that one day the longest cave will be 500 miles long.

Mark our words.

Roger W. Brucker and Richard A. Watson
12 May 1986

Index

Note: The abbreviations FCCC and CRF stand for Floyd Collins' Crystal Cave and the Cave Research Foundation, respectively. Page numbers in **boldface** *type refer to illustrations.*

A Note about the Authors

Roger W. Brucker is president of Odiorne Industrial Advertising, Inc., in Yellow Springs, Ohio. Past president of the Cave Research Foundation, he is an Honorary Life Fellow of the National Speleological Society. He has written adventure and technical articles related to caves and is the co-author of two other books on caving: *The Caves Beyond* (with Joe Lawrence) and *Trapped! The Story of Floyd Collins* (with Robert K. Murray). He has taught and written articles about marketing. He is the father of four children, Thomas, Ellen, Jane, and Emily. His wife, Lynn Brucker, is an electrical engineer and a caver.

Richard A. Watson is Professor of Philosophy at Washington University, St. Louis. He is past president of the Cave Research Foundation, a Fellow of the National Speleological Society, and director of Cave Books. He has participated as a geologist on paleoclimatological and archaeological expeditions in Kentucky, New Mexico, Iran, Turkey, and the Yukon. He is married to Patty Jo Watson (with whom he wrote *Man and Nature*) and has one child, Anna.